Critical Praise for INVISIBLE

"This is the best book written about the new age of biology....Mr. Hall has produced an authentic and fascinating account of the origins of bio-technology. His descriptions...capture the intensity and excitement as well as the social fabric of technology. I recommend it to everyone."

Phillip A. Sharp, Director
MIT Center for Cancer Research

"Hall is a first rate science writer....INVISIBLE FRONTIERS reminds me of Tracy Kidder's *The Soul of a New Machine*....Hall describes not only the science itself, but the scientists—their ambitions, jealousies, frustrations, paranoia, pride—and as a result the laboratories come alive. Hall has clearly done a lot of work, and he has produced a splendid piece of science writing."

Wray Herbert
Washington Post Book World

"Hall's account...is quite brilliant....He has found a ripping yarn to tell, and he explains the biology impressively clearly."

Maxine Clark
New Scientist

"A spirited account.... [Hall] succeeds marvelously in making the science accessible to the general reader, mixing straightforward description with inventive metaphors, similes, and analogies....An important and pioneer-ing book, dealing with events of high scientific and economic conse-quence. It is an absorbing narrative."

Daniel J. Kevles
New York Times Book Review

"Wonderfully instructive on the science—but especially on the scientists."

Kirkus Reviews

"Stephen S. Hall is one of the best of the new generation of science writers, and INVISIBLE FRONTIERS is one of the best of the new wave of science books....Superbly written and extraordinarily interesting....An extremely important book that deserves to grace the library of everyone who is interested in the future."

Allen L. Hammond
Washington Times

"If you only read one book about science this year, or this decade, it ought to be INVISIBLE FRONTIERS."

Susan Kelly
FAX on Books

"Stephen S. Hall has written a rare book. It is one of the very few scientific histories that I can wholeheartedly recommend to both my scientist and poet friends...perceptive, funny, and uncannily accurate.... [Hall] does an excellent job of explaining some rather esoteric molecular biology, while never forgetting to tell a terrific adventure story."

Harvey Bialy
Bio/Technology

"A suspenseful thriller...Hall paints a convincing picture of the people and forces that made human insulin the first recombinant DNA product used in humans. He sketches a scientist in a sentence or two, and proceeds to flesh out the drawing and give it breath and swagger and doubts and dreams."

David Ansley
San Jose Mercury-News

"With the page-turning tug of a thriller, INVISIBLE FRONTIERS presents a tale that most readers, especially scientists, will find irresistible....Hall is an excellent writer."

Rudy M. Baum
Chemical and Engineering News

INVISIBLE
FRONTIERS

The Race to Synthesize a Human Gene

Stephen S. Hall

T E M P U S ™

Tempus Books of Microsoft Press
A Division of Microsoft Corporation
16011 NE 36th Way, Box 97017, Redmond, Washington 98073-9717

Published by arrangement with The Atlantic Monthly Press.

Library of Congress Cataloging in Publication Data
Hall, Stephen S.
    Invisible frontiers.
    Reprint. Originally published: 1st ed. : New York :
Atlantic Monthly Press, c1987.
    Includes index.
    1. Genetic engineering—Popular works.   2. Recombinant
DNA—Popular works.   3. Insulin—Biotechnology—Popular
works.   I. Title.
TP248.6.H35   1988   615'.365   88-20042
ISBN 1-55615-172-1

Printed and bound in the United States of America.

1 2 3 4 5 6 7 8 9   HCHC   3 2 1 0 9 8

Distributed to the book trade in the United States by Harper & Row.

Distributed to the book trade in Canada by General Publishing Company, Ltd.

Tempus Books and the Tempus logo are trademarks of Microsoft Press.
Tempus Books is an imprint of Microsoft Press.

FOR MY PARENTS

Foreword and
Acknowledgments

· ·

When I started to write about science five years ago, two things immediately impressed me. One was how intrinsically exciting much of scientific research was, particularly in the field of molecular biology. The other was how rarely that excitement ever came across in writing about science.

These were admittedly presumptuous conclusions to reach for someone who came late and wandered sideways into the field, and whose last formal instruction in the biological sciences occurred during his freshman year of high school. In those days, we heard talk of Mendel, but the term "clone" hadn't even been coined. It was quite enough worry to keep a semi-pithed frog from fleeing the dissecting tray, its poor cranium dimpled by freshman angst.

I was the further (and thoroughly complicit) victim of a collegiate career in the early 1970s, a time when "liberal arts" was just undergrad code for "no science and math." Aside from an inconsequential course in physics, I happily volunteered for one of those wrong-headed experiments in education during which a small group of guinea pigs were sequestered in a dorm for eighteen straight days and asked to study nothing but chemistry. It dispatched a science credit in less than three weeks, and entailed nothing more onerous than taking water samples from rivers, creeks, and quarries in southern Wisconsin (followed by several hours of arduous sunbathing). At the end, a ten-question true-or-false test established our shaky competence in pH and valences. The professor was disgusted, the students were delighted, and that as well as anything conveys the ultimate educational value of the whole affair.

It thus comes as a surprise, especially to myself, that I have spent the last three years immersed in the world of molecular biology and, moreover, have chosen to write about it at a level of detail that may at first be as alarming to the general reader as those one-time college science courses were to me. At the same time, I have chosen to portray science as the profoundly human enterprise that it is, where scientists exclaim, curse, compete, get depressed, are swayed by rumors, make mistakes, luck out, fail miserably, pick themselves up and start over again. But to understand a scientist's emotional life, you have to understand at least a little of the work, and that is one reason I have tried to explain the science in some detail.

There are other reasons. In the course of writing this book and observing the periodic controversies that flare up about recombinant DNA, it cannot hurt to try to explain as much of the science as possible. A little knowledge can be terrifying, but a little more knowledge can be soothing (not without worry, certainly, but perhaps a bit more familiar). Nothing in our world is risk-free, including recombinant DNA, but we are helpless to weigh those risks until we begin to get a handle on the full dimensions of the problem. It is obvious that citizens need to make informed opinions about this technology and where it is taking us. Otherwise we are constantly beholden to the tyranny of experts, and as we learned in the early debates on recombinant DNA and in other areas of technological controversy, you can always trot out experts on either side of the question, which leaves everyone with an expert level of confusion and anxiety.

No one book can comfortably discuss all aspects of the revolution in biology, and I did not intend *Invisible Frontiers* to be a definitive history of science. It is not a rhetorical treatise on recombinant DNA, nor does it argue one way or another whether it is glorious or ominous. What it does attempt to do, however, is give a feel for science as it unfolds, haphazardly, during a historic period in the biological sciences, and it is a cautionary tale in the sense that in any human enterprise, of which biology is surely one, nothing is ever quite as anticipated and, as they like to say in the labs, Murphy—as in "Murphy's Law"—always catches up to you. In any event, I have attempted to keep the scientific explanations as clear and terminable as possible. Moreover, it is a dreary and hackneyed convention of science writing to imply that the scientist is a genius without ever explaining why. Mindful of Thomas Edison's famous observation, I have devoted much of my attention to perspiration in the scientific enterprise without, I hope, slighting that one percent of inspiration. And having spent several years in the occasional company of biologists, I am convinced that the only way to convey the resourcefulness of scientists is to do the only and obvious thing: talk about the science.

In attempting to do this, I am obviously indebted to all the biologists who agreed to discuss their work on the insulin project (a complete list appears at the rear of the book). In describing his policy of listing authors on scientific papers, Arthur Riggs of the City of Hope once mentioned that he tended to err on the side of including as many as possible, and the policy seems like a good idea for book acknowledgments as well.

This account would not have been possible without the generous and candid cooperation of the molecular biologists involved. I am deeply indebted to all those scientists who patiently submitted to interviews, second interviews, pestering phone calls, and hectoring letters. Among the Harvard researchers, I wish to express my thanks to Walter Gilbert, Argiris Efstratiadis, Lydia Villa-Komaroff, Stephanie Broome, Forrest Fuller, Peter Lomedico, William Chick, Allan Maxam, Karen Talmadge, Debra Peattie, Greg Sutcliffe, and Jeremy Knowles for consenting to interviews. Peter Feinstein and Robert Gottlieb of Biogen, Inc. provided useful assistance.

Among those affiliated with the University of California—San Francisco research effort, I would like to thank William Rutter, Axel Ullrich, Raymond Pictet, John Chirgwin, John Shine, Edmund Tischer, Fran DeNoto, Peter Seeburg, John Baxter, and Graeme Bell for submitting to interviews and sharing their impressions. I wish also to thank Michela Reichman and Roger Ditzel at UCSF for responding to questions and queries.

There was an inevitable overlap of personnel between UCSF and Genentech, but I wish here to acknowledge the help of those who have come to be associated most closely with Genentech during the time of this story. They include Herbert Boyer, Robert Swanson, Herbert Heyneker, David Goeddel, Dennis Kleid, Daniel Yansura, Francisco Bolivar, Roberto Crea, Ronald Wetzel, and William Young. At the City of Hope, Arthur Riggs, Keiichi Itakura,

Rachmiel Levine, Louise Shively, and Leonor Balce-Directo all generously agreed to interviews. Suzanne McKean and Debra Bannister at Genentech were most helpful in expediting matters there.

In the interests of fairness and accuracy, I asked one member of each of the three research groups discussed in the book to read, in confidence, a version of the manuscript; they know who they are, and I reiterate in public my deep gratitude for the time they took to perform this task. Any gaps, errors, or shortcomings that remain are solely of the author's stamp. In addition, several researchers donated time and assistance well beyond the bounds of polite cooperation, so it is a special pleasure to thank Argiris Efstratiadis, Raymond Pictet, Peter Lomedico, Karen Talmadge, Allan Maxam, Dennis Kleid, Axel Ullrich, and Lydia Villa-Komaroff for their not inconsiderable help at various stages of this project.

A great number of other people have contributed immeasurably to the preparation of this book by answering questions, making suggestions, and filling in the background in instructive, sometimes inspired ways. In particular I would like to thank Peter Greenaway for his recollections on the Harvard group's English expedition; Donald Steiner for his erudite and fascinating account of the history of insulin research (lamentably abbreviated here); Ruth Hubbard and Jonathan King for explanations of their opposition to recombinant DNA work in Cambridge; William Gartland of the National Institutes of Health; Ronald Cape and Thomas White of Cetus Corp.; and Irving Johnson of Eli Lilly & Co., who graciously consented to an interview after Lilly officials first agreed and then abruptly declined to provide interviews in connection with this project.

In terms of logistical assistance in research, I would like to thank Helen W. Samuels and Kathy Marquis of the MIT Special Archives' Recombinant DNA History Collection; Oscar Mastin, Ian Calvert, and Charles Van Horne of the U.S. Patent Office; Mary Grein and John Dolan-Heitlinger of Eli Lilly & Co.; Dorothy Born of the American Diabetes Association; and Carol Luczc of the Westinghouse Science Talent Search. My own research benefited from the assistance of Naomi Freundlich and Gerald Reagan, with a helping hand from Samuel Fromartz.

I was particularly fortunate, several years ago, to espy a New York Mets schedule taped to the office wall of Kevin Dolan, a molecular biologist at New York University. This assured we had at least two interests in common. In the ensuing period, we have attended baseball games and reviewed experiments (usually not simultaneously), and this account has benefited in ways too numerable to mention by Kevin's explanations, suggestions, and corrections. I am also indebted to Kevin's wife, Jo Alene Goidl, a diabetes researcher, who read an early draft of the manuscript, and to Kevin's colleagues at NYU, William Beers and Kathy Nelson, for letting me sit in on classes and labs. I remain eternally grateful to Kathy for letting me run my first gel.

Although mythology persists in characterizing the relationship between editors and writers as kin to that of lions and lambs, I have been very fortunate in working with a talented number of editors at several magazines who were cleverly duped into offering advice and assistance on various portions of this manuscript and, as always, the writer—and, more importantly, the reader—has benefited from their suggestions. At *Science 86*, I had the pleasure of working with Esther Mackintosh and Eric Schrier, both of whom provided cogent criticism and needed encouragement; I am also indebted to the former editor of *Science 86*, Allen Hammond, who generously allowed some of my magazine travels to take me, by prearranged coincidence, to the sites of interviews for this book. Joan Tapper of *National Geographic Traveler* was similarly supportive. John Tarkov made many edifying comments on an early draft, and I benefited as well from suggestions by readers Samuel Fromartz, Susan Ridge, and Nelson Smith. Charles Mann in particular proved to be an exemplary reader. Last but not least, my editor at Atlantic Monthly Press, Morgan Entrekin, not only penciled me toward readability, but has displayed boundless and infectious enthusiasm for this project from its very inception.

For spiritual as well as physical sustenance along the way, I am grateful to the friends who put me up as well as putting up with me, by which I mean that often self-absorbed world of writers muttering their way through a Work-in-Progress. In my mind, I associate them with happy moments and comforting memories in the midst of the deluge: a diffident iguana (Larry Sulkis and Margareta Schiappa); loveliest lanai (Brian and Peggy Sulkis); burritos in the Mission and the All-Star *Croix de Candlestick* (Jerry and Marilyn Burns); Slovenian wisdom and Jersey sunburns (Jerry and Linda Roberts); sundaes and pinball (Steve Rubenstein); best disc jockey, with consort (Charlie Haas and B.K. Moran); best music, original score (Joe Helguera); days in Maine (JoAnne Thompson and Tom Roos); mentor in a long-run continuing role (Robert Ray); the veal that could bring Skanderbeg back to life (Zef Albanese); my lunchtime gripe bunch (Thea Lurie and Mickey Friedman); best wedding, no guests (Tom and Stephanie O'Neill); best wedding, with guests (Eric Hall and

Dawn Seymour); the hippest Motown oral archivist (Barbara Ellmann); best common sense in a continuing role (Virginia Stern); the most difficult assignment of his life (Joe McNally); and the two cutest dogs you'd ever like to meet (Anne Friedberg and Howard Rodman).

To the two most important people in my corner, a special thanks. Melanie Jackson's unswerving loyalty and faith, to say nothing of good judgment and refreshing honesty, have been far more bountiful than any writer has a right to expect from a close friend, much less an agent. Joe McElroy has read, questioned, suggested, cajoled, encouraged, even bent spoons, and, perhaps more than the author, never doubted that this project was worth doing and would be completed.

Contents

CONTENTS

PART II

"... I must remind you of how I was made, a mass of protoplasm like a kind of pulpy dumpling with a nucleus in the middle. Now I'm not just trying to make myself sound interesting, but I must say that in that nucleus I led a very intense life."

—"Mitosis,"
Italo Calvino

"Biologists work very close to the frontier between bewilderment and understanding. Biology is complex, messy and richly various, like real life; it travels faster nowadays than physics or chemistry (which is just as well, because it has so much further to go), and it travels nearer to the ground. It should therefore give us a specially direct and immediate insight into science in the making."

—Pluto's Republic,
Peter Medawar

DRAMATIS PERSONAE

The Gilbert Group (Harvard/Biogen)

Walter Gilbert, *lab chief*
Argiris Efstratiadis
Lydia Villa-Komaroff
Forrest Fuller
Stephanie Broome

Greg Sutcliffe
Allan Maxam
Karen Talmadge
Debra Peattie
Peter Lomedico

The Rutter-Goodman Group (UCSF/Eli Lilly)

William Rutter, *lab chief*
Raymond Pictet
John Chirgwin
Graeme Bell

Howard Goodman, *lab chief*
Axel Ullrich
John Shine
Peter Seeburg
Barbara Cordell

The UCSF–City of Hope Group (Genentech/Eli Lilly)

CITY OF HOPE	GENENTECH	UCSF
Arthur Riggs, *lab chief*	Robert Swanson	Herb Boyer, *lab chief*
Keiichi Itakura, *lab chief*	Dennis Kleid	Herb Heyneker
Roberto Crea	David Goeddel	Francisco Bolivar
	Daniel Yansura	Raymond Rodriguez
	Michael Ross	Mary Betlach
	Ronald Wetzel	

PART I

One

Going for
the Gene

. . . .

It was not a seminal meeting, at least not in the usual sense of the word. No shattering new natural laws were proposed when a group of biologists and biochemists gathered in Indianapolis in the spring of 1976, no cherished long-standing dogmas were challenged. When this particular meeting began on May 24, the scientists did what they usually do at these affairs. They reported, one by one, on the latest results coming out of their labs, and each report provided a kind of pointillist dot of information; if you stood back and took it all in, you could begin to connect the dots, discern a pattern. And when the meeting concluded one day later, some of the scientists went home convinced that they could achieve something that had never been done before in the history of biology. With the eager and clumsy fingers of human curiosity, they planned to reach inside a living cell and tinker with its innermost mechanisms, manipulating the machinery of heredity in a way inconceivable even five years earlier.

Members of the general public had little reason to note the proceedings

. . . . . .

in Indianapolis. The sponsor of the event was Eli Lilly & Co., the large pharmaceutical company, and the ostensible subject of the meeting was insulin, not a terribly riveting topic unless you happened to be diabetic and needed insulin to survive. But after this meeting, the insulin molecule was destined to play a role with repercussions far beyond the area of diabetes. Biology—and the pharmaceutical commerce growing out of biology—was on the verge of a tremendous revolution. Biologists could see it in the distance, although they weren't quite sure how to get there. This revolution depended upon the laboratory demonstration that genetic material could be manipulated to produce useful new substances, and this manipulation depended upon a controversial new area of research known as recombinant DNA or, more popularly, "gene-splicing."

And Lilly was an appropriate place to discuss this coming revolution. Exactly one hundred years earlier, in May 1876, Lilly had entered the drug business with such stalwart products as Sea Wrack, Squaw Vine, Scullcup, and Wahoo. Since that time, it had saved and sedated lives with the Salk polio vaccine and the barbituate Seconal; its seasickness pills went with American GIs on D-Day, and its painkiller Darvon went to the moon with the astronauts. If major new drugs were to emerge from genetic engineering, as some observers predicted, Lilly wanted to get in on the ground floor. Especially if one of those drugs, as many suggested, would be genetically engineered insulin. That's what the scientists visiting Lilly came to discuss.

By its centennial year, Eli Lilly had long enjoyed its position as the dominant force in the U.S. insulin market; revenues from its several insulins accounted for $160 million annually and represented an estimated 80 to 85 percent of all American sales of insulin. But there was a chart making the rounds at Lilly that showed an ominous convergence of lines. One line represented the projected supply of beef and pork pancreases, from which all therapeutic insulins were derived to supply many of the nation's 1.5 million insulin-dependent diabetics; this line was beginning to lose its trajectory and level off as it headed into the 1980s. The other line showed the rate at which the diabetic population in the United States was expanding, and it made a steady, inexorable climb, a yearly increase of about 5 percent, into the 1980s. At some uncertain point in the future, somewhere off the graph, those two lines of medical supply and mortal demand would intersect, with what Lilly foresaw as dire implications for diabetics. "It was clear that if something didn't change, a shortage would develop," says Irving S. Johnson, the Lilly vice-president who organized the May meeting. "One could argue about when that would occur, whether it was going to be 1980 or 1990 or the year 2000. But the progression of lines—in terms of availability and requirement—those lines were clearly going to cross."

Lilly had plenty of economic motivation, and a good deal of proprietary pride, too, to prompt it to find a solution. It had already established an in-house group of scientists to study the possibilities of genetic engineering. But tinkering with the delicate genetic mechanisms of a cell seemed more like alchemy than biochemistry at that point. There were very few scientists who had mastered the freshly minted techniques that allowed one to manipulate the basic genetic material, DNA, and it was not easy to find, let alone hire, experts.

So Lilly did the next best thing: it organized one of its periodic insulin symposiums around the theme of making insulin by genetic engineering. And out went the invitations, not only to career researchers in the insulin field, but to prominent molecular biologists, who were leading the way in discovering and applying the new gene-splicing techniques. This commingling of academic and industrial interests may seem unusual, but Lilly had cultivated such ties since 1922, when the company originally reached a licensing agreement with University of Toronto researchers to produce insulin for diabetics; the May 1976 meeting was in fact Lilly's sixteenth such symposium.

And so scientists from all over the country converged on Indianapolis for the two-day meeting. They came from academic hotbeds on the East and West coasts where genetic engineering was in the process of being invented, and they came from the leading laboratories in insulin research. From the University of Chicago came Donald F. Steiner and his colleague Shu Jin Chan, two of the most respected researchers in the insulin field. From the University of Texas, Grady Saunders and his young assistant Peter Lomedico, also major players in insulin research. From the University of California–San Francisco, certainly among the pacesetters in recombinant DNA research, came William J. Rutter and Howard M. Goodman. And from Harvard University's Bio Labs, a late addition to the roster, came the noted molecular biologist Walter Gilbert, accompanied by an intense and talented young Greek scientist named Argiris Efstratiadis.

Efstratiadis took perverse pride in the fact that some people regarded his appearance as the stereotype of the mad scientist, with his wide eyes and wild hair and somewhat agitated manner. But it was probably the presence of

Gilbert, who enjoyed an enormous reputation for intellectual cunning and scientific success, that told the other participants in Indianapolis that this might be a meeting with more than modest implications. To characterize it in a more colloquial way, as scientists themselves often do, they realized that insulin had become a "sexy project."

When scientific meetings work well, individual research reports begin to fit together like fragments of a larger picture. Taken together, the research reports of the insulin people and the molecular biologists at the Lilly meeting became a mosaic of possibility. The scientists realized they could isolate the genetic information for insulin from the welter of genes in a mammalian cell. It was possible, too, to duplicate—or clone—the insulin gene by inserting it into bacteria. Finally, to the most ambitious and visionary, it even seemed feasible to get bacteria to make insulin—*human* insulin—as if each creature were a single-celled subsidiary of Lilly. The technical means were suddenly available, everyone realized, to go after the insulin gene. That was how biologists typically described it: going for the gene.

And as often happens in science, where the revolutions are inevitably silent and the lightning always comes before the thunder, it could be said that the age of genetic engineering began noiselessly in Indianapolis. Not in a laboratory, not in a test tube, but in the place where all scientific revolutions start to take shape, in the minds of a few scientists who weigh feasibility against failure, who have that special feel for the avoirdupois of doubt, and who then decide they have enough technology and enough intuitive know-how to press forward.

At this stage of the twentieth century, when almost everyone knows an insulin-dependent diabetic among family, friends, or colleagues, it is difficult to comprehend just how dramatic and life-restoring was the discovery of insulin in 1921–22 by Frederick Banting and Charles Best at the University of Toronto. At that time, severe cases of diabetes were invariably fatal; patients, sometimes maintained on diets as limited as 450 calories a day, were wraithlike shadows of their former selves. Those who witnessed the initial treatments were, writes Michael Bliss, author of the engrossing *The Discovery of Insulin,* "present at the closest approach to the resurrection of the body that our secular society can achieve."

Lack of insulin is the principal characteristic of diabetes. This disease has bedeviled mankind for many millennia. The ancient Egyptians did not understand the mechanics of the disease (just as we still remain uncertain about the causes), but they understood the symptoms only too well: a dramatic loss of weight, unquenchable thirst, persistent urination. It was the sugary content of the urine, giving rise to the name "sweetwater disease," that suggested the underlying physiological breakdown. Sugar built up in the blood and spilled over into the urine because it was not effectively metabolized. The missing metabolizing agent was insulin.

An estimated ten million Americans are diabetics, and about 15 percent of those suffer from Type I diabetes, the more severe form which usually develops during childhood and requires daily injections of insulin to sustain life. Type II diabetes, which normally develops in middle-aged adults, can usually be controlled by a prudent diet and exercise. Both forms of the disease, however, extort a terrible physiological toll on the body. Diabetes frequently leads to such complications as atherosclerosis, heart disease, kidney failure, blindness, and loss of limbs. Quietly, without fanfare, it kills about thirty-five thousand Americans each year, making it the seventh leading cause of death in this country. The disease is a complicating factor in another ninety-five thousand deaths.

Insulin, it turns out, functions as an indispensable middleman in metabolism. When we eat such carbohydrate foods as bread, vegetables, or fruit, a simple sugar called glucose is usually the end product of digestion, and this sugar provides energy to each living cell; insulin, in its turn, functions as the doorman to these cells, controlling the access of glucose molecules and other food sources such as protein and fat across the cell membrane and into each cell's interior. With insulin, metabolism is a finely tuned feedback mechanism. Without it, only a trickle of fuel leaks into the cells, hardly enough to stoke the great human metabolic furnace.

From 1922 to the time of the 1976 Lilly symposium, the only insulin available for diabetics came from the pancreases of slaughtered pigs and cows. Like all proteins, these animal insulins were formed of chains of subunits called amino acids, and neither the pork nor the beef version was identical to human insulin; pig insulin varied by one of fifty-one amino acids, cow insulin by three. The preparations weren't always totally pure, and some diabetics responded with allergic skin reactions, particularly to the bovine insulin injections. Clinicians also worried about the long-term effects of using animal proteins in the human body. Did they contribute to the complications—such as blindness and circulatory problems—that plagued elder diabetics? And then there was the problem of the chart developed by Lilly, of the two converging lines of supply and demand. In the 1970s, people were eating less meat, slaughterhouses were processing fewer cattle and swine, and consequently supplies of animal pancreases were beginning to level off. At the same time, the number of diabetics continued to climb.

Here, suddenly, was a situation where medical need could be married to the emerging biological technology. Just when long-term supplies of insulin could no longer be taken for granted, the promise of genetic engineering was rapidly maturing; it was reasonable, even prudent, to explore the possible production of human insulin. Here was a product that might eliminate the problem of allergic reactions. Here, from a manufacturer's point of view, was a way to control supply of the material. And here was a very seductive proposition for that intellectual recluse the research scientist.

Science, and molecular biology in particular, had become so reductionist, so focused on the barely visible integers of life processes, that biologists sometimes privately questioned the social value of their research. As Dennis Kleid, a West Coast biologist, later put it, "Usually scientists do experiments that are useful for other scientists. They make a discovery, publish it, and other scientists can use that information to do something that they're interested in. But this would be one of the few times a scientist really had an opportunity to do something for the general public." It was an opportunity many could not refuse.

And it was hard to ignore the glorious pedigree of those who studied insulin. Precisely because of its role in disease, insulin has always attracted more than its share of first-rank scientific attention, and indeed has been the silent partner to several Nobel Prizes—to Frederick Banting and J.J.R. Macleod in 1923 for its discovery, and to Frederick Sanger in 1958 for painstakingly determining its complicated structure, formed of two chains of amino acids; within eighteen months of the Lilly meeting, Rosalyn Yalow would receive a 1977 Nobel for discovering a technique, using insulin, to detect trifling amounts of any protein with a radioactive assay. It would be unfair to suggest that scientists flocked to insulin work in search of a Nobel Prize, but they were certainly aware the insulin molecule often accompanied groundbreaking, award-winning research.

As often happens in science, groundbreaking research becomes a matter of picking the right problem at the right time. In May 1976, it was the molecular biologists in particular who had truly sniffed the air and sensed the possibilities, and although scientists are viewed by our culture at large as dispassionate, cerebral sorts, they smell these moments as instinctively and fiercely as hounds. The molecular biologists, it became apparent, had picked up the scent and would lead the chase.

Who exactly were the molecular biologists? Molecular biologists devoted their attention to the smallest molecules of life: deoxyribonucleic acid (DNA), ribonucleic acid (RNA), proteins, enzymes. The activity of these molecules formed a biological vocabulary of transcriptions, translations, replications, catalyses, syntheses, and degradations; yet all that arch, latinate jargon added up to the supple, homeostatic throb of life inside each cell, as well as the perpetuation of life from one generation to the next through the mechanisms of heredity. Like the astronomers who inferred galaxies and quasars from invisible radio waves, molecular biologists inferred from the almost invisible life processes of bacteria and viruses the constellation of biochemical activities that applied to all living things, to human beings as well as to microscopic bugs.

They also had the reputation of being opportunistic, of trespassing onto other scientists' intellectual turf in search of answers to these central questions. And that was about to happen again in Indianapolis. They were extremely bright, extravagantly confident, and incredibly dedicated to their work. As is typical of revolutions in other spheres, it was the younger scientists in particular who understood the tremendous power of the new technology and embraced it with unconflicted fervor. Recombinant DNA promised a whole new view of the life processes, and in the spring of 1976 part of the seminal excitement running through the biological sciences reflected a kind of competitive anxiety. Richard Scheller, a young West Coast biologist at the time, recalls, "There was the thought that there were some real key questions and there were a few people who were going to answer them, and if you weren't one of them, then you were going to be left out."

One of those key questions was figuring out how a gene worked, and perhaps even getting a human gene to turn on and function in a bacterial cell. What was happening at the Lilly meeting, without anyone quite realizing it at the time, was that intellectual curiosity and applied possibilities were beginning to converge on this very question. And scientists of the first rank were converging on the insulin molecule in search of answers.

Nothing made that more apparent than the appearance of the delegation from Harvard. "That was the first time I knew that Wally Gilbert was interested as well," says William Rutter. Though the comment came eight years after the fact, his voice was still fresh with cynical surprise.

Gilbert headed a large and talented research group at the Harvard Biological Laboratories. Trained as a physicist, he had switched over to biology full-time in 1964, and his distinguished work had earned him the chair of American Cancer Society Professor at Harvard. He was widely regarded as one of the most enterprising and incisive biologists of his generation. A man of medium height, with curly dark hair, pale skin, and black functional horn-rims, he maintained unusually diverse research interests and addressed a broader range of scientific questions than most biologists. He was well known for solving difficult problems by patiently but aggressively attacking them. With his appetite for tough problems, Gilbert had left tracks all through molecular biology during his fifteen years in the field. Now those tracks led to Indianapolis and insulin, an area of research to which he had previously devoted not a single moment of his distinguished career.

Rutter headed a group from the University of California–San Francisco. Since becoming chairman of the biochemistry department at UCSF in 1969, Rutter had helped transform the somewhat moribund department into one of the leading recombinant DNA research groups in the country. With him was Howard Goodman, a molecular biologist whose UCSF laboratory had quietly but quickly moved to the forefront in the gene-splicing field. Reticent, intellectually intense, with a full beard and broad dome of forehead that gave him the unlikely look of a shaman, Goodman projected an air of tight-lipped religious asceticism. Rutter had been studying the pancreas for years and had a talent for incorporating new techniques into his research as soon as they became available. A short time prior to the Lilly meeting, the recombinant DNA specialist Goodman and the pancreas expert Rutter had joined forces to go after the insulin gene.

This collaboration was significant because Goodman had previously teamed up with another UCSF scientist, Herbert Boyer, on many projects. Boyer did not attend the Lilly meeting, but it would be fair to say that his spirit inhabited the proceedings. He had been among the inventors of the gene-splicing technology, and now he was busy back in San Francisco establishing an alliance, considered highly unholy at the time, of business and academia. Only a month earlier, in April 1976, a company called Genentech, which had been partly founded by Boyer, incorporated in California and began negotiating research agreements with UC–San Francisco and the City of Hope National Medical Center outside Los Angeles. Genentech's aim was to make a small human protein called somatostatin as a test project, then proceed on to insulin. Unlike the Gilbert and Rutter-Goodman groups, which were angling for a glimpse at the insulin gene and how it worked, the Genentech workers had from the start different intentions. They wanted to synthesize the human insulin gene and make human insulin. Then they intended to sell it.

The competitiveness that later caught up with all three groups unofficially commenced around four o'clock on the afternoon of May 24, 1976. That is when William Chick, a junior faculty member at Harvard Medical School in Boston, took the podium at the Indianapolis symposium. Tall, slightly burly, with light brown hair riding the crest of his head, Chick had about him an air of cheerful disorganization. He ran a lab at the Joslin Research Laboratories at Brigham Hospital in Boston, one of the premier research institutes in the country for the study of diabetes; he knew he had some interesting results to report that day, but nothing in the title of the talk—"Monolayer Culture of Beta Cells"—suggested the stir it would create among the molecular biologists in the audience.

The key to Chick's talk involved a very special and very rare type of tumor in certain pancreatic cells. The pancreas, a long slender gland tucked below and behind the stomach, possesses two different aggregates of tissues, with two radically different biological agendas. The majority of the pancreatic cells, about 99 percent, concern themselves with digestion; they manufacture and secrete the powerful enzymes that drip into the intestines and break down the food we eat into usable proteins. The other 1 percent of the cells produce hormones, principally insulin; these powerful chemical regulators, secreted not into the digestive tract but into the bloodstream, affect metabolism. To complicate matters further, these cells are all mixed together, looking to the unschooled eye like marbled fat in a cut of beef.

The endocrine (or secreting) cells were discovered in 1869 by the German medical student Paul Langerhans, and because they seemed to float like isolated entities within the ocean of pancreatic tissue, they became known as the islets of Langerhans. There are some five hundred thousand islets per human pancreas, and within each islet are arrayed some 1000 to 1500 beta cells. In the beta cells, and in these cells alone, insulin is manufactured.

Islets are rare. They are difficult to isolate, almost impossible to obtain in quantity, a headache to study. But Bill Chick and Shields Warren, a colleague at New England Deaconess Hospital, had almost by accident stumbled upon a way to get massive amounts of them. Warren had been bombarding laboratory rats with huge doses of radiation. The X-rays caused damage to the genetic wiring of cells in the rat pancreas, and that damage somehow flipped a cellular switch and provoked the cells to multiply rapidly—to become a tumor, in short. When a cell becomes tumorous, its internal cellular economy, normally measured and modest, suddenly embarks on wild and unregulated production. If it is the business of the particular cell to release insulin, for example, it will likely release insulin at a frenetic pace, around the clock, regardless of demand.

All this became apparent in 1975 when Chick ran tests on one of the tumorous rats. The glucose content of the blood was so low that Chick thought his testing machine was broken. He gave it a good whack and ran his tests again. "And it still came up with this very low number," he recalls. What he was discovering was an exceedingly rare tumor of the insulin-producing beta cells, referred to as an insulinoma. This insulinoma was churning out incredible amounts of insulin. The proliferating tumor cells and the beta cells were one and the same.

Chick and his colleagues set up a veritable tumor farm to harvest this precious tissue. They discovered that the tumors could be surgically excised and then transplanted into sibling rats, a special lab strain known as New England Deaconess rats. They further discovered that when they inserted bits of tumor into the space between the shoulder blades of the rats, the mutant cells actually took root and grew into benign nodules at the point of injection. Slowly but surely, a population of disabled hunchbacked rats were cultivated, with transplanted pancreatic tumors causing bulges in, of all places, their necks. Like plucking a tiny gray potato out of some furry plot of ground, the nodules could be harvested for scientific study.

Chick had barely stopped describing his work before Argiris Efstratiadis excitedly pulled Wally Gilbert aside. This insulinoma, they both realized, would be an extraordinarily rich source of raw material for the insulin experiment they were even now planning back at Harvard. They agreed to approach Chick later on and offer to make him a collaborator in the experiment to get access to the tumors.

The California researchers wasted no time at all. As soon as Chick finished his talk, William Rutter and Howard Goodman of the UC–San

Francisco group held a similarly excited conversation; their two labs had already embarked on virtually the same experiment to isolate the insulin gene. "Nobody was aware of William Chick's insulinoma prior to that meeting," recalls Raymond Pictet, a member of the UCSF team. "Goodman, Bill Rutter, myself—we were all surprised at the same time. And all three of us rushed to Bill Chick." Pictet, who knew Chick personally, expressed interest in using some of the insulinoma. Chick's reply, Pictet recalls, was "No problem."

Problems, as even Chick admits, began to develop later that day. Shortly after the main speech at Monday evening's dinner, the Harvard biologists Arg (rhymes with "sarge") Efstratiadis and Wally Gilbert made their approach to William Chick. Although he worked just across the Charles River from Cambridge and was affiliated with Harvard Medical School, Chick had never met Gilbert, and seemed barely to have heard of him. Efstratiadis explained to Chick that the Harvard group was interested in trying to isolate the insulin gene, and that the pancreatic tumors from his rats could prove immensely valuable; if Chick was willing to provide them with this very rare starting material, they would be pleased to have him as a collaborator. Chick, noncommittal but intrigued, said the idea sounded interesting. They would have to discuss it further.

It was becoming clear to Chick that this tumor was a hot item. Previously, a University of Chicago group headed by Donald Steiner had asked for the tumor to run preliminary tests on the tissue; Steiner, too, had thoughts of cloning the insulin gene. Now two high-powered research groups, the Harvard team and the California team, wanted his islet cells. But this sudden popularity, and the rivalry it reflected, did not go unnoticed. The possibility of a race against Wally Gilbert for the insulin gene was immediately brought to Bill Rutter's attention. Peter Lomedico, a young researcher from the University of Texas, recalls talking to Rutter during a break at the Lilly meeting. "Even though I was slightly removed from Gilbert's field I was still in awe of him," Lomedico recalls. "Everyone was. So I posed that to Rutter as being a problem."

Rutter didn't seem to think so. At forty-seven, the handsome Rutter had a wise and wary face with close-cropped sandy hair and sun-seasoned wrinkles around the eyes; he possessed a soft, deliberate, but unmistakably resolute voice that suggested he did not back down from challenges. "No," Lomedico remembers him saying. "There's nothing better than a good battle."

Rutter had every reason to be optimistic. When he left Indianapolis at the end of the Lilly meeting, he says, he was of the belief that William Chick had "virtually agreed" to provide the insulinoma to the UCSF group. It would give the California group a tremendous, perhaps insuperable, edge. They would have a very good shot at getting the rat insulin gene. And it was believed that once you got the gene from one species, it would be easy to track down other genes in the insulin family, including the human gene. In that sense, Chick's cells represented the biological gateway to all insulin genes. Now all they had to do was pick the lock.

T w o

"The Most Consequential Event of 1975"

The scientific value of William Chick's hunchbacked rats became clear to the biologists gathered in Indianapolis the following afternoon, when a researcher of unusually flamboyant demeanor addressed the group. He didn't look like a scientist and in fact had never wanted to be a scientist.

Born on the Greek island of Lesbos in 1941, Argiris Efstratiadis had acted in amateur theatrical productions in school and wanted to be a writer ("I still consider myself a poet who doesn't write poetry," he says of himself now). At his father's insistence, he instead attended medical school in Greece and reluctantly became an M.D. in 1966. After two years in the army, Efstratiadis began to work in a biochemistry lab in Athens and, through the efforts of a fellow Greek scientist, was invited to apply to Harvard University, where an informal "Greek mafia" of biologists received training in the laboratory of Fotis Kafatos, a fellow countryman from Crete. In 1971, at the rather advanced age of thirty, Efstratiadis commenced graduate studies at the Harvard Bio Labs. For the time being, the nucleic acids would be his muses.

Efstratiadis, as always, made a singular impression on his audience. Exaggerated in manner, his voice a deep, resonant instrument, expertly played, thick with Hellas and full of pregnant pauses, he sounded more like a thespian than a scientist. He was known for spicing up talks with jokes and humorous asides, and for dressing only in black (a color scheme that often extended to his dark-lenses glasses, the worry beads he habitually shifted in his hands, and the shadow of an unshaven face). He brought an obsessive, almost religious fervor to his work. Among the ranks of his more orthodox brethren, he roamed about like some manic Zorba. "I thrrriiivvvve on chaos," he would boast, turning each letter into a syllable.

Science does not disown theatricality, as long as the actor can deliver good scientific lines, and in that, Efstratiadis succeeded superbly. He opened his Lilly talk with a typically shaggy and arcane tale that involved the origin of life, the star Andromeda, and a play on words in four different languages, and there were the usual exchanged looks and raised eyebrows in the audience, but then he began to talk about his work. In 1975 and 1976, the Greek scientist formed part of a Harvard team which had fired one of the louder opening volleys in the recombinant DNA revolution. The implications of their work seemed so significant that pundit George Will concluded that "a century from now historians will identify the achievement of the Harvard scientists as the most consequential event of 1975." What they achieved, in effect, was a way to make a copy of a gene.

Working in the Harvard laboratory of Fotis Kafatos, and collaborating with Tom Maniatis and Allan Maxam (the latter of Walter Gilbert's lab), Efstratiadis had devised a way to make a genetic copy of a mammalian gene. The gene they worked on was not insulin but globin, a protein building block used to make red blood cells. That feat alone would have been highly significant, but more important, the procedure provided a biochemical skeleton key—a more or less universal method—for gaining access to any gene of interest, insulin included. Like illuminating an entire landscape with a penlight, it was one of those typical experiments in molecular biology that reductively focused on rabbit blood and ultimately had vast implications for studying how genes work in all organisms.

The key to the technique involves the unique interaction between DNA and RNA in the cells of all living things. Deoxyribonucleic acid—DNA for short—is the molecule that functions as the precious genetic blueprint for life, a biochemical inheritance locked and hidden, as in a safe, within the nucleus of every cell of the higher organisms known as eukaryotes. (Bacteria and algae, which have no nucleus and in which the DNA floats about naked in the cytoplasm, are called prokaryotes.)

The discovery of DNA's unique structure is known to all who have read James Watson's dramatic first-person account, *The Double Helix*. It is a tightly coiled double-stranded helix that carries coded instructions for the manufacture of the proteins necessary for living functions. If we think of DNA as a twisted rope ladder, as is often suggested, then these instructions are chemically encoded in the rungs of the ladder. All of life's proteins, all the constituent bricks, from the enzymes that digest our food to the proteins that color our eyes blue, can be spelled out by the astonishingly economical alphabet of biology, which consists of four letters: A, T, G, and C. They stand for four different chemical subunits, called "bases," that make up half a rung—adenine, thymine, guanine, and cytosine. These bases come together in pairs to form the rungs of the twisted ladder that is the DNA molecule. To get an idea of the incredible compression at work here, the reader is invited to consider for a moment a freckle on his or her arm. Each freckle may contain several thousand cells; each and every one of those cells contains an amount of supercoiled DNA which, if unraveled, would be about six feet long, yet be 550 times finer than the silk spun by a garden spider.

The bases form part of a nucleotide, which can be thought of as an L-shaped subunit of the ladder consisting of sugar, phosphate, and one of the four bases. These bases must join in the middle—adenine to thymine only, guanine to cytosine only—to connect the two outer spines, or backbones. These complementary chemical subunits, joined together in the middle as rungs, are called base-pairs. By biological convention, the length of any piece of DNA is measured not in millimeters but in the number of these base-pairs, and the obstacles confronted by the Indianapolis researchers can best be summarized within this context: the total amount of DNA in each rat cell was about 2.5 billion base-pairs, and they were looking for a small section of that—the insulin gene, measuring about four hundred base-pairs—without any clues.

When a carpenter builds a ladder, he would never put a joint in the middle of each rung, because that would be a structurally weak point. Nature, however, *does* put a joint in the middle of the rung, and for very good biological reasons. The base-pairs are held together by marvelously supple chemical bonds, which give the two strands of DNA the ability to unzip and reattach, depending on biological contingencies. In just such fashion, DNA's entire genetic manifest can separate, duplicate itself, and pass the information on to daughter cells; or a small section can be copied and communicated to the rest of the cell. This small section of DNA, which contains the coded instruction for a particular protein, is what we call a gene.

The first step in this process of communication, called transcription, occurs when the double helix of DNA uncoils and makes a kind of bowlegged separation—not all down its length, but for a discrete segment of base-pairs which roughly corresponds to the gene. With the help of special copying enzymes known as RNA polymerases, a molecule of RNA (or ribonucleic acid) begins to form adjacent to one of the now-separated strands of DNA. RNA uses the very same alphabet as DNA, with one important exception—in place of thymine, RNA uses uracil, so there is a U instead of a T. Since base-pairing is so specific (A to T, C to G), the emerging RNA contains a sequence exactly opposite the first DNA strand and therefore an exact replica of the second DNA strand. And because this RNA molecule, like a diplomatic courier, can pass freely beyond the boundaries of the nucleus into the cell proper, it carries the gene's biochemical instructions out to structures in the cell called ribosomes, where the code is "translated" into proteins. Hence this species of RNA is known as messenger RNA. Molecular biologists often speak simply of "the message."

Argiris Efstratiadis and his colleagues exploited the "message" quality of messenger RNA en route to their findings. In experiments over a two-year period, they isolated the message for globin from the red blood cells of rabbits. Red blood cells were a good place to look for this message: 90 percent of the cells' messenger RNA—an extraordinarily high amount—was earmarked solely for the production of globin.

Then, using an enzyme with the suggestive name "reverse transcriptase," the Harvard group in effect reversed the natural flow of information. They made a "backward" copy of the gene, from RNA back into DNA. It was a bit of biological wizardry not unlike going into the darkroom with a photographic print and, through a series of chemical manipulations, emerging with the original negative. Once they had this single strand of DNA, it required several more enzymatic manipulations to construct its complementary, second strand.

The end result, for all intents and purposes, was a recreated "gene"—a full-length, double-stranded piece of DNA that theoretically corresponded to the gene in rabbits that made globin. These putative genes, copied backward from message RNA, were called "complementary DNAs" (or simply cDNAs). The only way to find out if they were indeed functional in living cells was to splice this DNA into bacteria and see if they would start making globin. No one, however, had managed to get bacteria to read a mammalian gene and start making (or "expressing") the protein coded by the gene.

Efstratiadis's artful crafting of a gene for globin whetted Walter Gilbert's interest in the idea of gene-splicing. As a graduate student, Efstratiadis had taken courses from Gilbert, and "because I was bugging him all the time with questions, we started liking each other, I guess." As Efstratiadis achieved each step in recreating the globin gene, he would march upstairs to Gilbert's third-floor lab and wave the data excitedly in front of the older scientist. They both entertained thoughts of inserting this gene into bacteria to see if microbes could read the coded information. But globin did not have the medical impact of an insulin, nor for that matter its name recognition. As Gilbert recalls, "Somewhere very early on we switched to the idea of insulin, with the argument that it was a smaller protein, and actually a more useful one, to make in bacteria."

By Efstratiadis's method (one of several cDNA protocols simultaneously developed around that time), the essential starting material was lots of the right messenger RNA. That is why Bill Chick's work created such a stir. Most cells contain vanishingly small quantities of any given messenger RNA, and the molecules last only a few moments in the cell before being broken down; Chick's hunchbacked rats seemed to represent exceedingly rare cases where the tumors contained many more islets, and thus many more insulin "messages," than the average rat pancreas. It was, in lab parlance, an "enriched" source of RNA for insulin. It was the first step toward making an insulin cDNA. Whoever had access to it would enjoy a considerable advantage.

But what could they do with a gene? A gene by itself was like a light bulb still in its cardboard wrapper; until you very deliberately screw it into the proper genetic socket, there is no way to turn it on. That was the problem addressed by Wally Gilbert, who followed the voluble Efstratiadis with a few remarks. Gilbert, who had recently turned forty-four, had great scientific presence; when he spoke, the information tumbled out in a low, authoritative stream of facts. He wasn't listed on the speaking program and hadn't even been on the original list of invitees to the symposium until Efstratiadis mentioned his interest in insulin to the organizers. But he proceeded to describe a devilishly clever concept that had been experimentally pursued in the Harvard Bio Labs for the past six months by Gilbert and his graduate student Forrest Fuller, as well as by colleagues in the laboratory of Mark Ptashne.

They called it the "portable promoter," and it made genetic engineering sound potentially as easy as popping a cassette into a tape deck. In order to switch on a gene, you needed to line it up with an introductory piece of DNA that controls (or "promotes") the copying of the gene into RNA. This bit of control machinery is called the promoter; every gene in every organism comes with its own promoter. The Harvard biologists had figured out a way to excise the control machinery from a well-known bacterial gene and make it "portable"—that is, spliceable into other pieces of DNA useful in cloning. What's more, right next to this portable genetic control mechanism, which might be likened to the tape heads of a cassette recorder, they engineered the biochemical equivalent of a slot where a gene could be dropped in and "played" by the bacteria. The cassettes were the cDNAs, the product of Efstratiadis's methods; Gilbert and his colleagues were tinkering with the control mechanisms. No one at that point had succeeded in turning on a gene, but that was where things were headed. In a hurry.

Indeed, as Gilbert spoke, a grand scenario began to unfold, a progressive series of experiments that led inevitably toward the human gene for insulin. First you had to get your hands on source material to isolate some version of the insulin gene; that was the raw material provided by William Chick's insulinoma. Then you had to convert that into functional, double-stranded DNA; Efstratiadis's cDNA method produced "genes" in just such a way. Next, you had to *clone* that gene—slip it into bacteria and get it copied so that its A-T-G-C code could be properly analyzed. Finally, you had to *express* the gene—insert it so precisely into bacteria that the cells would be tricked into making insulin, perhaps in a model system like rat insulin first, and then with human insulin.

That was the journey, in four stages. Isolation of the message. Conversion into a "gene" (cDNA). Cloning. Expression. The first two steps were at least technically feasible; the last two were, as scientists like to say, on "the edge of doability." How long would it take? Optimists boasted that it could be done in two years; others predicted ten years of toil. It was one of those risky moments, in short, where scientists had to decide if the technology was ready, the project feasible, the problem soluble.

*

Revolutions are inevitably upsetting as well as exciting, and the Lilly meeting in a larger sense symbolized all the sociological turmoil that recombinant DNA brought to the biological sciences in the mid-1970s. Among scientists, it was cause for great public excitement and considerable private anguish, a schism reflected in the nomenclature of those who practiced recombinant DNA techniques (the "cloners") and those who didn't (the "noncloners"). By now the distinction has become meaningless, for nearly everyone is a cloner; at the time of the Lilly meeting, however, only a few had acquired the expertise. The cloners also quickly acquired a reputation for being opportunists, even among themselves. "If there's a field that's completely opportunistic," observes Brian McCarthy, formerly a faculty member at UC–San Francisco, "it's molecular biology. . . . You can't fault a guy for jumping onto a hot problem. That's what science is all about . . . it happens all the time. Some guy plodding away doesn't realize how important the problem is and doesn't know how to do it right, and somebody else jumps in and does it. Happens every day."

Recombinant DNA encouraged that kind of opportunism. It was as if the microscope had been reinvented. Everything had to be reexamined, and the molecular biologists roared like Huns through other scientists' turf, gathering key data and materials necessary for this new look at the world; to the traditionalists, it sometimes felt like the intellectual sacking of endocrinology, say, or hematology. "I was standing up there talking about my work," recalls Alan Permutt of Washington University, who presented data at the Lilly meeting, "and I was just so stupid because I was really surprised to look down and see Wally Gilbert taking notes about what I was saying, not realizing that he was getting ready to use everything I had to make some really great discovery." Permutt thought to himself, "Oh my God, here I've told them everything I know!" He returned home, he admits, a very disconsolate researcher.

The mid-1970s were thick with discomfited biologists. Donald Steiner had devoted a long and distinguished scientific career at the University of Chicago to insulin research; if long and distinguished research alone conferred special merit, he by rights should have been the scientist to lift the veil from the insulin gene and solve the mystery of its inner organization. The Lilly meeting made it clear, however, that a high-speed chase was about to get underway, and it was the upstart cloners, with their mastery of the new recombinant DNA techniques, who were in the driver's seat.

"It left us a little bit . . . in a peculiar position," Steiner would recall later, selecting his words with gentlemanly Midwestern discretion. "That what we had been working on for years suddenly became of interest to the molecular biologists, and they suddenly jumped in. And as often happens in science, we were working in a methodical way; we were taking one step at a time. These guys decided to leapfrog everything and just go for the gene. . . . When that happens in science, it's important because it's the way real progress occurs. You need people to do that. On the other hand," he added with a rueful laugh, "if you're one of the people working in the field, you always feel a little bit put upon by it."

The cloners, for their part, felt a little bit put upon at the Lilly meeting when Steiner raised a question that had by that time attracted considerable public interest as well. Following Efstratiadis's presentation, Steiner wanted to know if this kind of work, the splicing together of genes, needed to be regulated? Indeed, was it even safe?

Within a month, the National Institutes of Health would issue a long-awaited set of guidelines for recombinant DNA research, and only then could the scientists proceed with the advanced cloning experiments, such as the insulin project, that they wanted to do. But just as the Lilly discussion was, in the words of one observer, "spirited," so too was the debate raging around the research in places other than Indiana. Indeed, while Walter Gilbert and Arg Efstratiadis were explaining how one could isolate and clone the insulin gene to researchers in Indianapolis, colleagues back in Cambridge were mounting a campaign to assure that those types of experiments would never take place in the Harvard Bio Labs.

The dispute did not arise in a vacuum. The scientific community had become sensitized to (not to say traumatized by) the issue of recombinant DNA several years earlier, when the safety of that type of research had first been questioned. In a series of events that have been well documented elsewhere (see John Lear's *Recombinant DNA: The Untold Story*, Michael Rogers's *Biohazard*, or Nicholas Wade's *The Ultimate Experiment* for lively accounts), scientists brought these concerns to public attention in 1973, shortly after the pioneering gene-splicing experiments in California by Herbert Boyer and Stanley Cohen were reported that summer at sessions of the Gordon Conference in New Hampshire. Maxine Singer of the National Institutes of Health and Dieter Söll of Yale University then sent an open letter to the journal *Science*, detailing some of the scientific concerns about gene-splicing. Within a year, in a missive known ever after as "the Berg letter," a

22

committee headed by Stanford biologist Paul Berg called upon scientific colleagues around the world to desist voluntarily from certain types of recombinant DNA experiments until questions of safety could be addressed.

This extraordinary self-imposed moratorium, and the process of scientific self-examination it triggered, led in February 1975 to a historic three-day meeting in Asilomar, California, where the issues were hotly debated. Those discussions culminated, about a year later, in a series of guidelines drafted by the National Institutes of Health. Recombinant DNA had already attracted considerable press and public attention; what's more, it had become associated, among skeptics, with the notion of narrow-minded, technology-loving scientists who refused to look beyond their petri dishes to the broader social and ethical implications of their research.

Into these treacherous political waters, then, ventured the Harvard scientists. The Harvard dispute had begun innocently enough about a year before the Lilly meeting, in May 1975, when biology department professor Mark Ptashne applied to the National Cancer Institute for funding for a new laboratory on the fourth floor of the Bio Labs building. Three rooms were to be upgraded in the process; one of the three new labs would be a so-called P3 containment facility, a workplace with special physical-safety features to reduce the possibility of disease-causing organisms escaping (physical containment proceeded through four stages, from P1 to P4). In all, Ptashne sought $500,000 in federal funds to refurbish the labs, a total that would be matched by Harvard. Only later did it occur to researchers that recombinant DNA work could take place in the lab as well.

As required by law (but with unseemly discretion, given the public uproar), Harvard placed a legal notice in the *Boston Globe* and other local newspapers in mid-February 1976, announcing plans for its P3 lab and inviting the public to attend a February 26 meeting to discuss aspects of the plan. Few people saw the notice, of course; fewer still felt sufficiently alarmed to attend. "A couple of administrators and I were there to answer questions," Ptashne recalled in an interview with archivists from the Recombinant DNA History Collection at MIT, "but nobody came."

When certain faculty members got wind of the revised plans, however, reaction was swift and fierce. Ruth Hubbard, a spry and lively eminence in the biology department, and several other professors viewed the proposed experiments with alarm. The concern, as Hubbard recalls, involved the possible spread of genetically engineered organisms in a building where students and employees—not to mention a seemingly uneradicable colony of pharaoh ants—roamed at will. Hubbard expressed particular concern about the fact that recombinant DNA researchers used the weakened K-12 strain of *Escherichia coli*, a common bacterium related to *E. coli* strains inhabiting the human gut. "We were thinking at the time," Hubbard says, "that since this was unknown territory, *ideally* one would want to have an isolated laboratory where one could, in fact, look at what the health histories [were] of the people who worked there, and who [came] in contact with the people who work there."

Hubbard's opposition was not limited to questions of safety in recombinant DNA work. She had shifted interests from active research to the history of science, and part of her concern reflected deeper philosophical questions about the headlong rush into the future that typified modern biological research. Inextricably bound to this scientific juggernaut, in Ruth Hubbard's mind, was the engine of competition driving the enterprise. "It's not that there aren't enough questions to go around," she would remark later. "There are lots. But there are the *hot* questions. And this kind of competition to get the hot answers to the hot questions has been fostered very much in this whole molecular biology thing. I really feel that the image of the race for the double helix—I mean, that whole story and the paradigmatic character of that story—has done enormous damage, because [it has created] the notion that that's how the best science gets done, that's how the best science *must* be done, by racing for the prize. And that's how young scientists are trained." The advent of industrial applications was unlikely to change matters, either. "Both in science and in the commercial thing," she believed, "the second person to get there might as well stay home."

A number of junior faculty shared Hubbard's reservations, and a biology department meeting on April 14, 1976, brought the dispute out in the open. The powers in molecular biology at Harvard were intent on getting their laboratory; faculty members against the lab were equally adamant in their opposition. Proponents of the technique professed sympathy for the public debate, but their comments also betray great exasperation. Arg Efstratiadis would later complain of the fuss, "Americans have seen *too* many monster movies, okay? It's the classic theme of all horror movies that the monster is going to attack back. Here was the monster, being created in test tubes. It was something concrete."

Both proponents and opponents held meetings to plot strategy. Ptashne, along with Matthew Meselson and Daniel Branton, formed the core group arguing in favor of the lab, although many graduate students and postdocs shared their position. Opposing the laboratory were a number of Cambridge scientists identified with a public-interest group known as Science for the People; prominent among the critics were Hubbard, her husband George Wald (a Nobel laureate), biology department professor Carroll Williams, Richard Goldstein and Jonathan Beckwith of the Harvard Medical School, and several MIT faculty members, including Jonathan King and Ethan Signer. This cast, with occasional understudies, would appear again and again in the next few years, a small but vocal minority raising unpopular questions about the safety and wisdom of recombinant DNA work.

The dispute also represented a traumatic break from a tradition of political solidarity in the Cambridge scientific community. During the political storms of the 1960s and 1970s, most scientists had shared liberal sentiments on a variety of causes. There was general unity in opposing the Vietnam War. Ptashne had paid a sympathetic visit to North Vietnam, and Meselson had been a major figure curtailing biological warfare testing. Suddenly colleagues and collaborators found themselves painfully in opposition. Recombinant DNA seemed to represent a different sort of political fork in the road, and this time not everyone took the same path.

Indeed, the two camps moved farther apart with each encounter. A campus-wide meeting was scheduled for May 28. Hubbard circulated a memo protesting the lab and sent it to Henry Rosovsky, Harvard's dean of the Faculty of Arts and Sciences. "A final showdown," as Ptashne characterized it, was scheduled for May 27 in Rosovsky's office. Walter Gilbert, due back from the Lilly symposium by then, would attend the meeting. He too would need the P3 lab, now that he planned to do the insulin work.

Those who opposed the new lab felt outnumbered and exposed. Untenured faculty members who had been outspoken against the lab in April became mute by May. "Untenured people very quickly got the message ..." recalls Hubbard. "Given the degree of vehemence that even got into the debate between colleagues on the same level—that is, tenured faculty—it was obvious that it was not a healthy thing for an untenured person to do." The message, as Hubbard saw it, was "that the people with power in molecular biology were all on one side, and that if the untenured people wanted to go somewhere with their careers, this was a good subject to stay out of." Opposition began to dwindle—the result, according to some, of subtle intimidation.

It was finally decided at a meeting among critics to take the debate outside Harvard. It was a momentous decision—"the sin," as Hubbard later termed it, "for which there is no forgiveness." In mid-May, around eight o'clock in the morning, Ruth Hubbard picked up the telephone and made a call.

On the receiving end was a woman named Barbara Ackermann. Former mayor of Cambridge, now on the nine-member Cambridge City Council, once employed by the scientific publisher John Wiley & Sons, Ackermann had been active in protests against the Vietnam War, which is how Hubbard had originally made her acquaintance. On that May morning, Hubbard described Harvard's plans for the new laboratory and revealed her fears about recombinant DNA. She suggested that Ackermann might want to drop in on the May 28 meeting. Ackermann was intrigued by the idea. "But I remember saying to her on that morning, 'What difference does it make what *we* do?' " Ackermann later recalled. 'How can you stop them from doing it? Now that they've thought of it, how can you stop it?' And I also said, 'What difference does it make if we stop it in Cambridge?' " Those questions, reflections in a private conversation, were destined to be answered in an exceedingly public manner.

Unbeknownst to Ruth Hubbard, it would have been difficult to imagine a more propitious moment for making that fateful call; indeed, it was Arg Efstratiadis's worst fear come to life. The previous evening, Barbara Ackermann had stayed up to watch the late show on television. The film happened to be the biological science fiction thriller *The Andromeda Strain.* Her view of the issue, she would admit later in an interview with MIT archivists, "was slightly colored in my mind" by having seen the 1971 movie. It tells the story of a lethal organism that comes from outer space, kills humans almost instantly, and, despite the most stringent containment conditions imaginable (at least to fiction writers and Hollywood fantasists), even escapes from a top-security government laboratory.

Like the fictional germ in the movie, the scientific dispute over Harvard's new facility had escaped from the laboratory and was about to spread through the community. Barbara Ackermann made plans to attend the May 28 meeting.

T h r e e

In the Land of
Invisible Frontiers

The Harvard Biological Laboratories are in a U-shaped building dressed up in Cambridge's hoary and ubiquitous red brick, set back from the street on Divinity Avenue. Two huge bronze rhinoceroses, as massive and inexorable as scientific progress itself is presumed to be, flank the main entrance. In their immense bulk, their aggressively homely and stolid appearance, they make a blunt statement about the changing nature of biological inquiry. At the time the labs were built, in 1931, it was still possible to indulge in the illusion that biology was the province of rhinoceroses and redwoods, whole beasts and intact plants. Since that time, it has become increasingly clear that biology's prized game is much smaller, for all practical purposes unseeable. At the present scale of inquiry, even the most fundamental biological frontiers become invisible. Is a virus—little more than a piece of nucleic lint sheathed in protein—a living thing or not? No one can say for sure. This is the shadowy domain where molecular biologists work. Their most persistently difficult task, in many ways, is making the invisible visible.

Inside the Harvard Biological Labs, despite recent renovations, the air has a musty, palpably historic feel to it. In its old wooden labs, on its venerable blackboards, with its bulky Tinkertoy-like constructs of molecules, postwar generations of biologists attacked the molecular identity of the gene; it was a campaign that, with the identification and characterization of DNA as the likely genetic blueprint in 1953, became increasingly sophisticated, increasingly molecular, and increasingly revelatory in the subsequent generation. James D. Watson and Francis Crick described DNA's unique double-helix structure in a 1953 *Nature* article as brief as a thunderbolt, and that report rattled every window in the house of biology. Watson had ruled a third-floor wing of the Bio Labs since 1955, and it was to this very same lab that Walter Gilbert returned after the Indianapolis meeting.

Gilbert returned just in time for the May 27 "showdown" in Dean Henry Rosovsky's office over the P3 lab. It was a fractious meeting in which a divided faculty aired its positions with the kind of heat that spilled out into more public discussions. The laboratory dispute had become a campus-wide issue, and would be the subject of a debate at the Harvard Science Center the following day, with all the potential for attendant publicity. Unbeknownst to Rosovsky, Hubbard had already contacted Barbara Ackermann, and the dispute was headed for still wider public exposure.

The gathering in the dean's office was relatively small. Only two non-tenured biology faculty members attended. Rosovsky expressed surprise at this, Hubbard recalls, going so far as to say, "It's very strange. I have these letters of opposition from a lot of the junior faculty, and none of them is here." Ptashne, for his part, recalls contradicting one of the opponents of the lab in the discussion; in reaction, he says, the disgruntled researcher "threw down whatever he was holding and stormed out of the room." What impressed Walter Gilbert most, however, was the philosophical nature of the confrontation. After hearing arguments that recombinant DNA was a scientific unknown, possibly a dangerous unknown, Rosovsky finally turned to Hubbard and Wald and said: "What? You're telling me it's new? You're telling me you're afraid of the unknown? I thought the purpose of science was to explore the unknown." Gilbert recalls it as a "wonderful moment."

The movable debate shifted to the May 28 meeting in the Science Center. There, a parade of advocates, pro and con, argued the safety of the lab and the propriety of the techniques. But perhaps the most significant

28
. . . . . .

speech was one of the shortest. At the conclusion of the three-hour meeting, Barbara Ackermann stood up and identified herself as a Cambridge city councillor. The moment she began to speak, the issue was no longer an internecine biology department fray nor even a university matter that Harvard sages could decide in decorous privacy. As if to underscore the point, Ackermann had noticed a reporter in the audience taking notes of the proceeding. She learned that he was from the *Boston Phoenix,* an alternative weekly newspaper, and she recalls thinking, "We're going to be spending our summer on this."

Rosovsky deferred his decision on the P3 lab for several weeks, although at the time he felt inclined to approve the laboratory, in part because the majority of Harvard scientists seemed to feel the work was safe. But there was another compelling factor. "I always had a great deal of faith in Wally Gilbert himself," Rosovsky said later. The fact that Gilbert backed the lab, he recalls, carried a lot of weight.

Walter Gilbert inspired that kind of faith. Testaments to his scientific style and brilliance come in almost embarrassing abundance. Phillip Sharp, a researcher at MIT (and himself one of the bright lights of molecular biology), once remarked, "If I had a problem and had one person to take it to, I would take it to Wally. I consider him the brightest man I ever met." A former postdoctoral fellow, Peter Lomedico, recalls, "He could go to a meeting and not know anything about the subject, but he'd come back and give the impression that he knew more about the subject than the people who'd given the talk. And he did. He could sense the important points and focus on them better than the people who were working on the problem."

To his peers, Walter Gilbert possessed a most desirable array of scientific traits: great intellectual curiosity, rigorous scientific standards, a rich imagination, and a lust for understanding the way life worked in its most microscopic and, in many respects, most intricately beautiful manifestation. "I'm driven by a . . . just an intense curiosity," he would say, his very self-explanation riven by a kind of driving impatience. "I love new things, new ideas, new facts. It goes along with a tremendous impatience. It's very nice to have the old things, but a week or so later, they're all old hat, and you want something new." He could be arrogant, in the view of some, and aggressive. His was a pursuit of correctness, a kind of intellectual high ground, so focused and astute that temporal distractions—like a caviling colleague—intruded at some peril. As former graduate student Karen Talmadge once put it: "Nothing—*nothing*—daunts Wally."

Yet he also possessed the deeper, serene patience essential for long-haul work—a profoundly optimistic self-confidence. Solutions *could* be found. Gilbert had the knack of peering over the scientific horizon and spotting distant but attainable destinations—what's more, he had a knack for figuring out the steps necessary to get there. He likens his own career in molecular biology to a meandering path. "You wander quite freely from question to question," he says. "The basic attitude I always took was that the field was so *productive*, you see. You looked in front of you, and you stumbled over something interesting. You'd go and pursue that until you'd stumble over something else that would raise some other question, and you'd pursue that."

The pursuit began at an early age. There were shades of the master scientist even in the precocious young child. Born in Boston on the vernal equinox in 1932, Gilbert's lifelong ties to Cambridge came almost as a birthright. His father, Richard Gilbert, was one of America's early Keynesian economists and taught at Harvard University from 1924 to 1939; his mother, Emma, trained as a child psychologist at Radcliffe. The grandparents on both sides were Russian Jews, and there was a strong intellectual and anti-authoritarian heritage in the family roots. One grandfather was an anarchist printer. Once in America, Gilbert's maternal grandfather founded an anarchist commune in Stelton, New Jersey, where Gilbert's mother lived from 1913 until 1921 and where young Wally often spent time. "The community," said Emma Gilbert, "was built around the idea that in order to change the world, you had to educate your children." Walter Gilbert's education ranged from the nontraditional to the quintessentially establishment.

He grew up with an early taste for astronomy and chemistry. When Richard Gilbert went to Washington, D.C., in 1939 to join Harry Hopkins's brain trust and work in the Office of Price Administration during World War II, it represented one of the few times Wally strayed much beyond Cambridge. Diligent as a schoolboy, he ground his own mirrors for telescopes and, at the age of nine, joined his first science society—a graybeard group of rockhounds in Washington.

There is a story from that Washington period which, in retrospect, suggests a lot of the qualities that distinguished Gilbert's mature scientific career. He had set up a laboratory in the pantry of his parents' suburban Virginia home and one day was absorbed in an experiment that required dropping sulfuric acid on a piece of zinc in a flask, a reaction which produces hydrogen gas. When the experiment is done correctly, the hydrogen can be ignited and burns with a low steady flame. Twelve-year-old Wally lit a match and instead triggered the *bête noire* of boy scientists: an explosion. Flying glass from the broken flask slashed into his wrist, causing a nasty cut.

Gilbert bled profusely from the wound, but showed no apparent sign of upset. He calmly coached his mother in the proper application of a tourniquet, and then the two of them hopped in a car and rushed off to the hospital. Wally didn't have much to say on the ride over, but finally he turned to his mother as they neared the hospital and delivered his first comment on the episode, which was "I know what I did wrong." It is not untypical that when Gilbert is reminded of the incident forty years later, he remains fuzzy on the details of the ride to the hospital, but instantly remembers where the experiment went bad. "Didn't flush all the oxygen out," he recalls with a chuckle. This obsession with being right, accompanied by an almost acquisitive passion for facts and understanding, have been trademarks of his work. In the first three decades of his scientific career, Gilbert's name appeared on about 75 papers, by most standards a remarkably low number. "Yeah," agrees one former lab member. "But every one of them is dynamite."

Despite this bent for scientific rectitude, Gilbert could also thumb his nose at convention, particularly when he was bored. His parents sent him to a Quaker high school in Washington, but he did not feel obliged to attend classes all the time, and at one point the school principal finally felt moved to send Gilbert a note reminding him that Sidwell Friends was not a correspondence school. Not the type to loiter on street corners, though, he spent his senior year at the Library of Congress, doing research in preparation for the annual Westinghouse Science Talent Search for high school science students (one of which he won in 1949). He also found time to attempt to desegregate a city-wide high school radio club in Washington. It was a fellow member of that club, Celia Stone (the daughter of writer I. F. Stone), who became his wife. She would go on to become a poet, and he would go on to become not a biologist, but a physicist.

Gilbert elected to study chemistry and physics at Harvard. Almost every friend or mentor of his was on the fastest of fast tracks: headed for Nobel laureateship. His physics adviser at Harvard was Julian Schwinger (Nobel laureate in physics, 1965), and in 1954, after graduating from Harvard, he went to Cambridge, England, where he worked on his Ph.D. in theoretical physics with Abdus Salam (Nobel laureate in physics, 1979). It was in this other Cambridge, at a party thrown by some colleagues, that he happened to meet a young, skinny American named Jim Watson (Nobel laureate in physiology/medicine, 1962).

At that point, molecular biology was just getting off the ground; it was Watson and Crick's elucidation of DNA's structure, in fact, which had given the field its trademark molecule. It was the missing piece of a puzzle that began almost a century earlier. The nucleic acid DNA was first discovered in 1869, but no one knew its biological significance. In the 1860s, meanwhile, a monk named Gregor Mendel first observed the basic principle of heredity, that certain traits are passed on from parents to offspring. It was not until 1944, however, that Oswald Avery and his colleagues at Rockefeller University brought these two distant insights together by proving that DNA controlled the process of heredity. Genes, which passed along the traits first identified by Mendel, were simply segments of DNA. Learned men since Aristotle had wondered about the nature of heredity, but now, finally, there was a molecule that might provide some answers. How did DNA work? How did it chemically interact with proteins and other workhorse molecules of the cell? What qualities about its biochemistry allowed it to control life processes? An army of biologists set out to answer these questions.

Gilbert, however, was not one of them. By 1957, he was back at Harvard, but still a physicist. And a very good one, too. One of his fellow faculty members and close friends at that time, Sheldon Glashow (Nobel laureate in physics, 1979), recalls that "he was one of the most abstract of physicists. The work he did was very mathematical and very far removed from experiment. Maybe he was unsatisfied by that. In biology, theories and experiments are much closer together, theories he could justify with his own fingers."

In any event, Gilbert had begun to be "seduced" by the biological sciences. Glashow recalls watching Gilbert's "eyes glow" as he described experiments. It was a seduction skillfully orchestrated by James Watson, whose most singular talent may well be his gift for recognizing talent in others. At the time, Gilbert's wife, Celia, was Watson's first lab technician, and the physicist saw a lot of the biologist. Then, probably over dinner, one of those dinners the young bachelor was endlessly showing up for, Watson urged

Gilbert to drop by the biology labs. "There's something very exciting going on," he promised.

What they were doing, in the spring of 1960, was looking for something called the "messenger." It was a short-lived, transient molecule that seemed to ferry messages from the DNA to outposts in the cell where proteins were manufactured. Gilbert read about six papers to learn what was going on in the field, and then he showed up in Watson's lab. The French scientist François Gros was spending the year at Harvard, and Gilbert recalls, "I remember getting curious about it and going over and just talking to Jim and François, following them around as they were doing experiments one day, starting to do experiments with them the next."

This was Gilbert's first formal introduction to the nucleic acids, DNA and RNA. The species of RNA that carried these instructions out to the cell became known as messenger RNA (or mRNA). This was the molecule that Gilbert and his colleagues tracked down; by 1961, Gilbert's first paper on messenger RNA appeared in the British journal *Nature*, and his RNA work was subsequently hailed for the elegance of its experimental approach.

After that, there was no turning back. Gilbert began to spend more and more time—first evenings, then weekends, then holidays—in the Bio Labs. In 1964, the last of his eight papers in theoretical physics—"Broken Symmetries and Massless Particles"—was published. In 1964, he officially became a tenured biophysicist.

Around this time, Gilbert tackled the preeminent mystery of molecular biology of that era. The French geneticists François Jacob and Jacques Monod had proposed that unknown agents in the cell could somehow block information in certain genes, so that the genetic instructions for making proteins remained unread and unheeded. It was a theory of surpassing importance, for it might help explain why, for example, cells in the human pancreas produce the hormone insulin while brain cells do not, even though the nucleus of both cells contains *exactly* the same information in their DNA.

Gilbert, studying the common bacterium *Escherichia coli* with Benno Müller-Hill, broke this intriguing theoretical question down to a specific, testable hypothesis. In order to digest the milk-sugar lactose, *E. coli* manufactures an enzyme called beta-galactosidase. But a living cell, like any other energy-dependent entity, must use its resources efficiently, so the microbes manufacture this enzyme only when it is necessary, *only* when lactose is there to be digested. Otherwise, something efficiently turns off, or represses, the *lac* gene, as it became known. How did this gene switch on and off? Thus began the search for the repressor.

While Gilbert and Müller-Hill sought the *lac* repressor, a postdoctoral fellow in the same lab, Mark Ptashne, looked for an analogous repressor agent in a virus known as lambda phage. It has been suggested that Watson shrewdly pitted two exceedingly bright Harvard researchers against each other by giving them a conceptually similar problem to solve; and in this manipulated atmosphere of creative tension, a competition that had moments of both collegiality and intense rivalry developed between Ptashne and Gilbert. Both, ultimately, were successful, but only after several lean years of research marked by many false starts. Gilbert's young children, Kate and John, knowing only that their father was "looking for" something, would occasionally stand in the grassy courtyard of the Bio Lab and shout up in the direction of Room 371, "Did you find it yet, Daddy?"

Gilbert and Müller-Hill did find their repressor, eventually, as did Ptashne. Their discoveries revealed a vastly more complicated picture of gene control. The bacterial work, in addition, unintentionally paved the way for biotechnology's first breakthroughs. Gilbert and Müller-Hill showed that the *lac* repressor turned out to be a protein molecule, ten or twenty molecules per bacterium. Like all compounds, the repressor had its own gene, which was located cheek-to-jowl with the gene for beta-galactosidase. Each one of these repressor molecules functioned like a speck of dirt in highly sensitive machinery; it fit snugly into a prefatory segment of the *lac* gene known as the "operator" and clung to this portion of the bacterium's DNA. This molecular grit was sufficient to interfere with the *lac* gene's start-up signal; in effect, the repressor prevented the enzyme that copies DNA into a RNA "message" from getting close enough to the DNA to do its job. As a result, beta-galactosidase production never got underway.

When the bacteria grazed in a lactose-rich pasture of growth nutrients, however, an entirely different scenario unfolded. Those ten to twenty repressor molecules tended to get snarled up biochemically with lactose-related sugars. Entangled, as it were, in another reaction, they were no longer free to seize onto the operator DNA and thus no longer could interfere with the gene's start-up signal. "De-repressed," as the scientists called it, the bacteria proceeded to make beta-galactosidase.

In light of subsequent milestones in molecular biology, it is difficult now to convey the historic importance of this work. Gilbert and Ptashne, using two different systems, had shown the world for the first time how the hierarchical genes proposed by Jacob and Monod controlled other genes and how these genes could be switched on and off. What began as rather forlorn and lonely scientific paths opened up broad avenues of research that, twenty years later, are still among the most heavily traveled of the biological sciences: the control and regulation of gene expression. The field has proved to be immensely more complicated than the simple *lac* repressor system would suggest, but that was the system that got things started, and that was the system that attracted huge numbers of researchers throughout the world.

The discovery of the *lac* repressor earned Gilbert a place on biology's international stage, and his frequent journeys to meetings meant frequent absences from his lab group. It was upon his return from one of these many scientific forays that the eminent Dr. Gilbert discovered his lab coat had been dyed purple by a prankster. Unfazed, he wore it, and in that simple gesture personified the tone of a very special lab group. Frontier science and social informality seemed to nourish each other. "He did not pull rank or assert his authority in arbitrary sorts of ways," recalls one graduate student. "The authority would come strictly in terms of the science."

In 1976, Gilbert had about fifteen graduate students, postdoctoral fellows, and technicians in the laboratory. The principal area of study was regulatory mechanisms in bacterial genes, but the Gilbert lab was notoriously catholic in its pursuits. To be a graduate student in Gilbert's lab sometimes seemed a mixed blessing. Gilbert's scientific reputation was so towering that mere association with his laboratory virtually assured good connections in one's subsequent scientific career. On the other hand, the standards were severe and some students wondered whether they would ever be perceived as independent scientists with enterprising ideas of their own when the world suspected that many of their projects originated with the man some referred to as "the guru of the Bio Labs." As in similar relationships between highly reputed mentor and high-powered protégé, the students vied to win the lab leader's attention and approval while at the same time striving against dependence on just such approval. It was like a second adolescence, and declaring one's intellectual independence was essential to becoming a mature scientist.

To be a student in molecular biology in the mid-1970s was to gain admittance to an esoteric, high-spirited, driven fraternity that, by good fortune, was coming into maturity just as biology itself was about to enter a gloriously fertile period. "We *knew,*" says one, "that we were on the threshold of the vault." The atmosphere in Gilbert's lab reflected the personality of the leader in two important respects: the craving for information was immense, something akin to physical need, and the tone of the place was casual, almost fiercely informal. Graduate students would drift into the lab around noon or shortly after, and often work until the wee hours of the morning or on through into the next day. There would be mass excursions to the local Szechwan restaurant for meals, or sandwiches would be grabbed on the fly. DNA would be chopped and mixed and analyzed to the sounds of Joni Mitchell and the Rolling Stones. "At about three or four A.M.," recalls former graduate student Karen Talmadge, "the stereos would turn up very loud. People would be working madly."

The Gilbert group occupied part of the third floor of the Bio Labs, an area shared with Professor David Dressler. Each individual laboratory was furnished with two workstations—"benches," as they are called—and was shared by two students; so much effort and thought and emotion were expended in those little rooms that, in the words of one student, "it was so much more than work. It was *life!*" At the very end of the hallway, past the "Tea Room" on the left where group seminars were held, past the elevator and the darkroom, with a view looking back at all the corridor traffic, was Gilbert's office. Small, narrow, piled high with papers and scientific journals, it contained a desk to one side, an easy chair in the corner, and a blackboard famous for its colored-chalk renderings of molecular interactions. The pipes in the corner had been painted purple. Another prankster, another trip.

Like any organization, the Gilbert lab took on the personality of its leader. "The work was *always* urgent in Wally's lab," says Talmadge, who worked in the lab from 1975 to 1980. "It was too boring not to have results, so everything went as fast as possible. It was a great training for how to have everything ready, and how to interweave things. Never let a night go by without bacteria growing on a plate or something growing up in culture or a reaction going overnight if it had to. Things like that. Because otherwise life is *boring!*"

No one wanted to be boring, of course; indeed, the Gilbert lab cultivated a collective sense of humor both witty and raucous to balance that sense of urgency, and it had the psychologically salutary effect of making peers of everyone. If there was any one person who, despite the democratic spirit, lent the place a kind of specific social gravity, it was probably a talented biochemist named Allan Maxam, who had joined the lab as a technician in 1971 and did sufficiently advanced work to receive a Ph.D. from Harvard. Heavyset, light-featured, with a round and immensely candid face, Maxam was known as "Maine Man"—an acknowledgment of his native state, but also of his social eminence in the lab. In the wee hours of the morning, he could usually be found in Room 388, discussing biochemical tricks or ruminating on the sociology of science.

It was the lab group of the mid-1970s which had, in his words, "an atmosphere, a style, a point of view." It incorporated the open, egalitarian spirit of the '60s, and pointedly tried to eliminate all the unnecessary obstacles to making discoveries, which the students interpreted to include authority, formality, rules, financial problems, and people problems. "The philosophy," says Maxam, "was that we didn't want to be pretentious or formal because that can be inhibiting." The scientific upshot, he says, was a little bit like a gourmet potluck dinner, a small private banquet with course after course of interesting work. "We had the liberty to expand our research in various directions," Maxam recalls. "There wasn't a czar at the top. Definitely not that type of arrangement. Everybody there had the liberty to pursue unexpected, new observations." From the retrospective view of the 1980s, Maxam described it as "the best arrangement I've seen—before, during, and since."

Emblematic of that spirit was a most unusual, and often very funny, photocopied lab newspaper known as the *Midnight Hustler*. It was Maxam's brainchild, but a number of other members of the Gilbert group—Karen Talmadge, Lorraine Johnsrud, Philip Farabaugh, Debra Peattie, Winship Herr, Gregor Sutcliffe, and fellow travelers Arg Efstratiadis and Nadia Rosenthal, to name a few—were frequent contributors. Its issues were as irregular as the hours of its contributors, but it was a devastatingly democratic organ. No one, not even Wally Gilbert, was spared. Sometimes the tweaks were gentle, sometimes they drew blood. They made fun of Gilbert's wardrobe, with its purple and orange turtlenecks and screaming scarlet cravattes. There were parodies of Hunter Thompson on gonzo biology, self-deprecating jokes about "Big Guy Science," and tongue-in-cheek ads. "Not enough of that 'belly-up-to-the-bench' action to get those exciting results that impress the other Big Shots at meetings?" read one ad for clamps designed to affix graduate students to the research bench. "You need BENCH BONDAGE!"

One of the graduate students who took to this casual atmosphere with great relish was a self-described "young buck" out of Salt Lake City named Forrest Fuller. An affable fellow with sandy shoulder-length hair, he wore cowboy boots, drove a broken-down purple Mercedes, and had a relaxed storyteller's gift of gab. After completing an undergraduate degree in chemistry at the University of Utah, he'd gone east to Harvard in the fall of 1974 for his graduate education. When asked what scientific area he wanted to investigate, he replied with naive directness that he was interested in working with "RNA and DNA." Just that: RNA and DNA. His attention was directed to Walter Gilbert.

Fuller's initial impression of Gilbert was of "a very bright person, kind of aloof in his brightness." He found him "somewhat impatient with people talking to him if they weren't saying what he wanted them to say in a concise manner," and also "appeared to have a lot of chutzpah." Like many of the graduate students, Fuller regarded Gilbert with complete awe; at the same time, however, he had a disarming talent for finding himself in the thick of interesting projects, aspiring to a kind of peerage based not so much on experience as enthusiasm. Fuller the hero-worshiping student once tried to align bacterial colonies on an agar plate so that certain positive colonies, the ones that triggered chemicals in the agar to turn blue, would bloom into a big blue W for Wally (an experiment that didn't, alas, work). Fuller the ambitious scientist, on the other hand, felt quite at home with leading-edge projects. "I viewed Wally as really doing this New Science and these new things, and doing them in recipe-type witchcraft, almost, and I called him the Sorcerer . . ." Fuller remembers. "So I called myself the Sorcerer's Apprentice, and Wally began, I think, to call me that occasionally, too."

As early as 1975, the Sorcerer's Apprentice had what he considered "a great idea." Fuller had just learned about Arg Efstratiadis's work down in Fotis Kafatos's lab on the first floor. The Greek scientist had figured out a way to make a copy of the globin gene. Inspired by this work, Fuller walked into Gilbert's lab, cheerful and typically confident, and spelled out the plan for an ambitious new experiment. Not only would they make a copy of the gene for rabbit globin, but they would induce bacteria to make this rabbit protein. "The next thing I knew," Fuller recalls, "I was collaborating with Arg on getting globin." It was clear from the outset, too, that globin would be an obvious stepping-stone to insulin. "Insulin was always in the back of our minds," Fuller says, "because that would be the big thing to do with recombinant DNA."

It was certainly ambitious and visible, and those were two of the prerequisites for what the Gilbert group had mockingly christened Big Astounding Science. Fuller became more and more fascinated with the idea. It would be his graduate school project, he decided. He would work at it painstakingly, methodically, as was his style, do all the controls, know why everything worked or didn't work. After two years or so, he figured he would have the globin gene and could move on to insulin.

What he didn't figure was that, in the fields of Big Astounding Science, where "Big Guys" like Walter Gilbert played, other scientists often had the very same ideas. And as the recombinant DNA gold rush began to get underway, some of the play could get a little uncollegial. Gilbert's attitude was typically aggressive. Says one of his graduate students, apropos of the project: "It reflected Wally's approach to any system or any experiment. He goes for the kill immediately."

The University of California researchers in Howard Goodman's and William Rutter's labs were pursuing the same project, but the West Coast biologists were about to learn one of those rude, unexpected lessons in the new scientific etiquette. Shortly after the Lilly meeting, Goodman had plans to be in Boston; it had been agreed upon, according to the UCSF group, that he would drop by William Chick's lab, learn how to dissect the tumors, and come back to San Francisco with both frozen tumor and some rats. Then, in the Californian's view, the plans changed rather abruptly. As Raymond Pictet, one of the UCSF researchers, describes it, "Goodman told us that he went there, and Bill Chick said that they had problems and that he couldn't do it, that he couldn't give anything now." Chick's reason, he would explain later, was that there simply wasn't enough material to spare. The important thing was that suddenly the UCSF researchers were cloners left out in the cold. They suddenly found themselves deprived of the rare starting material for their experiment.

Not long after that, Argiris Efstratiadis and Forrest Fuller hailed a cab and headed from Cambridge over to William Chick's lab in Boston to pick up some of the insulinoma tissue. The tumors promised to be solid masses of islet cells, impossible to isolate otherwise and rich in messenger RNA for insulin. This, they knew, would give them the inside track for the first step, making a cDNA copy of the gene. They were poised to move on from bacterial genes to explore the intricate, complicated beauty of the eukaryotic gene—the kind found in human beings and other higher organisms.

But the California group did have a last laugh of sorts. No sooner had the Harvard scientists begun the work than they ran into an unexpected and sobering obstacle. The problem had nothing to do with biochemistry, however, and everything to do with the controversial nature of recombinant DNA work. The citizens of Cambridge simply didn't want them to do those kinds of experiments. "We were not unaware of their problems," said one of the San Francisco researchers, with a wry little smile.

F o u r

''Fire, Too, Is a Vague Hazard''

. . . .

The *Boston Phoenix* article, when it came out in early June, minced no words about the imminent danger soon to be loosed on Cambridge. "No one knows what a recombinant DNA microorganism will do—how it will affect man, how it will affect the environment," read the article.

The type of hazard that recombinant DNA research poses is thus different from any other type—different, in fact, from any hazard previously known to mankind. If General Motors manufactures a defective steering column, the company can recall the automobiles. If the Food and Drug Administration discovers that a drug has harmful side effects, the agency can ban it from the marketplace. If a vial of cancer virus spills in a laboratory, scientists can scrub it up. And if one is working with a known carcinogenic like the chemical polyvinylchloride (PVC) or asbestos, one knows not to ingest it because it may cause cancer.

But how do you defend yourself against a microorganism which has never existed before, whose effects are completely unknown, which cannot be seen, which reproduces itself, and which scientists may not even know has escaped?

"This is a biohazard," explains Harvard biologist Ruth Hubbard. "It's not enough to stop doing the experiments. Once it's out, it's impossible to shut off. It's worse than radiation danger."

As of June 8, 1976, when the *Phoenix* hit the streets, the proposed P3 laboratory at Harvard officially became a public hazard. All the experiments underway or anticipated, the insulin work included, immediately bore the stigma of being dangerous, unpredictable, possibly lethal. The lengthy article in the *Phoenix* relied heavily on the testimony of critics within Harvard's biology department and cited several disquieting "scenarios." The technology, the article suggested, could produce novel diseases, a slow-acting cancer, or epidemics of flulike contagion. There was even speculation about an insulin-producing recombinant organism escaping into the water supply.

These were all hypothetical dangers, and clearly stated as such. But the mere suggestion of microbes run amok, of the irrevocable laboratory blunder, touched a sensitive nerve, and the *Phoenix* article did little to minimize the hypothetical aspect, suggesting analogies to corporate arrogance and insensitivity (General Motors and pharmaceutical companies), to all the more demonstrably worrisome environmental hazards of the era (asbestos, PVCs, carcinogens), and to ill-fated technological imperatives (the nuclear energy industry). These guilt-by-association analogies had an alarmist elegance to them, but was recombinant DNA inherently hazardous? Did the technology cross some invisible threshold of danger? Could recombinant organisms, like fibers of asbestos, lay silent like microscopic time-bombs and reveal their deleterious effects only decades after exposure? There was scant scientific reason to think they would be more harmful than ferociously lethal smallpox viruses or the bacteria that cause botulism, but there was no conclusive way to prove otherwise, either.

Of all the passages in the article to which the Harvard researchers could point with dread or alarm, however, perhaps the most damaging was among the more subtle. ". . . no representative of the public," authors Charles Gottlieb and Ross Jerome wrote, "has participated in a decision which some Harvard officials concede may affect the health and safety of the general public." All politicians abhor political vacuums. To the great dismay of Harvard's molecular biologists, the politician who roared into the void was the mayor of Cambridge, Alfred Vellucci.

Vellucci, a native of Cambridge, had been a city councillor since 1956 and had gained political prominence in the community in the 1960s when he led a highly publicized fight to oppose a NASA research facility in South Cambridge. A crusty blue-collar conservative in a town where Harvard took a notoriously aloof attitude toward the local community, he worked town-gown tensions to flamboyant political effect; he was a self-described "ordinary citizen" who never had a college education, and he seemed to enjoy tweaking the university, like the time he proposed paving over Harvard Yard to ease local parking congestion. Just as Harvard sought to lay the issue to rest, in stepped Vellucci to stir things up.

On June 13, 1976, Harvard dean Henry Rosovsky announced the university's official verdict on the controversial recombinant DNA laboratory. Construction of the P3 facility in the Bio Labs building would proceed according to plan. In a statement accompanying the decision, and quoted in the *New York Times*, Rosovsky took pains to reassure members of the college community that rigorous safety steps would be taken. In a curious aside that seems never to have been clarified or explored, the dean added, "I say this with the knowledge that our safety record has not always been perfect." Walter Gilbert, cited in a *Harvard Crimson* article, suggested that the likelihood of creating a virulent pathogen in the P3 lab was about as farfetched as a meteorite devastating Cambridge the following day.

Something very much like a meteorite, a political meteorite, did in fact strike the following day, in the form of Vellucci. Shortly after the *Phoenix* article was published, Ruth Hubbard had prepared information packets for Cambridge's nine city councillors. It was June 14, a sunny Monday afternoon, she recalls, and on the spur of the moment, her husband, George Wald, offered to join her on the walk to Cambridge's imposing City Hall, a stodgily elegant gabled brownstone structure on Massachusetts Avenue.

No sooner had they arrived and their identities been determined than they were asked to join the mayor in his chambers for an informal conference (Vellucci maintains that Wald and Hubbard had made an appointment). The fuming Vellucci was not happy to have learned about the Harvard lab uproar from the *Phoenix* article and asked Hubbard and Wald questions about the situation at Harvard. The biologists revealed their fears about safety and contamination; Vellucci recalls, "They even talked about ants getting into people's lunches." He asked if they would be willing to make statements at a hearing before the city council. The Harvard scientists assured the mayor they would be delighted. Seizing the opportunity, armed now with the political legitimacy a Nobel laureate conferred, Vellucci announced that the Cambridge City Council would hold hearings on Harvard's controversial P3 laboratory the following week.

Among the scientists interested in recombinant DNA work, the reaction ranged from joking dismissal to a more sobering, and unpleasant, surge of *déjà vu*. By the spring of 1976, just when many scientists had concluded three tumultuous years of debate with the belief that the technology was safe, the public and the politicians became aroused and interested. There were Ruth Hubbard's public-health fears that novel pathogens might be created, and George Wald's moral caveats that billions of years of evolution would be abrogated by the new gene-splicing techniques; perhaps even more troubling was the fact that Jonathan Beckwith of Harvard Medical School and Jonathan King of MIT, both biologists of the first rank, insisted the techniques were potentially harmful.

With scientific expertise lined up on both sides, how could lay citizens hope to make crucial distinctions on safety? It was not easy material to convey or master. As Barbara Ackermann later remarked in an interview, "Very rarely does an issue come up where you have to learn as much." Rent control, she noted by comparison, "wasn't such an abstract issue."

When the debate finally went public, it began with a local high school choir filling the City Council chambers with adolescent strains of "This Land Is Your Land" and a sign reading "No Recombination Without Representation." Onlookers and kibitzers packed the gallery, jammed the balcony, spilled out into the hallway on the evening of June 23, 1976. Along a row of chairs to one side, shoulder to shoulder, sat Mark Ptashne, Matthew Meselson, David Baltimore, and Wally Gilbert—molecular biology's equivalent of baseball's Murderers' Row. Television lights and cameras, for both television news and archival footage, were trained on the councillors and on the witness table. Having gaveled the meeting to order, Al Vellucci basked in the glow.

It is difficult to imagine a city council meeting, in any place, at any time, creating more hullaballoo and attracting more national attention and generating more ink and comment than the Cambridge DNA debates. Rare are the town meetings where the proceedings are treated as national events, attracting the attention of the *New York Times* and other national publications, beamed around the world by the wire services. "Creation of life" experiments held sensational fascination for people far beyond Cambridge. There was also appealing confrontational value: the intellectual titans of Harvard and MIT would be forced to explain themselves and their work before a committee of lay citizens. "At times unnervingly like an inquisition" in the opinion of one chronicler, "a very good thing" in the words of a participant, it was an extraordinary confrontation of science and society. And it caught the biologists by complete surprise. "Everybody assumed the whole thing would just sort of go away," Walter Gilbert recalls. "Once the city councillors discovered that television cameras would be called in, and the whole meeting televised, then it all went to pot quickly." "If the first hearing on the recombinant DNA debate had had a hundred people from Harvard supporting it and a hundred people opposing it, and *didn't* have 60-some television cameras and radio stations and reporters, it probably would have died a natural death that night," says a former City Council member.

Mayor Vellucci set the tone, for better or worse, at the outset when he issued the following warning to the assembled scientists: "Refrain from using the alphabet. Most of us in this room, including myself, are lay people. We don't understand your alphabet. So you will spell it out for us so we'll know exactly what you're talking about, because we are here to listen." Vellucci was not referring to the genetic alphabet of A, T, C, and G; rather, this was an invitation to the scientists to keep it simple and straightforward.

A limit of ten minutes was imposed on each speaker, but the first two witnesses—Mark Ptashne of Harvard and Maxine Singer, a representative of the National Institutes of Health—spent more than two hours fielding questions. Much of the flavor and substance of the dispute emerged in those first few hours.

Ptashne explained the background of the P3 lab at Harvard, citing National Cancer Institute figures that there were already hundreds of P3 laboratories in use in the country, including one at the Harvard Medical School across the river in Boston and another at MIT, about a mile down Massachusetts Avenue from City Hall in Cambridge. "Most molecular biologists—not all, but most—expect that the information being derived from these experiments that are now ongoing will profoundly advance our understanding of life processes," he said. Moments later, he added, "We must realize that unlike other real risks involved in experimentation, the risks in this case are purely hypothetical." In perhaps the most aggressive assertion on behalf of recombinant DNA made that evening, Ptashne went on to add:

"Nevertheless, we cannot say that there is absolutely no risk involved in these experiments. But then, Mr. Mayor, I ask you to consider that [that] statement—little risk—can be made about few human activities. It certainly cannot be made for many of the experiments performed every day in biological and chemical laboratories. The degree of risk involved in carefully regulated recombinant DNA experiments is almost virtually, in my estimation, less than that in maintaining a household pet." In an audience stacked with partisans of both stripes, the comment was greeted with hisses of derision and swells of applause. What was stunning, at least to Councillor Ackermann, was the unusual political schism revealed in this partisanship. "There were more students for it," she recalled in surprise, "than against it." One more reminder that this wasn't like Vietnam.

If the atmosphere occasionally tilted toward the carnivalesque, the impresario of the hour was Vellucci. Flamboyant and grave, shrewd and bumpkinish, good cop and bad all at once, by sheer dint of style and personality he wrestled the agenda onto turf of his liking and scored political points with calculated yelps of Everyman outrage. He chided Daniel Branton, head of Harvard's biosafety committee, for failing to contact public health officials about the Harvard meetings on the lab, and reacted with chagrin and anger when Matthew Meselson, then chairman of Harvard's biochemistry and molecular biology department, conceded that chemicals from experiments were sometimes flushed down the Bio Lab drains. After the attractive Maxine Singer introduced herself, Vellucci inquired for her phone number, a request overlooked for other witnesses; at another point, he observed to no one in particular, "There are sharp minds in this audience," and then added in an aside to city councillor Leonard Russell, "You and me are a couple of sharp minds."

Singer explained the new NIH guidelines, terming the likelihood "extraordinarily low" that P3 experiments could cause serious trouble. Vellucci then weighed in with his "statements of questions." It was a masterful populist *tour de force*, not so much for its logic or rhetoric (or grammar, for that matter), but for its unabashed, bald-faced exercise of power in setting the terms of the public debate.

"One," Vellucci began. "Did anyone of this group bother at any time to write to the mayor and the City Council to inform us you intended to carry out these experiments in the City of Cambridge, and you just said that you had public hearings.

"You plan to use *E. coli* in your experiments. Do I have *E. coli* inside my body right now? That's a question. Don't answer, but you may, as you go along.

"Does everyone in this room have *E. coli* inside their bodies right now?

"Can you make an absolute, one hundred percent guarantee that there is no possible risk which might arise from this experimentation? Is there zero risk of danger? Answer that question later, too, please.

"Would recombinant DNA experiments be safer if they were done in a maximum security lab, a P4 lab, in an isolated, nonpopulated area of the country? Question.

"Would this be safer than using a P3 lab in one of the most densely populated cities in the nation? Question.

"Is it true that in the history of science mistakes have been made, or known to happen? Question.

"Do scientists ever exercise poor judgment? Question.

"Do they ever have accidents? Question.

"Do you possess enough foresight and wisdom to decide which direction the future of mankind should take? Question.

"The great war poet Joyce Kilmer once wrote, 'Poems are made by fools like me, but only God can make a tree.' I have made references to Frankenstein over the past week, and some people think this is all a big joke. That was my way of describing what happens when genes are put together in a new way. . . ."

At this point, a flicker of a smirk spread on Mark Ptashne's face. Vellucci's radar picked it up immediately. "This is a deadly matter, sir," he said sharply. "Ma'am, sir. Harvard University. This is a serious matter. It is not a laughing matter, please believe me. It is not a laughing matter, and this is for the National Institutes [Maxine Singer]: this is not a laughing matter. If worse comes to worse, we could have a major disaster on our hands. I guarantee everyone in this room that if that happens no one will be laughing then.

"Protecting the health and safety of the people of Cambridge is a solemn trust," the mayor intoned. "I intend to treat that trust with complete dedication, and, madam, it was only twelve years ago that I sat in that seat with the

City Council full to capacity, as full as it is tonight, fighting the coming of the NASA site in Cambridge. And I predicted that that whole thing would collapse, and it collapsed.

"And now," Vellucci vowed, "tonight I come here with the same fight in me."

Vellucci was not the only city official to vent his spleen that evening. One can infer from the comments of other city councillors an uncomfortable mix of bewilderment, suspicion, resentment, paranoia, and alarm over the entire issue. After two hours of testimony, Barbara Ackermann made the following observation to Daniel Branton of Harvard's biosafety committee: "I would say that a committee that was dealing in matters which are even one millionth of one percent dangerous ought to involve at least half public representation." If risk is as much a social calculation as a statistical assertion, this was the new math of public hazard.

The prevailing fears were perhaps best articulated by a tall, bearded councillor named David Clem, who found it paradoxical that the NIH, the federal agency charged with funding recombinant DNA research and therefore "promulgating" it, now assumed the mantle of regulator as well. "I really don't give a damn about a P3 laboratory at Harvard University," he said, "because I can't visit that laboratory and discover whether it's P1, P2, or P4 or whatever. I don't have the expertise to analyze or investigate any type of laboratory facility at Harvard University. But," he added, indignation growing in his voice, "it strikes me as very to the point that there is an important principle in this country that the people who have a vested self-interest in certain types of activities should not be the ones who are charged not only with promulgating it but regulating it." To a huge outburst of applause, Clem continued, "This country missed the boat with nuclear research and the Atomic Energy Commission and we're going to find ourselves in one hell of a bind because we are allowing one agency with a vested interest to initiate, fund, and encourage research and yet we are assuming that they are nonbiased and have the ability to regulate that, and more importantly, to enforce their regulation."

Soon after, to competing choruses of applause and boos, Alfred Vellucci—who was just there "to listen"—introduced a resolution. "Whereas," he read, "there is still considerable doubt concerning the safety of experimentation dealing with recombinant DNA, therefore be it resolved: that the Cambridge City Council insists that no experimentation involving recombinant DNA should be done within the City of Cambridge for at least two years, and be it further resolved: that the Cambridge City Council will do anything and everything in its power to enforce this resolution."

Ptashne's reaction was instantaneous. "If you pass that resolution," he said, visibly agitated, "virtually every experiment done by members of the biochemistry department at Harvard will have to stop and virtually every experiment done by about half the members of the biology department would have to stop, including experiments that no one, sir, *no one,* has ever claimed had the slightest danger inherent in them—namely recombinant experiments done under P1 conditions. And so on such an important issue, it seems to me you have to clarify much more clearly what the issues are before proposing such a resolution."

"What I want to alert you to," Alfred Vellucci replied, "when I was a little boy I used to fish in the Charles River and I woke up one morning and found millions of fish dead in the Charles River, and you tonight tell me that you've dumped chemicals into the sewer system of Cambridge and the sewer system overflows into the Charles River." Peremptory and dismissive, he grumbled, "Carry on."

Round one to Vellucci.

Not all citizens shared Alfred Vellucci's grumpy indignation. As one reader complained in a letter to the June 24 issue of the weekly *Cambridge Chronicle,* "We are amazed that anyone should express concern about the creation of a laboratory at Harvard to experiment with new life forms. A look around Harvard Square at nearly any time of day or night reveals life forms sufficiently grotesque to convince us it is already too late for such protest."

Humor, however marginal, was rare as the debate heated up. Writing in the *New England Journal of Medicine* some months later, essayist Lewis Thomas noted that the recombinant DNA debate "has become an emotional issue, with too many irretrievably lost tempers on both sides. It has lost the sound of a discussion of technologic safety, and begins now to sound like something else, almost like a religious controversy, and here it is moving toward the central issue: are there some things in science we should not be learning about? There is an inevitably long list of hard questions to follow this one, beginning with the one that asks whether the mayor of Cambridge should be the one to decide, first off."

Many biologists at Harvard and MIT were asking the same question, but without Thomas's evenhanded elegance. The first Cambridge hearing provoked utter shock and disbelief at Harvard and MIT. It is unlikely that any of the Harvard molecular biologists fully anticipated the threat which public debate represented to their ongoing research until June 23. Only then did the fervor and threat of opposition become all too clear.

The citizenry had lambasted the NIH for its dual (and obviously awkward) role of bursar and auditor. Vellucci had maneuvered the safety debate onto impossible terrain; no one could guarantee 100 percent safety crossing the street, much less doing an experiment. And as Lewis Thomas noted, the underlying tone of the conflict was one of intellectual freedom: "Are there some kinds of information leading to some sorts of knowledge that human beings are really better off not having?" Scientists work and thrive in that conceptual twilight where knowledge dissolves into the unknown, so restrictions on their intellectual curiosity created very real feelings of psychological violation. The whole experience was frustrating and alien to the scientists. Not only were these eminent biologists not accorded the deference and respect to which they were accustomed in the academic milieu, but they discovered to their chagrin that attempts to make "reasonable" arguments in no way put a stop to what they viewed as irrational, hysterical opposition. It was as if "reason," the *lingua franca* of science, common coin of the realm, suddenly lost its buying power in the public arena. The journey from meritocracy down to democracy was, in some cases, sufficiently steep and swift to provoke cases of the ivory tower bends.

The gloom of the Gilbert group reflected the general dismay in the Harvard Bio Labs. Graduate student Debra Peattie would later recall, "It was the first time I had ever seen general public reaction to something that reflected on what I was interested in. And I think in some ways it alarmed me because it indicated that the lay public do carry an awful lot of clout." Another graduate student, Karen Talmadge, found the hearings to be "quite a zoo, quite horrendous," and became active in the debate because "I think that the scientific arguments against recombinant DNA were very weak." Argiris Efstratiadis, about to embark on the insulin cloning and already engaged in other recombinant DNA projects, lost a valued collaborator to the controversy. Because Harvard's P3 lab was tied up in the imbroglio, Tom Maniatis left Harvard to pursue his work at Cold Spring Harbor Laboratory in New York and later took a faculty position at Cal Tech. Rarely at a loss for words, Efstratiadis suffered this time in silence. "I didn't even have a green card then," he says, "so I didn't want to say anything."

At that time, the Gilbert lab became one of the primary meeting places to discuss the issue. Talmadge and Peattie virtually abandoned their scientific work—no small concession in as competitive an atmosphere as Harvard's—to devote their energies to the DNA debate. Along with Gilbert, they drafted an information sheet that attempted to explain recombinant DNA technology in layman's terms; they invited people to call up the lab if they had questions; they testified before the city council.

Wally Gilbert rarely appeared in the front lines during the public debates, but he was steadily active behind the scenes. This was probably due as much to temperament as strategy: for all his scientific aggressiveness, Gilbert often struck students and colleagues as socially reserved, even shy, and Talmadge observes that "he didn't really want to be out there and be pegged as one of the very pro-recombinant-DNA people." But he had several projects underway—insulin among them—which relied on recombinant DNA technology, and he clearly advocated use of the techniques. To the *Real Paper*, a Cambridge weekly, he commented, "There are two kinds of benefits to be accrued from this kind of research: abstract knowledge and long-term practical benefits. Recombination of DNA is a new technology—the closest parallel I can think of is the role of fire in metallurgy. Fire, too, is a vague hazard."

If anything, Gilbert managed to lend advocacy without creating enmity, which was not easy in those days. "I didn't have any terribly public role in it," he says now. "I testified occasionally because I was interested in doing insulin, doing this sort of project, and was a spokesman for the general use of the technology during that period." Amid the "irretrievably lost tempers," his was a low-key, noninflammatory presence. He testified at the hearings, helped draft the information sheet, debated Ruth Hubbard before the Cambridge Rotary Club, stood in a booth in Cambridge's Kendall Square during a Saturday street fair and answered questions about the issue, and talked to all reporters, from the *Boston Globe* to the student newspaper.

In retrospect, Gilbert considers the debate a useful exercise in democracy. Undoubtedly, though, he simply wanted to get on with the experiments. The public clamor, the extreme suggestion that tampering with DNA was "sacrilegious"—as MIT's Jonathan King had testified—made work virtually impossible. "There's an underlying fear of knowledge and fear of technology which is what was really catalyzing the public reaction there," Gilbert says in retrospect. "That's in fact one of the themes that always goes beyond this debate, because the actual debate was about a totally irrational thing. When the debate involves words like 'sacrilege,' you know something pretty weird is going on. And it goes back to underlying emotional attitudes toward genetic material. It's one thing to look upon DNA as a chemical entity, but then suddenly people are saying, 'My God, if you cut the DNA molecule apart by chemical means, you've done something horrifying and dramatic. . . .' " That same level of genetic manipulation, he points out, "goes on all the time naturally."

Gilbert even believed that in a subtle psychological way, the biologists fathered part of the controversy themselves with their penchant for self-aggrandizement. "In another sense I think that we—and here I speak for myself as one of the biologists—are responsible, and it's the macho impulse side. 'Stop us, we're doing something dangerous. We're as good as the physicists. We are handling these great, dangerous forces.' "

If the message was "Stop us," the City Council got it.

The Cambridge hearings resumed on the evening of July 7, 1976. Some twenty-two witnesses lined up to testify, and the session was only slightly less tumultuous than the first. Late in the evening, when many spectators had already left and the city councillors were restlessly pacing behind their desks, Walter Gilbert took his place at the table before the council and began his testimony.

In a low, somewhat tentative voice, Gilbert introduced himself and then said, "I think it's very useful for the public to get involved in these questions to educate themselves in science, because I, as a scientist, find science the most exciting thing going and I'm delighted when anybody else is interested." Having said that, Gilbert went on to talk about an experiment he wanted to do using recombinant DNA techniques. He spoke, of course, about insulin.

"One of the experiments we're interested in is putting the gene for human insulin into a bacterium," he told the councillors. "The purpose of that experiment is to make human insulin, a specific hormone, available. You probably know that insulin is used to treat diabetics. You may or may not know that the source of that insulin is from pigs or from cows, and that molecule is not identical to the one that is in all of our bodies." Commercially available insulins, Gilbert explained, were "not perfect" because they differed from the human hormone. He proposed instead to insert the human gene for insulin into bacteria in such a way that the bacteria would make human insulin.

"We are making something which is, in a sense, beyond price," he continued. "It is not that we are making cheaper medication. We are making something which we cannot *get* by other means. It is not necessarily the case that we can succeed in such an experiment. It is not the case that we can succeed in this experiment tomorrow. The actual time scale, if we are thinking of time, is [on] the order of a couple of years." The preliminary work would take place in a P3 laboratory, he explained, and the final experiments in a P4 facility. Such work, he suggested, would be a way of testing the real, not hypothetical, hazardousness of this type of experimentation. "You can learn if it is or is not hazardous," he said. "You can therefore devise the safety measures that are required."

Gilbert's testimony, it must be said, had absolutely no bearing on the opinion of the city councillors. Several hours earlier, they had voted five to three, with one abstention, to approve an amended version of the mayor's resolution; it called for a "good faith" three-month moratorium on recombinant DNA experiments in Cambridge. Specifically, P3 and P4 work—the very level of work Gilbert proposed to use—was enjoined while a special citizens' review panel and committee attempted to establish local guidelines. In a revealing comment to the journal *Science,* councillor David Clem—who introduced the amended resolution—explained, "I tried to understand the science, but I decided I couldn't make a legitimate assessment of the risk. When I realized I couldn't decide to vote for or against a moratorium on scientific grounds, I shifted to the political."

Although few lay people in the audience could appreciate it, Walter Gilbert had done something that scientists rarely do: reveal in public an ambitious project, risking all the perils such disclosure entailed. Gilbert had publicly spelled out the scenario for the insulin work, right down to the timetable. In a scientific sense, he had put himself on the line. From the very beginning, he had had designs on the human gene. He adjudged it feasible within two years. And in a nod to scientific realism that might easily have been overlooked, he even admitted the possibility of failure.

Moreover, Gilbert's comments rendered very explicit the idea that medical applications justified the use of recombinant DNA research without delay. Insulin had been invoked as a potential benefit of recombinant DNA work almost as soon as that work became controversial. Now Gilbert, a scientist of world stature, repeatedly identified human insulin as one of the practical benefits of the research, tantalizingly within reach. He testified to that effect, gave interviews to that effect, participated in public debates to that effect. And as other communities and states pondered moratoriums or regulations following the Cambridge precedent, the swift delivery of those benefits became a matter not only of medical but of political importance.

So even while workers in Massachusetts and California had already commenced the effort to genetically engineer insulin, the hypothetical advantages and risks of just such a product became a matter of public conjecture. In fact, the insulin argument formed a kind of social and political counterpoint to the scientific work going on simultaneously in several labs. As the debate heated up, competing exaggerations hardened into polarized dogmas. Critics invoked fears about man-made epidemics and genetically engineered human clones, perhaps inflating the scientific dangers in order to attract attention to their political points. Unable to explain the enormous yet complicated intellectual rewards of studying immunoglobulin or histone or ovalbumin genes to the lay public, defenders of recombinant DNA predicted dramatic medical advances. If you closed your eyes and just listened, it sounded as if the future offered either genetic monsters or wonder drugs, with little in between.

The bureaucratic tangle in Cambridge did not make the work any easier. The three-month moratorium ultimately lasted seven months, until February 1977, and for all the public testimony and fact-gathering and explication, one wonders just how successful the process of public education had been when no less a model of quotidian restraint than the *Times* of London reported the end of Cambridge's moratorium with the headline " 'Frankenstein' project given go-ahead in US."

F i v e

Sticky Ends

. . . .

As the occasional "Frankenstein" headlines suggest, the scientific realities diverged substantially from popular imagination. Nowhere was this more apparent than in the feelings roused by a single word. To the biologists, the word fairly throbbed with power and promise; to lay people, it provoked queasy ethical questions and vague dread. That word was "cloning."

The citizens' committee set up by the Cambridge City Council spent months wrestling with this term, and it is clear that as the science became better understood, the notion of creating Frankensteins in a test tube by cloning, while fanciful and amusing, was seen to be hardly feasible. The work of this board was an early and encouraging sign that lay citizens could not only come to grips with thorny technological issues, but could also achieve competence in regulating it. That competence, it seems clear, came with greater familiarity with the science itself.

Recombinant DNA has sometimes been likened to a kind of editorial enterprise, and this is probably the most useful analogy for thinking not only about cloning, but about the whole range of biological manipulations that became possible: scientists could cut and paste genes, make photocopies of particular genes in great number, and read genes as sequences of biochemical words and sentences. To perform all these tasks (and to maintain metaphorical consistency), you need scissors and glue, a good Xerox machine, a dependable dictionary, a primer of biochemical syntax, and a very, very fine magnifying glass, so that you can see exactly what you are doing on texts that, page by page, twist by twist of the double helix, take up little more space than a fraction of a wavelength of light. DNA, in that sense, is heredity's fine print. In the years between 1968 and 1976, all these tools became available to molecular biologists.

Writers often wax rhapsodic about DNA, likening it to blueprints, masterplans, rope ladders, and whatnot, but biochemists are mostly impressed by its physical qualities. To them, DNA is a biochemical entity, no more intrinsically wonderful than a two-by-four is to a carpenter. It is amazingly tractable, and the ease of working with the molecule made learning about it much easier. DNA can be frozen, boiled, dried, bounced around, gouged by chemicals, whirred about in centrifuges, stripped apart, and reassembled and still maintain its informational integrity. That information is stored in the sequence of bases climbing like steps along the backbone, the As and Ts and Cs and Gs.

Beginning in the late 1950s, the basic syntax of genetic grammar has become well defined. The sequences of bases within a gene has an inner organization. Each group of three letters, known as a "codon," corresponds to an amino acid, which might be thought of as a biochemical word. To use an example more fancifully illustrative than technically accurate, one might encounter a series of three codons that "make sense" in English, such as CAT ATE RAT. Three bases make up each word; in a chain of amino acids, these words are joined together, often in sentences extending for hundreds of words, to spell out a protein. There needs to be a way of getting this information from the DNA to the rest of the cell, and that's where messenger RNA comes in.

The message passes out of the nucleus and into the cell proper and proceeds to the cell's manufacturing district, the ribosomes. Each ribosome, like a small factory, processes the message as if it were a punched computer tape; it recognizes within the genetic sentence the three-letter words, and, in this sense, life's dictionary is disarmingly economical. With an alphabet of only four letters and only three letters per word, it follows that there are only sixty-four possible three-letter combinations that can form a word (four times four times four). With some built-in redundancies and a few punctuation marks, these sixty-four possible words account for all twenty amino acids used to build and maintain the microscopic infrastructures of life. In DNA code, CAT actually stands for an amino acid named histidine.

Word by word, amino acid by amino acid, long strings of proteins—enzymes, hormones, hemoglobin, neurotransmitters, insulin—are thus created in the cells; our genes contain sentences for enzymes that digest and enzymes that build, hormones that control other molecules, structural proteins that link up to become the mortar and brick of cells, of organs, of bones, of organisms. It is the correct assembly of all these sentences into paragraphs and chapters that forms the rich, self-renewing text of life. And in biology, all creatures (with a few rare exceptions) speak the same language, a kind of biochemical Esperanto; a single-celled blue-green alga is as literate when it comes to reading CAT ATE RAT as a Rhodes scholar.

This, then, is life's basic grammar. Nature tends to be far less tolerant of grammatical mistakes than the average teacher or editor; such errors in DNA can prove debilitating, even fatal, to an organism. Since nucleotides are read in threes, the insertion or deletion of a single letter—referred to as insertion or deletion mutations—can change the entire meaning of a sentence and can spell out a kind of nonsense protein. Inserting one letter, R, at the end of the first word of the CAT ATE RAT sentence, for example, has a domino effect. All the ensuing words are thrown out of what is called the "reading frame," and CAR TAT ERA T is the result. Similarly, if the letter C is deleted from CAT, the sentence now reads ATA TER AT. Or if one letter were to change, say the T in CAT, the sentence's meaning might become CAR ATE RAT. In biochemical code, those words still spell out amino acids, but almost invariably they spell out the wrong ones. In order to be successful, molecular biologists could not be haphazard in their manipulations; a single misspelling, one out-of-place letter, could throw off an entire experiment.

Some of biology's syntax is a little more complicated. The gene, it turns out, is more than just a sentence or declaration of information. As the work by Wally Gilbert and others on the *lac* gene indicated, it resembles a line in a script, a core of coded information flanked by stage directions. Stuck on one flank are instructions on how quickly or often the sentence should be read (the "promoter"), a start signal akin to a capital letter to mark the beginning (the "start codon"), and a stop signal like a period (the "termination sequence") at the other end. Just as an actor can deliver the same sentence in a great many ways, the inflection given a genetic sentence varies with the directions surrounding it. The genetic sentence for insulin, for example, is never read in brain cells at all; something in the stage directions around this gene render it mute, unenunciated.

Biologists obviously considered it of profound interest to figure out why some genes were silent and others not, an area of research known as gene regulation and expression. A route of attack began to take shape in the 1960s, and it grew out of an area of bacterial research so basic and narrow that it seemed destined to reveal nothing whatsoever of practical relevance. It had to do with a curiously violent little sideshow of bacterial life.

Bacteria, in a cunning way, had evolved a class of ferocious enzymes which recognized short "phrases" of invader DNA and chopped right through them. When bacteria were infected by viruses, these so-called restriction enzymes would wheel into action, dicing up the viral DNA wherever the offending phrase appeared and thus preventing the viruses from taking over the cell. "Restriction," in short, could be considered a kind of biochemical xenophobia.

The isolation and purification of the first restriction enzyme was reported in 1968 by Harvard professor Matthew Meselson and his postdoc Robert Yuan. Two years later, Hamilton Smith and Kent Wilcox of the Johns Hopkins University School of Medicine reported restriction activity in a bacteria known as *Haemophilus influenzae* strain Rd. This enzyme ultimately became known as *Hin*d II (the etymology of these enzymes derived from the organism and strain from which they came).

With the keys provided by Meselson and Smith, biologists began to unlock hundreds of these enzymes in other bacteria. The names, to the uninitiated, may read like sound effects in a comic book brawl: *Bam, Sma, Pst, Sal, Hae, Hha,* and *Eco,* to name a few (they joined a pantheon of fanciful enzyme names such as decapitase, transferase, cocoonase, and the wonderful swivelase). But these restriction enzymes were to the biological revolution what machine tools were to the industrial revolution: precise, extremely powerful means for cutting and shaping and manipulating raw materials. The raw material, in this case, was DNA.

Of all the restriction enzymes to hit the scene during this period, no pair of scissors found more immediate or widespread or more dramatic use than an enzyme isolated from the common bacterium *E. coli.* It was discovered by

Herbert W. Boyer, a microbiologist at the University of California–San Francisco, and was so identified with its discoverer that other researchers, in lapses of scientific neutrality, sometimes refer to *Eco* RI as "his" enzyme. With this one enzyme, Boyer stood at the crossroads of enzymology and genetics. It was like having a front-row seat to a revolution.

Herb Boyer achieved prominence the hard way. He had toiled in the relative obscurity of bacterial genetics through the late 1960s and early 1970s, maintaining a small, overcrowded, chronically underfunded laboratory; Boyer rarely paid his postdocs out of his own grant money, instead attracting people who had already arranged funding from other sources. While President Nixon declared war on cancer, and scientists migrated en masse to cancer research in search of funding, Boyer continued to study the lumpen prokaryotes, those simple single-celled organisms like bacteria that browsed through life without so much as a nucleus.

The work seemed off the beaten path, scientifically interesting but of no medical import, and funding was tight. But, as often happens in science, yesterday's overgrown, undertraveled path becomes today's four-lane highway, crowded with researchers in search of scientific answers (and public funding). Boyer's enzyme joined several others to become the invisible workhorses of the recombinant DNA revolution. Herb Boyer didn't plan on being one of the founding fathers of genetic engineering, but the dramatic power of the restriction enzymes thrust him into the forefront, and his University of California laboratory became overnight one of a handful of labs on the leading edge.

In a way, Boyer looked more like a railroad engineer than a genetic engineer. He had a cheeky, guileless face framed by a wild nimbus of permed brown curls, dark eyebrows, a thick mustache, and a small mouth. Chastened by years of publicity (some of it adverse), he rarely unfurls much of his personality in the few interviews he concedes these days, but there are flashes of a nimble intelligence and a puckish sense of humor. While it seems like his entire generation of contemporaries headed radio clubs or physics clubs in high school, Boyer was a football player, a lineman even, and there is a down-to-earth quality to him, as if he's spent his share of Saturday afternoons picking divots out of his faceguard. "He can step back and laugh at himself," says a scientist friend, "which is not that common in the business."

That adolescent athleticism remained with him; his colleagues knew him both as a runner and as a football fan. And although he was raised in the East, Boyer seemed perfectly at home on the West Coast. "He pretends to be more relaxed and laid-back than he really is, of course," says a frequent collaborator. "But his style is to be laid-back and sort of relaxed, everything's okay." That became the demeanor of his laboratory as well.

His background was lower-middle-class, his intellectual development neither precocious nor conspicuously brilliant. He was born in 1936 in Pittsburgh, Pennsylvania and raised in the area around Latrobe. The elder Boyer worked in the coal mines and on the railroads, which, as his son once recalled, was "about all you could do in western Pennsylvania in those days." Boyer grew up in the town of Derry, population 3,000, and went to one of those small high schools where a single teacher handled chemistry, physics, mathematics, and biology; it was fortuitous that at this particular school, the science teacher happened also to be the Derry Borough High School football coach, because in those days, Boyer was a guard on the football team. While western Pennsylvania has produced a disproportionate share of good football players, it is not normally considered a hotbed of molecular biologists. Boyer enjoyed football, but the science courses were beginning to interest him, too. "I guess it's that they were orderly subjects, logical; they made sense," he once put it.

When he entered St. Vincent College in Latrobe in 1954, Boyer's initial impulse was to follow a premed curriculum. St. Vincent was a small Benedictine liberal arts school that is probably best known, certainly in western Pennsylvania, as the summer training camp site of the Pittsburgh Steelers. Boyer lived at home during those years, taking the bus or hitchhiking to classes. By his junior year, the prospects for a career in medicine seemed less appealing, perhaps diminished a bit by a summer job when he worked in the clinical laboratory of a nearby mental hospital. One gets the impression that Boyer's initial interest in bacterial genetics and DNA was, unusual in his field, the intellectual equivalent of a found object. During a cell physiology course, he was assigned to give a lecture on DNA and independently became very excited by this enigmatic but clearly important molecule. "To hell with medicine," he remembers thinking at the time, with typical collegiate impetuousness. "Who needs all these sick people to take care of? I want to do something that's interesting!" Not long after, he was raising a pair of pet Siamese cats named Watson and Crick.

Boyer graduated from St. Vincent's in 1958 and married a local girl named Grace in 1959, and then this self-described "small-town boy" experienced his scientific coming of age studying bacterial genetics at the University of Pittsburgh. He spent three more years as a postdoctoral fellow at

Yale, where he also participated in the burgeoning civil rights movement. Having picked up some additional expertise in biochemistry, protein chemistry, and enzymology, Boyer was poised to explore that strange mechanism by which bacteria were able to chop up and disable foreign DNA when he reported to the University of California–San Francisco in 1966 as an assistant professor.

By 1969, Boyer and his colleagues at UCSF were well along in identifying two restriction enzymes from *E. coli* bacteria. Another young assistant professor named Howard Goodman arrived from a postdoctoral fellowship in Switzerland in 1970, and the two men embarked on an immensely productive collaboration. Goodman, trained at Williams College and MIT, had early on established a reputation as an outstanding scientist. Boyer and Goodman prepared grant requests together and shared federal funding; their two labs interacted, according to one researcher, like "members of the same family," and their research headed into promising territory.

The properties of these fascinating enzymes, as they became better defined, caused great excitement. Wherever *Eco* RI found the genetic phrase GAATTC, for example, it cut through. Moreover, as chemical scissors go, *Eco* RI made a very special kind of cut. Whereas many restriction enzymes cut flush through DNA, leaving blunt ends, *Eco* RI left a mortiselike incision. As a result, two single-stranded flaps of DNA were created wherever the enzyme made a cut—"sticky ends," as they became called in the business. And since opposite bases attract and form bonds, it was as if these overlapping flaps were like tiny fingers of Velcro which would rejoin if they came into proximity. This is one of the most fundamentally important characteristics of DNA: whenever one strand of DNA encounters another with the precise complementary (or opposite) sequence of As and Ts and so on, they tend to come together and attach as if magnetically attracted. The chemical joints between the rejoined flaps can then be sealed by another enzyme. Since *Eco* RI both cut DNA and left sticky ends, it was a very powerful enzyme: in terms of cutting and pasting, *Eco* RI offered scissors and pastepot all rolled into one.

None of this was lost on a Stanford University researcher named Stanley N. Cohen when he heard Boyer describe his recent work in November 1972 at a scientific meeting in Hawaii. Cohen had been studying little ringlets of DNA in bacteria known as plasmids. These are free-floating hoops of DNA that drift through the cytoplasm of a bacterium and replicate independently of the bacterium's chromosomal DNA. Cohen had developed a method of removing these DNA ringlets from bacteria and then reinserting them. In the hands of genetic engineers, which Boyer and Cohen were about to become, plasmids became the biological equivalent of a smuggler's bible—a device for smuggling bits of DNA from other organisms into bacteria. That was exactly the purpose Cohen had in mind.

When Cohen heard Boyer describe his work with restriction enzymes, he realized that any piece of DNA cut with *Eco* RI could be recombined with any similarly cut piece, regardless of species difference. You could join the DNA from two different plasmids, for example. You could just as easily join the DNA from cobra and mongoose, sloth and gazelle, fruit fly and rhino (not that you'd necessarily want to), because in all cases Boyer's enzyme left identical sticky ends. Cohen acted almost immediately on his insight. "That evening," he recalled later, "at a delicatessen across from Waikiki Beach, I proposed a collaboration with Boyer."

This historic collaboration began shortly after they returned to California. Cohen's research assistant, Annie Chang, lived in San Francisco, so she became a shuttle service for materials going back and forth between San Francisco and Cohen's lab in Palo Alto. Given the revolutionary nature of the experiment, splicing together DNA from two different bacterial plasmids, the experiment worked smoothly and the data were in place by March 1973. Many cloners informally date the birth of recombinant DNA—or "gene-splicing," as it popularly became known—to the Cohen-Boyer experiment, which was published in the November 1973 issue of the *Proceedings of the National Academy of Sciences* (or *PNAS*).

The Cohen-Boyer team had taken texts from two different bacteria, recombined them, and slipped this composite plasmid into *E. coli*. That was where the matter rested. It was a tool with enormous potential. If one could insert a specific gene, a gene from one of the higher organisms, into bacteria, the experimental worth vastly increased. Plasmids replicated along with bacteria every time the bacterial cells divided, which occurred at an astonishingly rapid pace, every twenty minutes or so. After a day of these serial replications, one could harvest a considerable amount of the spliced-in DNA. That is exactly what a Boyer and Cohen team, led by postdoctoral fellow John Morrow at Stanford, did in 1973. The end result demonstrated the true value of cloning.

Morrow and his colleagues combined DNA from species on either side of the prokaryote-eukaryote divide, an extremely important demarcation to biologists; the researchers plucked a single biochemical sentence out of the chromosome of the African clawed toad *(Xenopus laevis)*, using the cutting ability of the restriction enzymes, and then inserted it, using the sticky-ended glueing ability of those same enzymes, into the long genetic text of *E. coli.* Second, the researchers were able to determine that this amphibian clause was copied and carried by subsequent generations of bacteria as they divided. The inserted toad sentence had been incorporated into the bacterial DNA and passed on to each succeeding generation. In short, it had been cloned.

When Boyer called up a scientific colleague, Stanley Falkow, to relay the exciting news, Falkow naturally wanted to know how Boyer's team had been able to identify the few bacterial colonies—literally one in a million—containing the toad DNA. "Herb just said he kissed every colony on the plate," Falkow later remembered, "until one turned into a prince."

At the time, it was fashionable to call such hybrids "chimeras" or "chimeric molecules," a term coined by Stanley Cohen in acknowledgement of the mythical Greek beast that was part lion, part goat, and part dragon. There is more jargon and less literary allusion in molecular biology these days, and such hybrids are usually known as recombinants or, even more commonly, clones.

With all the sinisterly dull, *Brave New World* thud of the term and its social echoes of genetic tampering and biological imperialism, the key advantage of cloning probably hasn't been sufficiently understood by nonscientists. In standard laboratory applications, the ultimate goal of cloning is not, as Lewis Thomas once wryly observed, to replicate "prominent politicians from bits of their own eminent tissue." Nor is it, as Woody Allen comically implied in his film *Sleeper*, to reproduce entire human beings from a random scrap of tissue—a nose, say. The aim, pure and simple, is quantity. Scientists exploit bacteria, with their prodigious replicative properties, as copying machines. Every time bacterial cells replicate, the DNA inside makes a copy of itself, too; if a biologist manages to smuggle his plasmid into a single cell only, there will be millions of copies of it within eight hours. Then the goal is to extract the desired DNA segments.

Cloning, like photocopying, is intended to produce multiple copies from a single original. And just as photocopying produces copies of a particular document, cloning produces copies of a discrete segment of DNA—a particular genetic document. Unfortunately, at the time that Boyer and Cohen did their initial cloning experiments, there were no means available to read these genetic documents.

But at this critical juncture, Wally Gilbert's lab at Harvard provided a dramatically useful new tool. The continuing work on the *lac* repressor led, "by accident" in Gilbert's estimation, to one of the most powerful techniques at the disposal of today's molecular biologists. Allan Maxam and Gilbert— with crucial kibitzing from Soviet scientist Andrei Mirzabekov—developed a quick and reliable method of reading the sequence of nucleotides, or base-pairs, in any given snippet of DNA.

"Rapid sequencing" involved a sophisticated series of chemical manipulations. They were performed on small amounts of DNA that sound astronomical, a million billion identical fragments—just the kind of identical copies produced by cloning. By tagging the end of each molecule with a radioactive marker and then chopping them with specialized chemical shears, Maxam and Gilbert designed a way in which randomly generated lengths of this DNA would sort themselves out by letter—A or T or C or G—as they were pushed by electric current through a stiff, thick gel of acrylamide, a process known as gel electrophoresis. It meant that each organism's hereditary library was theoretically readable and open to inspection. British biochemist Frederick Sanger, a 1958 Nobel laureate, independently developed and reported a different method for the rapid sequencing of DNA at about the same time. Thus two techniques were suddenly available to read what had previously been an invisible and inaccessible biological transcript.

In terms of speed, the Maxam-Gilbert technique represented the difference between copying a book out by longhand and making photocopies of it. In terms of impact, it joined other implements in the biochemist's burgeoning tool bag, tools that allowed scientists to cut specific segments of genetic material, splice them into other pieces of DNA, insert the recombined DNA into organisms like *E. coli,* and make multiple copies (or clones) of the genes. The orchestrated manipulation of this material provided a wonderful means for studying gene regulation. With the development of the sequencing techniques in 1975 and 1976, virtually all the basic tools had been assembled. Gilbert looks back upon it as "one of the most exciting moments in science."

Indeed, by the mid-1970s, the molecular biologists had their enzyme scissors and their enzyme pastepots, a well-defined dictionary of genetic usage, and an expanding grammar to explain the tricky syntax of DNA; they had a dependable means of copying the material, and a powerful lens that allowed the text to be read accurately and rapidly. They talked about isolating specific human genes and inserting them into bacteria. They talked about peeking "over the wall," the invisible biological partition that separated bacteria, with their free-floating cellular DNA, from the higher organisms, whose genes resided in a well-defined nucleus and whose DNA seemed peculiarly complex and highly regulated. And in science, where "visionary" may just be another word for acute optimism, some biologists even talked about making things like hormones in bacteria using these recombinant DNA techniques. Herb Boyer was among the earliest.

In May 1975, in an interview with historian of science Rae Goodell of the MIT Recombinant DNA History Collection, Boyer talked about inserting the gene of a higher organism into bacteria and getting the bacteria to express it. Other people were thinking similar thoughts, of course, but Boyer put a different spin on the concept. He said, "I think this has a lot of implications for utilizing the technology in a commercial sense; that is, one could get bacteria to make hormones, etc., etc. So that's one project that we're involved with."

It was an appropriate sentiment for the fellow who, as a graduating senior from Derry Area Senior High School, confessed his ambition to become "a successful businessman."

Given these entrepreneurial impulses, Herb Boyer was a company waiting to happen. He dutifully approached several companies about exploiting the technology in 1975 and, like many lone inventors with an idea ahead of its time, was dutifully discouraged. The discussions never went beyond "very general terms," he recalls, and there were no takers.

At roughly the same time, a young venture capitalist named Robert Swanson was sniffing around the edges of the West Coast molecular biology milieu, likewise convinced that recombinant DNA represented a technology with vast industrial implications. With that in mind, he tried to interest a California company called Cetus and telephoned a number of prominent molecular biologists, such as Stanford's Paul Berg, sounding them out on the possibility of commercializing the technology. No one had a good word to say about the idea. Molecular biology at that time had a long and illustrious tradition of basic research—untainted, as the biologists themselves would put it, by commercial interests. Venture capitalists were held in about as high esteem as ambulance chasers.

Given the similarity and singularity of their interests, it was almost inevitable that Boyer and Swanson, like a pair of sticky ends, would ultimately seize onto each other. Indeed, Boyer made for an especially sanguine and optimistic listener when he received a phone call in January 1976 from Swanson. Swanson, with a background in biochemistry and an M.B.A. from MIT, had worked at the venture capital firm of Kleiner & Perkins in San Francisco. By his own definition, he was "pretty driven to succeed"; he was even prepared to dump his job and stake his career to a start-up company. Boyer agreed over the phone to meet with Swanson for a few minutes on a Friday afternoon.

Swanson hadn't really done his homework, though. He had no idea that in Boyer he was talking to a coinventor of the very gene-splicing technique he sought to exploit commercially. He was excited, he recalls, that anyone sounded even vaguely encouraging. "Here was someone," Swanson says, "who was clearly a leader in the field, although when I talked to him the first time, I didn't know about the Cohen-Boyer experiments. So I got even luckier in a sense. But here was someone who said, 'Gee, we can do that.' And nobody else had said that!"

On the agreed-upon Friday afternoon, Swanson dropped by the lab. His appearance inspired immediate comment, for only an outsider would wear a suit in a molecular biology lab in those days. The two men retired to Boyer's office, apparently liked what they heard from each other, and continued the discussion over beer at Churchill's, a local bar. "After that meeting," Swanson says, "we did some thinking, him on the technology side, me on the business side, to see what was possible. We started out with a list of known proteins and looked at what the markets were, which were the most interesting." The decision to plunge ahead, in those heady days, does not seem in retrospect to have been horribly costly. The businessman and the biologist each ponied up a modest $500 (Boyer had to borrow his stake), and that $1,000 became the initial operating capital for the new company. For the real work, the research, they would of course use other people's money.

Swanson's cheerful, frat-brother bonhomie masked an edgy, impatient ambition and considerable organizational skills for a twenty-eight-year-old. He hewed to a well-defined list of economic prerogatives. All of them sounded crass and heretic to most academic ears, but not to Boyer's. Swanson's ambition was nothing less than to establish a new major pharmaceutical company, with research, development, manufacturing, and marketing all under one roof. In so competitive a field, the selection of the initial products could make or break the company. "They had to have an existing market, because we didn't want to do missionary marketing," Swanson recalls. "It had to be a significant market. We had to see a way clear to have the right economics in terms of manufacturing a product. . . . In theory, if we could come close, then it was a possibility. [And] we looked for a product that could have a relatively short time through the regulatory hurdles. That was a key factor."

But perhaps the most important goal in the beginning was an intangible: credibility. The company needed to prove that the technology worked, that somehow bacteria could be coaxed into making useful human drugs. Since recombinant DNA was so new, and the investment community so green to its promise, it would not hurt if the company's first product had a little name recognition in the general public. These were difficult first decisions: on the one hand, you wanted to select a biological product that would impress venture capitalists and loosen the purse strings of private investors; on the other hand, you didn't want to embark on a project so technically ambitious and difficult that precious amounts of time, money, and credibility went up in smoke.

At that first meeting, Swanson did not have a particular protein in mind, but after talking to Boyer, human insulin emerged as one of several candidates. Shortly after their meeting, Swanson was off doing some legwork, contacting the American Meat Institute, getting numbers on the price of pancreases, figuring out the economics of making and selling insulin produced by genetic engineering. "The idea to do insulin," says Swanson with surprising assertiveness, "was mine."

The idea about *how* to do insulin, however, fell to Boyer. In many ways, it was the more determining of decisions. The selection of a scientific strategy can, in a way as determining as destiny, give a group an exceptional long-term advantage or doom it to futility. Boyer knew exactly how he wanted to proceed on the work, and he also knew that there were only a few scientists in the country with the expertise to help him out. That same January of 1976, Boyer picked up the phone and called his friend Arthur Riggs, a respected biologist at the City of Hope National Medical Center, just east of Los Angeles in Duarte, California. As Riggs recalls, the conversation was short and to the point.

"Hey, Art," Boyer said. "I have this businessman friend of mine who thinks he can raise money to do insulin."

"Oh, that's interesting," replied Riggs. A native Californian, Riggs tended to be both modest and deferential (traits not always in conspicuous abundance among molecular biologists), but he reverted to a kind of polite tenacity when it came to his pet scientific interests. Insulin was not at that point among them.

"Would you be willing," Boyer continued, "to accept the contract to make the gene for insulin?"

"Well, that's interesting," Riggs replied again, "but there are a couple of things. Itakura and I are just in the middle of writing a grant to do somatostatin. Could I interest you in somatostatin?"

Boyer's call came at a time when Riggs and his colleague, Keiichi Itakura, already had several other irons in the fire. They planned nothing less than to *synthesize* a human gene—that is, build the correct sequence of DNA from scratch in a test tube, using chemicals off the shelf. If that worked, they planned to plug this man-made gene—for a hormone called somatostatin—into bacteria and see if it could be cloned. And if *that* worked, they even had hopes of getting the bacteria to *express* this synthetic human gene—that is, read the genetic instructions and make a human brain hormone in a petri dish. Riggs argued that somatostatin would be a good dry run—a "model system," in lab lingo—for insulin.

So young Bob Swanson, not even a month on the job, faced his first tough business decision. "As a philosophy," he says, "I've never liked the idea of model systems. It seems to me that if you can work on a real system, you're better off doing it that way." But Riggs was pretty firm about somatostatin. While Swanson ruminated, Riggs and his colleagues proceeded on another little experiment they'd cooked up. They planned, in the words of one collaborator, to "immortalize" a man-made piece of DNA.

s i x

What the
Blue Colonies
Meant

Arthur Riggs was daring the way scientists tend to be, in an esoteric and nearly invisible way. In Riggs's case, the invisibility was as much a matter of temperament as of profession. As one former colleague notes, "Art is one of those people who, almost as a matter of principle, won't toot his own horn." All the scientific questions he investigated reflected his basic scientific philosophy, which he describes, with typical self-deprecation, as the ability "to flounder around in the right direction." The man could flounder.

He roamed the precincts of molecular biology as a gentle anomaly. He was a family man who actually left the lab early enough to spend time with his family; in a field populated with giant egos and abrasive personalities, he was reserved in manner, almost visibly self-effacing. His lone concession to flamboyance in appearance or demeanor was a runaway mustache that spread uninterrupted into a pair of conflagratory sideburns. Born in Modesto, California, son of a pipefitter and a nurse, Riggs grew up in San Bernardino, devouring science fiction as a youth and dreaming about being a spaceman. He studied chemistry at UC-Riverside, earned his Ph.D. at Cal Tech, did a postdoctoral stint at the Salk Institute outside San Diego, and since 1969 had been lodged at the City of Hope—virtually his entire scientific career maturing within a hundred-mile radius of his childhood home.

His scientific imagination, however, ranged considerably farther afield. A colleague recalls how Riggs once devoted a nonstop half-hour monologue to ravaging a proposed experiment as absurd and impossible, concluding his remarks by saying, "But it's so exciting! What the heck, let's try it." He had won distinction for his work on the *lac* repressor, being the first scientist to obtain pure repressor protein in a test tube. Herb Boyer considered him "a very creative and imaginative guy," and another collaborator described him as "amazing," but in a surprised way, as if to say, "You'd never think it, but the guy is amazing." Louise Shively, his longtime research technician, says, "Once Riggs convinces himself that something should work, he won't give up. He's sure of it. He's astonishing that way."

Herb Boyer decided to call up this stubbornly optimistic fellow in January 1976 because "it didn't make sense to look anywhere else." Riggs in fact would turn out to be one of the primary architects of the first two experiments conducted by Boyer's new San Francisco company. But Boyer's interest extended beyond Riggs alone. Riggs had a colleague who was a superb organic chemist, and he represented a crucial piece of the collaboration Boyer was trying to put together.

What happened next is a reminder that science moves forward as often on the basis of coincidence and misstep as design and intent. While the merits of insulin versus somatostatin were being argued up north, Riggs and a group at Cal Tech found themselves mired in another of those absurd and impossible and exciting experiments. Struggling to extricate themselves, Riggs and his colleague got sidetracked onto another project. This second experiment proved to be dramatically important to both molecular biology and its commercial offspring, genetic engineering.

It was an obscure, complicated project called the synthetic *lac* operator experiment. Of all the experiments undertaken during the mid-1970s, it may have been the most symbolic of that era—a project that bridged the gap between the previous decade's basic research on the *lac* gene and the next decade's rapid explosion into genetic engineering. It provided the first conclusive evidence that man could manipulate genes virtually at will.

*

The technical temerity of this landmark experiment reflected, at least in part, a letter-writing campaign initiated around 1974 by a young Japanese scientist working at the National Research Council of Canada in Ottawa. His name was Keiichi Itakura, and he toiled in a particularly abstract, and to many minds generally impractical, branch of organic chemistry that had to do with the synthesis of nucleic acids. "The one goal of a chemist—and it's always a goal—is to synthesize some chemicals which function better than the natural product," he would say later. Usually when chemists say things like this, they are talking about plastics or deodorants or artificial sweeteners. The chemical Itakura chose to improve was DNA. He planned to make the "molecule of life" with ingredients you could buy over the counter at a chemical supply house.

At the time, Itakura was thirty-two years old and a long way from home, literally and figuratively. Born in Tokyo, he had at first studied pharmaceutical science, acquiring a Ph.D. at the Tokyo College of Pharmacy; in 1971, resolved to make his career and fortune far from home, he began a postdoctoral fellowship in the Ottawa laboratory of Saran A. Narang. A small, lean, athletically compact man, Itakura could be sober-faced one minute, cheeks furrowed by high-pitched laughter the next. When asked to describe him, colleagues invariably refer to his quiet ambition. That was what had made him an expatriate scientist in the first place.

Itakura went to Canada, he says, because the Japanese culture of science mitigated "against doing what you want to do." Like many of the other foreign scientists who gravitated toward America, Itakura thrived in the wide-open, highly competitive culture of science in the United States. "In the field of science [in Japan]," he explains, "there is not as much freedom. The system is completely different from that in the U.S. In a scientific group, you have a professor and an assistant professor and three or four assistants. The professor says, 'Do this,' and you have to do it. There's almost no freedom to do what you want to do."

Although his English was not spectacular, Itakura drafted a letter and sent it out to a select fellowship of scientists. "I was looking," he says now, "for some molecular biologist who was working on cloning." He recalls sending out only four or five letters (Itakura did not keep copies of his correspondence, so the precise wording is not known), to such scientists as Paul Berg at Stanford University, Herb Boyer at UCSF, and Susumu Ohno at the City of Hope. The letter revealed a degree of immodesty more occidental than oriental, and on some level it must have astonished its recipients. Itakura asserted that he could build functioning genes out of synthetic DNA; what's more, he claimed he could do it ten times faster than by the so-called "diester" method currently in favor, which happened to have been developed by the 1968 Nobel laureate in physiology/medicine, Har Gobind Khorana.

Itakura's boast was enormous. In the mid-1970s, only a handful of scientists indulged in the difficult art of DNA synthesis. More significant, it was well known within molecular biology circles—indeed, it was almost viewed as a kind of long-running scientific saga—that a group of researchers had dedicated years to the construction of a functioning gene. The effort was headed by the Indian-born Khorana, who began this epic biochemical crusade while at the University of Wisconsin and continued the work when he moved to MIT in 1971.

The task was considered so immense, so preposterously difficult at the time, that Khorana's group had no ambitions whatsoever to build the gene for a miracle drug or a super enzyme. Instead, they applied their considerable energies to a relatively small, well-known gene for a nucleic acid known as transfer RNA. Eventually Khorana got his gene. But it was, to say the least, a labor-intensive exercise. By later estimate, it took about two dozen hardworking and exceedingly bright young scientists about nine years to build the gene—something like two hundred man-years in all to string together two strands of DNA totaling 207 base-pairs. When the team announced that the gene was biologically functional, in August 1976, there was a significant and well-deserved rustle of interest, including a front-page story in the *New York Times*. But, given the pace of the work, most everyone concluded that synthetic DNA chemistry represented a tedious and time-consuming way to get at genes for study. While it took Khorana's team about a decade to build its gene from scratch, Efstratiadis and his Harvard coworkers by comparison spent only about two years to obtain a gene twice that size. Thus the cDNA approach developed by the Harvard group and others—getting a gene by enzymatically copying its messenger RNA back into DNA—still seemed the way to go.

Then, onto the desks of some of the leading biologists in the country fell this unsolicited letter from a Japanese scientist based in Canada, claiming to be able to do the unreal: synthesize genes *ten times faster* than Khorana. And, perhaps more important, he claimed he could make DNA as *pure* as

Khorana's. Itakura received polite but unencouraging responses from the scientists he'd written to. All except one. His letter to the City of Hope's Susumu Oho had been passed on to Art Riggs. Having already spent some time in Herb Boyer's lab in San Francisco, Riggs knew that cloning represented the wave of the future in molecular biology. With the Maxam-Gilbert rapid sequencing method and the Cohen-Boyer cloning techniques, synthetic DNA would form the third leg in a triad of technical tricks with great potential. Riggs reviewed Itakura's work and said to himself, "We've got to get Itakura here."

The machinery was set in motion to bring Itakura out to the City of Hope. The Japanese scientist first settled into a position at nearby California Institute of Technology, in the laboratory of Richard Dickerson, in the summer of 1975. But the position was a joint appointment, kind of a one-year holding pattern until Itakura could set up permanent shop at the City of Hope. "We became convinced," says Riggs, "that he was a very good chemist. I was also convinced that it didn't matter if the techniques were perfect, because we'd get something interesting out of it anyway."

Even from the normally understated Dr. Riggs, that turned out to be an understatement of historic proportion.

The future of biotechnology, for anyone who cared to notice, began to take shape during the first few months of 1976 in an experiment that, on the surface, had nothing whatever to do with insulin or pharmaceutical products. While the faculty of Harvard and the citizens of Cambridge bickered about the direction in which biology should move, in fact, a giant step toward the uncertain recombinant future was being confirmed and written up on the other side of the continent. It was an experiment that gave Herb Boyer the confidence to proceed with the formation of Genentech, yet it was his East Coast rival, the Harvard molecular biologist Wally Gilbert, who ironically provided the crucial tip that got the West Coast experiment untracked.

Around the beginning of 1976, Gilbert was on a West Coast seminar tour, spreading the word about the Maxam-Gilbert sequencing technique. In late January, he happened to be visiting Cal Tech and dropped in by invitation on the laboratory of Richard Dickerson. As it turns out, Gilbert had wandered into a spontaneous and somewhat gloomy lab meeting. Dickerson and his collaborators—Art Riggs, Keiichi Itakura, a postdoc named John Rosenberg, and a young graduate student named Richard Scheller—had reached a go or no-go point of decision in a complicated, costly, and time-consuming experiment. They wanted to take an X-ray picture showing the interaction of the *lac* repressor and operator, to see exactly how the repressor protein attached to DNA. But in order to do that, they needed to crystallize large amounts of the protein and DNA. And that in turn posed severe problems: it would take Itakura the rest of his life to synthesize enough DNA for the experiment, and for complicated reasons, it was impossible to clone it, too. "That was a terrible crisis," one participant, John Rosenberg, recalls. "We would need millions of dollars and an army of people to make enough of this to purify. So this project was threatened with the end."

At first, Gilbert's appearance was a "source of concern" to some of the younger scientists. He was considered a competitor on this particular project and, in the words of one participant, "definitely not a competitor to be taken lightly." It could have been a terribly awkward moment, a duel of mumbled homilies and evasive courtesies. Instead, the Cal Tech group frankly described their problem and asked Gilbert if little tails of DNA could be glued onto their operator DNA to allow it to be cloned. "Well," he said, "T4 ligase will do it." The effect was electric. "I will always remember it," says Scheller. "Really an exciting moment!" Rosenberg confirms.

Possessed of a prodigious recall for data and trivial details, Gilbert in 1975 managed to recall a relatively obscure paper by Vittorio Sgaramella that had come out of Khorana's laboratory in 1970. It described some fascinating and, at the time, unappreciated qualities of an enzymatic glue known as T4 polynucleotide ligase. A ligase, as the name implies, is an enzyme that ligates or ties together molecules; this particular ligase conveniently glued together pieces of DNA. It was the *way* this enzyme did its glueing that held revolutionary promise for genetic engineers. It allowed efficient end-to-end fusions of DNA, like glueing dowels together. Toward the end of 1975, Gilbert tested the principle himself and discovered that it "worked like a charm."

"The moment we did that experiment," Gilbert recalls, "it became obvious that one could tie together arbitrary pieces of DNA. Since we had the sequencing technique, you could know everything you did. You could put together promoters, tack bits on, put DNA pieces together in total freedom. As opposed to all the previous strictures, where you had to do things with sticky ends, and there wasn't very much you could do."

With typical conciseness, Gilbert described the unique qualities of this "rediscovered" enzyme to his hosts at Cal Tech, and the scales fell from everyone's eyes. Gilbert's lab did not have a Keiichi Itakura and could not synthesize DNA; the West Coast group did. And synthetic DNA added a new dimension to the scientific possibilities. Art Riggs realized that the combination of this blunt-end glueing and synthetic DNA chemistry could revolutionize cloning. With Itakura, they could scratch-build a short piece of DNA to use as a kind of coupler on either end of a gene. This DNA would contain the sequence recognized and chopped by a restriction enzyme—*Eco* RI, for example. With Gilbert's "glue," they could jam these short, bookend segments on either end of a gene. They could create their gene of choice, put couplers on either end, drop it with incredible precision into a plasmid that had the same sticky ends, pluck it out with unprecedented ease.

Riggs talked out these ideas aloud with Wally Gilbert, as the scientists sat around in the Dickerson lab, and vividly recalls Gilbert's reaction. The Harvard scientist, after thinking over the matter for a long time, said simply, "Yes, it would probably work."

Almost from the instant that Gilbert gave this collegial affirmation, Riggs and Itakura began to bail out of the experiment that had brought them to this impasse in the first place. The "Magnificent Failure," as Riggs called it, was shelved; their attention shifted to a different, more brazen project. Indeed, Riggs and Itakura—with the help of Dickerson and Rosenberg and others in Canada and San Francisco—saw a dramatic opportunity to show the world, and many skeptics in the scientific community, that synthetic DNA could change the way molecular biology was practiced. One of Itakura's last projects with his mentor Saran Narang in Canada had been to synthesize a small piece of DNA that duplicated the twenty-one-base-pair *lac* operator. As Gilbert had shown a decade earlier, this was the segment to which repressor proteins seized, and so it had a clearcut and detectable function in the cell. It would provide a perfect test strip for these new ideas.

All they needed, then, was a way to splice it into a cell. Itakura recalled that Herb Boyer had a piece of DNA that fit the requirements for those bookend segments—a tiny scrap recognized by *Eco* RI. With Gilbert's key suggestion, they now had a way of glueing Boyer's pieces, like adaptors, on either end of Itakura's DNA piece. Efforts were initiated to "wine and dine" Herb Boyer, as one participant put it, to get access to his DNA.

*

The collaboration turned out to be one of those increasingly typical platoons of scientists, an armylike assault that was about as far removed as biology gets from the monkish nineteenth-century ruminations of Gregor Mendel, the solitary gardener who tended his peas and quietly laid the groundwork for modern genetics. Four labs, four different cities, two countries, a total of ten scientists. And yet the experiment itself had an exquisite simplicity.

The experimental strategy owed all its elegance to years of basic research. Itakura provided the main piece of DNA—the twenty-one-base-pair *lac* operator segment. Meanwhile, Herb Boyer, successfully wined and dined, donated the couplers, and these were promptly glued on either end of Itakura's DNA. By such patchwork stitching, they took this DNA and inserted it via a plasmid into the ever-obliging *E. coli*. But how could they tell if the man-made DNA worked inside the cell?

The answer rested with a young Dutch biochemist in Herb Boyer's lab, Herb Heyneker, who had volunteered to handle the project. Heyneker knew that if the synthetic operator pieces worked like their natural counterparts in bacteria, they would soak up the repressor molecules cruising around the interior of the bacteria. Heyneker shrewdly used a plasmid that made about thirty copies of itself, including the operator segment, per cell—well in excess of the ten to twenty repressor molecules per bacterium. It was inevitable that these "impostor" operator segments would biochemically distract the repressor molecules from their appointed task; in some cases, the *lac* gene—the "real" gene—would not get blocked by the repressor. Rather, the gene would turn on and the bacteria would commence to make the enzyme beta-galactosidase. And biologists had devised a simple, color-coded signal to detect the presence of this enzyme; by mixing into the growth media a chemical fortunately abbreviated X-gal (the full name is 5-bromo-4-chloro-3-indolyl-beta-D-galactoside), colonies would turn blue wherever bacteria were making beta-galactosidase. "Blue colonies," as the scientists called them, would be a sign of success.

The sign, when it came late in February 1976, did not flash with the brilliance of neon. "I can remember," says Boyer, "that we were in the lab late one night, around nine or ten o'clock, and Herb Heyneker had done the experiment that morning. He brought plates in to look at. We knew there were clones if the colonies turned blue. Herb said, 'It didn't work.' I said, 'Let's look a little bit more closely.' If you looked *real* closely, it was very obvious that they were turning blue. They were very faint, but there were quite a few blue colonies on the plate." Heyneker, who recalls it as an early-morning experience, concedes, "I was expecting something more dramatic."

However faint, blue colonies meant that the bacteria were making beta-gal. And that meant Itakura's synthetic DNA had fooled a living cell. Man-made DNA functioned in a living organism.

To Heyneker, who was first author on the ensuing paper, the result was nothing short of "mind-boggling." "What was really the most exciting part was," he recalls, "here we were, the first to immortalize a piece of chemically synthesized molecule. . . . I mean, we were able to add to the genetic load using off-the-shelf chemicals!"

The experiment gave an enormous psychological boost to the idea of a biotechnology company. Boyer now regards it as "a very key bit of experience on our part, that was a contributing factor to my positive response to Bob [Swanson]." Just as Boyer had spotted the blue colonies missed by others, he clearly saw the implications of this experiment where others didn't: DNA *could* be manipulated; genes *could* be engineered. The experiment bolstered Boyer's faith in pursuing the synthetic DNA strategy, which would pay tremendous economic dividends for Genentech two years down the road.

The *Nature* publication that resulted from this work, "Synthetic *lac* operator DNA is functional *in vivo*," gave crucial legitimacy to several emerging ideas at the time. First and foremost, the collaborators—Heyneker, John Shine, Howard Goodman, and Boyer in San Francisco, Rosenberg and Dickerson of Cal Tech, Narang in Ottawa, and Itakura, Riggs, and Syr-yaung Lin at City of Hope—had demonstrated that man-made DNA could be inserted into a bacterium and become part of the organism's functioning genetic repertoire. They had shown, in addition, that blunt-end ligation and linkers (as those small couplers came to be called) added incredible precision to cloning. The *Nature* paper, in fact, introduced the word "linkers" to the scientific literature, and soon Richard Scheller was synthesizing and freely distributing these DNA couplers to all who asked, merely dotting a drop of solution on a piece of paper and mailing it in an envelope.

As the authors noted in the penultimate paragraph of the *Nature* paper, "The combination of chemical synthesis technology with cloning technology will provide a powerful approach to many problems, including the chemical synthesis of long DNA sequences."

"Long DNA sequences" was rather coy phrasing. What they were talking about was genes.

The operator experiment had the unintended effect of beginning the drumroll for cloned insulin. The University of California–San Francisco, in a press release about the work, noted that the experimental advance "for the first time could make possible the economical production of scarce biological substances such as human insulin and pituitary growth hormones." That view was echoed in an article in the *New York Times*, which spoke of "revolutionary practical applications—for example, inducing bacteria to grow scarce substances such as insulin."

Indeed, biologists were coming face to face with what Peter Medawar, the British Nobel laureate and essayist, once referred to as the "dire equation" of basic research—*Uselessness = Good*. Recombinant DNA now offered an opportunity to rewrite the equation, giving greater weight to practicality. And of all the scientists in the country who could exploit these possibilities, no one was better situated than Herb Boyer. "He was in the position," says one former lab member, "to look out into his lab and see some of the latest developments in terms of techniques, so he was really in a unique situation. In that two- or three-year period, from 1975 to 1977, the departments of biochemistry and microbiology at UCSF were essentially *where* genetic engineering was happening." Indeed, Boyer did not have to look far for the best and the brightest in the field. They all seemed to shuffle in, in their running shoes and blue jeans, each morning at UCSF.

It was a small group of researchers who collectively were inventing an entire technology as they went along. If technology provides the tools for scientific revolutions, Boyer's modest fourth-floor laboratory at UC–San Francisco was a kind of high-tech blacksmith's shop where the finest implements were crafted. Five floors up, in Howard Goodman's laboratory, these same tools were tempered and tested in state-of-the-art cloning experiments.

The two laboratories, although in separate departments of the university, worked hand in hand, prospering intellectually from the people who had assembled there. And since part of the talent of running a laboratory is identifying and attracting good researchers, both Boyer and Goodman made

78
. . . . . .

some blue-chip recruits in the mid-1970s. "We all came at the same time, more or less, and we all came from different parts of the world, and we all brought different expertises," recalls Herb Heyneker. "That is really, in a nutshell, what made this so *extremely* successful."

In Heyneker, the Boyer lab possessed an extremely gifted biochemist, a vintner of enzymes so extraordinarily pure that they were as prized as a rare Bordeaux. A slender Dutchman with a long, narrow face, high forehead, and animated blue eyes, Heyneker had a cheery yet brisk air about him. He had been working at the University of Leiden, where he had isolated and purified several "repair enzymes," which proofread and repair DNA damaged by radiation.

Heyneker met Boyer at a scientific meeting in Belgium. Boyer described the early Cohen-Boyer experiments, and Heyneker, excited by the implications of gene-splicing, boldly approached him and asked to do a postdoc in California. Boyer surprised Heyneker by simply looking at him and saying, "Sure." That was in 1973. Heyneker had doubts that Boyer had taken him seriously, but the laconic invitation was formalized a year later. Armed with a one-year grant from the NATO-funded Netherlands Organization for the Advancement of Pure Research, Heyneker arrived in San Francisco in the fall of 1975. More important, he arrived clutching a thermos, and in that thermos were vials of terribly precious enzymes he had made and purified with his own hands. At that time, chemical supply houses had not yet begun to make and market restriction enzymes, ligases, and other condiments of the cloning profession. "In those very early days," Boyer explains, "getting the enzymes was not easy. You had to make your own, or you begged, borrowed, or stole them." Indeed, the sociology of collaboration during the early days of cloning is full of bartering for enzymes, wining and dining for bits of synthetic DNA, transcontinental swaps of plasmids.

Heyneker joined other key people in the lab. Francisco "Paco" Bolivar was a bearded, rail-thin Mexican postdoc who was experimenting with the development of plasmids. Raymond Rodriguez, a native Californian, had joined the Boyer lab in 1974 and was also working with plasmids. Postdoc Patricia Greene made exceptionally pure enzymes and technician Mary Betlach fashioned state-of-the-art plasmids. Self-starters all, they plunged into the work, and, to hear the postdocs tell it, Boyer was hard pressed to keep up. "He was trying to ride herd," recalls one former postdoc, "on a lot of very willful horses."

Up on the ninth floor, meanwhile, three talented foreign postdocs set up benches in the Goodman lab—Peter Seeburg, John Shine, and Axel Ullrich. Peter Seeburg was a tall, lanky German with a wispy mustache; his scientific resourcefulness was widely admired, even though his independent streak was an occasional source of friction. The soft-spoken and astute John Shine arrived from Canberra in the fall of 1975 and was perhaps the most scientifically innovative of the group; as one researcher later remarked, "Whatever John Shine worked on *moved.*" Axel Ullrich came by way of the University of Heidelberg. He had expressed interest, quite ignorant of Genentech's intentions, in going after the insulin gene. As one researcher recalls of the two German researchers, "They were good. And they knew it. They were very pushy, very aggressive, very ambitious, and very smart."

The three Goodman postdocs were highly motivated and highly independent, which had both scientific and nonscientific ramifications during what turned out to be a brief but historic period of work. Working closely with the Boyer postdocs, they formed UCSF's critical mass. "Whatever we did turned into gold," Heyneker recalls. "It *all* worked. It was absolutely amazing."

Heyneker settled into the routine right away. The Boyer lab was small and crowded; Heyneker's first "desk" was a clipboard placed flat over a sink. But the intellectual environment was in full flower, with a constant cross-pollination of ideas with the Goodman group. In the corridors outside the laboratory, where perhaps the most substantive exchanges of scientific ideas take place in contemporary institutions, the two groups supplied each other with materials, ideas, mutual enthusiasm, and a little friendly competition. A fair amount of social cogitation was also accomplished at the Chelsea Pub, a few blocks down Irving Street. To the young scientists who lived it, it was a magic period.

Boyer's lab early on earned the reputation as the leading design center for versatile, imaginative new cloning vectors. At that time, its particular métier was the design of plasmids. Plasmids are to genetic engineering what chips are to computers: they can be removed from a larger system (in this case a bacterium rather than a computer), "wired" with a particular genetic program, and reintegrated into a living biological machine. And yet they were simply hooplets of bacterial DNA that floated around in the jelly of the cell, quite apart from *E. coli*'s main DNA. Like artists signing their canvases with large letters, molecular biologists christened each plasmid they created with their initials, giving it a name and number of its own. Mary Betlach originated the pMB series, and Bolivar and Rodriguez produced the pBR series. The plasmids were not living things; they were merely loops of DNA, so stable and resilient that they could be dried and shipped in test tubes around the world. Boyer's laboratory routinely shared its creations with hundreds of other laboratories, a form of scientific magnanimity which later caused Boyer considerable political distress.

With everything falling into place, Boyer was convinced the timing was right to go commercial. Boyer's business input was essentially technical, but it was also momentous in a way that few people in the business world were equipped to appreciate. Among all the decisions made by all the lab-coat executives at all the biotechnology companies in the 1970s and 1980s, probably the most implausible (then) and most admired (now, at least by scientists) was the decision Herb Boyer made to use synthetic DNA. The first step in any attempt to make insulin was to get your hands on the insulin gene itself. You could build it from scratch (the synthetic DNA approach), or you could isolate the messenger RNA for the gene and make complementary DNA (the cDNA approach). Wally Gilbert's lab and the Rutter-Goodman group at UCSF were taking the cDNA approach. Boyer, against the prevailing wisdom of the day, opted for synthetic DNA.

"I feel terribly immodest saying this," Boyer admits now, "but the chemical synthesis of DNA was waiting for the cloning techniques to make its usefulness known. . . . I felt that its real value was being able to manipulate DNA. You could synthesize a gene or modify a gene or make adaptors to a gene. I think people did not have the means to utilize all the terrific things it could do."

Opposition to recombinant DNA work had somewhat influenced the decision, too. Boyer feared, correctly as it turned out, that the hostile political climate and regulatory uncertainty might ultimately interfere with the work; the NIH guidelines, issued in June 1976, in fact stipulated that any recombinant work with human genetic material would have to be conducted in the equivalent of a germ-warfare laboratory. At the time, there was not a single such laboratory in the United States easily accessible to researchers for such work. The original NIH guidelines, however, said nothing about synthetic DNA; it had never been considered a feasible alternative to the cDNA approach. With their synthetic DNA, therefore, Boyer's troops could march through a loophole in the guidelines and avoid some truly nightmarish logistical constraints.

In those days, choice of strategy almost automatically dictated choice of personnel. Having decided on the synthetic DNA route, Genentech chose Keiichi Itakura to be its chemist. And having considered the various markets, the company chose insulin as the most promising target. So insulin it would be, and by the synthetic DNA route. That was why Boyer decided to call Art Riggs in January 1976. And that was when Riggs rocked the boat by saying, "Could I interest you in doing somatostatin?"

Although he is loath to betray the slightest intimation of discord in his organization, Bob Swanson clearly did not like the idea. It probably doesn't stretch things to say that the company's very first business disagreement began when Swanson heard that funny word. Somatostatin.

It was sometime toward the end of 1975 that Riggs and Itakura sat down with a piece of paper and designed a human gene. Consulting the genetic code as if it were a foreign dictionary, they created "words" of DNA that coded for somatostatin.

As a scientific target, somatostatin was of obvious interest; as a commercial target, no one heard cash registers jangling. In the case of Genentech, there was only one cash register that counted. His name was Thomas Perkins, and he headed the successful venture capital firm Kleiner & Perkins, where Swanson had cut his entrepreneurial teeth. Handsome, wealthy, with a reputation for gilding new high-tech companies with his financier's Midas touch, Perkins was expected to arrange the initial funding for Genentech—*if* the project sounded feasible and convincing. Art Riggs's first hard sell was somatostatin.

This protein had been discovered only in 1973. Like insulin, somatostatin was a hormone—a powerful chemical secreted by the hypothalamus, at the base of the brain. Somatostatin seemed to inhibit the release of growth hormone; in cases where somatostatin is absent, the result is a relatively rare disorder known as gigantism, a form of unbridled growth. Medical demand for the hormone was not great, but then neither was the supply—milligram amounts were recovered, at great expense, from the brains of hundreds of thousands of freshly slaughtered sheep and pigs. The hormone could be chemically synthesized, but the cost was $55 per milligram. Riggs and Itakura predicted they could get bacteria to produce the same amount for about a penny.

Somatostatin's small size made it particularly appealing to Riggs and Itakura. At fourteen amino acids, it was a very small molecule, which meant it would be feasible for Itakura to synthesize the gene for it. Moreover, Riggs knew that there was a laboratory assay for somatostatin so sensitive that it could detect the presence of a mere molecule or two per bacterium. Hence they could prove the existence of even scant amounts. Itakura's gene was different from nature's gene. He used the same four letters of the genetic alphabet to build the fourteen words, or codons, that spell somatostatin. But using a kind of biochemical thesaurus, he selected particular synonyms—different codons for the same amino acids—that he knew were likely to be more easily read by bacterial enzymes (bacteria, oddly enough, tended to "stutter" on certain combinations, for example).

In a forty-page grant proposal, which was submitted to the National Institutes of Health for review on February 26, 1976, Riggs and Itakura asked the federal government for $401,426 to fund the somatostatin project. Whether it had any commercial value or not, Riggs felt committed to the project. "Itakura and I were going to do it, just as a scientific project," Riggs recalls, "to show that we did understand enough about the workings of the bacterial cell to go to the genetic code tables, write out a sequence of a gene, make the gene, put it in the bacteria, and have it function. So that was a scientific project which, quite frankly, I thought was of the highest order—totally independent of whether it would ever have a practical application or not." That was the argument Riggs waged, and he finally won over the businessmen. "Somatostatin could be interesting in and of itself, and it was argued that way," Swanson recalls. "And the technology needed to take one step at a time. Finally Art convinced us that this was the right way to go." Swanson was persuaded, and so was Perkins.

The decision also established an important precedent for the young company. "I think one of the things that has driven Genentech is a philosophy that maybe came out of the resolution of the somatostatin-insulin thing," Swanson says now, "and that is to make sure that different elements of the technology that you are putting together for an achievable result are all doable. If you need five different pieces of technology in order to achieve a result, you have to make sure that each of them is doable, rather than require one or two breakthroughs along the way. There was a clear desire to make sure we could *do* what we said we could do."

On April 7, 1976, just weeks after the synthetic operator experiment, the prospects for "doability" looked sufficiently rosy for Boyer and Swanson to incorporate. Perkins launched the venture with $100,000 in seed money. After a few false starts in the name category, including the wisely jettisoned Her-Bob, they settled on the name Genentech (accent on the second syllable), which stood for genetic engineering technology. The first project on Genentech's agenda was the somatostatin gene. Keiichi Itakura and Art Riggs, informally considered cofounders, received the first contract issued by the young company, worth $300,000. But there was no question that somatostatin was a guinea pig molecule. "Insulin was definitely part of the contract," Riggs says. "The plan was just to demonstrate feasibility, and obtain patents, with somatostatin, then immediately go for insulin."

With Swanson and Boyer's commitment to the project, Riggs and Itakura decided they could not accept any NIH grant money, but they didn't withdraw their application, because they were curious to see what the response would be. As it turns out, Itakura received word that the NIH declined to fund the somatostatin work. "The most significant criticism by the NIH study section was that it was an intellectual exercise," Itakura recalls with a laugh. "I didn't understand [it] at that time. Well, I *still* don't understand it. Because this was a very practical approach to produce peptide hormones, and even proteins for use in the treatment of patients. . . . The other criticism was that they didn't think I could make a gene in three years." The ambitious Itakura set out to prove them wrong on both counts.

Riggs, who had submitted his share of grant proposals, could read between the lines of the rejection. "They just didn't think we could do it," he says. "They just didn't believe in DNA chemistry. They didn't believe it could be done."

In fact, the scientists weren't all that sure, either. Riggs remembers that the scientists developed a "syndrome" of saying, "Yes, of course," whenever Swanson asked, "Can you make somatostatin?" or "Can you make insulin?"

"There's a time to be optimistic, and a time to be pessimistic," Riggs says. "This was a time to be optimistic. We were never sure it could be done, but we didn't want to admit that."

All this took place behind the scenes, and it wasn't until October 28, 1976, that the results of the synthetic *lac* operator experiment appeared in the scientific literature. By then, Genentech's scientists had fitfully started out trying to build and clone their human gene. Almost inevitably, they ran into a few snags along the way, and Swanson naturally repeated the crucial question: "Are you sure you can make it?"

And the scientists replied, "Yes, of course."

S e v e n

"A Mad Dash
with Odd Bedfellows"

When Axel Ullrich arrived from West Germany to begin his postdoctoral fellowship at the University of California–San Francisco on October 17, 1975, he was about to turn thirty-two years old and came with aspirations that were ambitious, but not immodestly so. He wanted, in his own words, "to learn English, to do some interesting work that would get me a job, and to go back to Germany after two years."

The work he had in mind was a good deal more than merely interesting. He wanted to clone the insulin gene. His doctoral work at Heidelberg University had been headed in that direction. "At that time," he says, "the whole idea that you could clone something, combine two different pieces of DNA, was of course already available. But all the ideas related to applications, even to any major scientific application, had not been done. People had pieces of DNA, but didn't know what they were and didn't know how to approach this problem."

Ullrich could not have picked a more auspicious time to show up in San

Francisco to attack that type of problem. In the mid-1970s, the biochemistry and biophysics department of UCSF held a position second to none in terms of recombinant DNA research. The department laboratories, located in the Medical Center on Parnassus Avenue, not only provided a stimulating scientific ambiance, but boasted splendid physical environs as well. The laboratory of Howard Goodman, where Ullrich worked, was located on the ninth floor of Health Sciences West, one of a pair of research towers planted on the north slope of Mount Sutro, and Ullrich's bench was a few steps away from a spectacular panorama. To the north, overlooking Golden Gate Park and San Francisco's Richmond district, the stunning view took in the towers of the Golden Gate Bridge poking up like two rusty wickets above the trees, and the dramatic Marin headlands and Mount Tamalpais farther north; to the west, one could follow the line of Judah Street as it ran out of continent and into the gray Pacific. Like the earlier explorer Sir Francis Drake, who had poked into nameless bays and waterways within view of these ninth-floor windows, the UC–San Francisco researchers found themselves poking around the edges and inlets of biological unknowns. "We didn't even know what we were doing in those days," laughs one member of the Goodman lab, "but I guess that's why it's called research."

In another sense, Ullrich could not have picked a worse time to set up shop at UCSF. What Ullrich did not know, what no one at UCSF could have foreseen at that point, unfortunately, was that important and ambitious recombinant DNA projects like the cloning of insulin would initiate, for a complex mix of reasons, a period of great turmoil at UCSF. Unexpected celebrity, ill-advised rivalries, disputes over priority and credit, personal jealousies, the possibility of great wealth—all enveloped the twin towers on the hill like the swift-moving fogs off the ocean. No one could have quite prepared for it, and the UCSF biochemistry department became, unwittingly and rather traumatically, a social laboratory for all the growing pains associated with fame and achievement in genetic engineering. A member of the Goodman lab remembers, "Everyone knew the discoveries were right around the corner, here and all over the world. Being at San Francisco and having the technology here, people were racing for the finish line. They wanted to be first."

The first inklings of trouble appeared as early as 1976. The fact that Genentech had contracted for work in Herb Boyer's lab inspired both concern and controversy. Almost simultaneously, the familylike collaboration between

Boyer and Howard Goodman came to an abrupt conclusion, amid much speculation by people in the department. The split, some said, reflected the excitement and ambition coursing through the department. "I think they both thought that in the early days of cloning, they had struck a gold mine," theorizes a faculty colleague, "and they both wanted to exploit it independently." They were hardly alone.

These competitive ambitions tended to complicate an already difficult scientific relationship, that between lab chief and postdoctoral fellow, as Ullrich would soon learn. Postdocs are young scientists, fresh out of graduate school, Ph.D.s in hand, who sign up for fellowships and pursue pure research. While the lab chief is busy procuring supplies, juggling administrative chores, applying for grants, attending faculty senate meetings and committees, traveling to meetings, quelling mutinies, recruiting next year's postdocs, and in general imbuing each lab with its intellectual identity and spiritual élan, the postdoc slaves away in a frenzy of pure science. The postdoc has no meetings to attend, no bureaucratic duties to perform. His or her lone responsibility, for an annual stipend that in the 1970s was little more than $10,000, is to be at the bench. The postdoc lives and breathes and dreams science.

If the conceptual outlines of a project originate with the lab chief, the hands-on and day-to-day problem-solving research is actually performed by the postdocs, and the closer the work is to the leading edge, the likelier it is that the postdocs, in solving those day-to-day problems, are doing considerable trail-blazing on their own. Sometimes lab chiefs discourage postdocs from tackling particularly difficult projects. This can be a psychological ploy to fire up the postdocs and goad them into defying authority, or it can simply reflect caution on the part of the lab chief, who worries about galloping budgets as well as errant scientific imaginations. It is an area where faith and doubt often rub against each other, and the friction is almost guaranteed to generate a lot of interpersonal heat.

The biochemistry department of UCSF, Ullrich discovered, was highly competitive, crowded with postdoctoral fellows all scrambling to produce an eye-catching piece of work that would earn them a faculty position somewhere. The researchers themselves referred to it as a "postdoc mill." The atmosphere created by this overcrowded, underpaid, teeming, overachieving mass of talent, Ullrich quickly realized, was "one of the secrets of the success of American science—that you, as a postdoc, have this pressure to produce something that will get you a job."

With ample amounts of both scientific naiveté and personal faith, Ullrich advanced his case for working on the insulin gene. Even he concedes that the younger scientists like himself gravitated to projects like this "without realizing the magnitude of the technical problems." Goodman suggested that his German postdoc work instead on another project involving yeast cells. Ullrich, without much enthusiasm, started to work with yeast. It was one of those classic conflicts of faith and doubt.

Postdocs do not routinely give up on a good idea simply because it seems futuristic and difficult. They tend to strike off in their own direction; indeed, it appears part of a risky and often surreptitious ritual passage toward independence. Ullrich followed this time-honored tradition. With the insulin idea foremost in his mind, with Howard Goodman by all accounts less than enthusiastic, they approached another professor at UCSF named Gerald Grodsky, a pancreas researcher, and plotted out a collaboration on insulin together. Grodsky, with his expertise in pancreatic physiology, would try to isolate those scarce islet cells where insulin was made; using those cells, Ullrich could attempt to fish out the messenger RNA for insulin and clone the gene, using essentially the same approach that Argiris Efstratiadis and his Harvard colleagues had worked out for globin.

They had hardly begun, however, before they ran into obstacles of a nonscientific sort. The problem surfaced in late 1975, Ullrich recalls, when Donald Steiner came to UCSF and gave a talk about the recent work coming out of his University of Chicago laboratory.

One of the most attentive listeners at the seminar was Ullrich. He had already started working with Gerald Grodsky and was moving forward with his insulin plans. William Rutter, the chairman of the biochemistry department, was also there, along with one of his postdocs, John Chirgwin. Shortly after Donald Steiner finished his talk, Ullrich recalls, Grodsky introduced the German scientist to Steiner with the words "This is Axel Ullrich. He is working on translation and attempting to clone the rat insulin cDNA." At that particular moment, Bill Rutter happened to be standing within earshot with Chirgwin. Rutter promptly said, "Well, here's John Chirgwin. He's doing the same thing!"

It was an embarrassingly public way to establish that you were standing on a colleague's toes, and vice versa. "Everyone," Ullrich admits, "felt a little strange."

Since it was the end of the day, they all rode down to the ground floor in the same elevator. Ullrich did not speak English very well at the time, but he knew enough to convey how stunned he was to learn that someone else in the same department was working on the same problem—an unanticipated rivalry that could potentially result in a waste of time, energy, and money, to say nothing of intramural goodwill. "Well," he recalls telling Rutter, "I'm really shocked."

Rutter was called off guard, too. "The reason we were surprised to learn that [Howard's group] was interested," he would say later, "was that he hadn't had any background interest in that area, and although he was in the same department, he'd previously been looking at—in collaboration with a number of people, essentially at the genetic level—viral and microbial systems."

As the elevator continued downward, Rutter replied to Ullrich, "Well, we have to get together and talk about this."

In its own ironic, microcosmic way, this little turf skirmish on the ninth floor perfectly captured the scientific clashes of the recombinant DNA era, when cloners from the molecular biology side stepped on the toes of scientists in their rush to clone important genes. Although intradepartmental competitions on scientific problems are not unheard of, they are indulged in at considerable risk to general morale. They can be divisive, expensive, redundant, and, worst of all, redundantly inconclusive. The issue was further complicated by the proprietary way in which scientists view their systems and their research. Although neither researcher chooses to characterize it this way, one can easily imagine the awkwardness when one of Goodman's postdocs, a newly arrived foreigner in the department, was discovered to be poaching on the department chairman's pet research interest.

Rutter happened to be the most powerful member of the UCSF biochemistry department, and in the view of some the person most responsible for shaping it into a world-class group. The department had been in decline only a few years earlier—had, in the words of one faculty member, "sort of decayed away." For a number of years in the 1960s, it did not even have a regular chairman. Into that rudderless turmoil stepped Rutter, who had been persuaded to leave the University of Washington in 1969 to take over the chairmanship of the department.

Rutter hailed originally from Malad City, Idaho, a small town about ninety miles north of Salt Lake City, where he was undoubtedly the only student in his high school class to express interest in attending the School of Tropical Medicine in Calcutta. That unusual ambition stemmed from his grandfather, who had served as an officer in the British army in India and had intrigued his grandson with tales of the exotic tropical diseases found there. Rutter traveled east only as far as Harvard for his undergraduate education, however, and then went on to get a Ph.D. at the University of Illinois.

He quickly established a reputation as an able administrator. He approached any job with a tremendous amount of energy and conscientiousness, and he was not afraid to manage people; "Bill uses people," says a former associate, "but he uses them well." When there was competition, he rose to the challenge. When there was a problem, he would step forward to put out the fire. He had "a great sense of responsibility for things that happen in his department," according to a former faculty member. He typically took on a lot of commitments. He always found time for one more committee membership at the National Academy of Sciences, one more charity foundation to assist, one more editorial board to serve on, one more industrial consultancy to fill. And one more mountain to scale or ski—those were his recreations.

Rutter engineered a massive overhaul as soon as he took over as department chairman in 1969. Unlike many university laboratories which were dedicated to studying the genetics of bacteria (or prokaryotes), Rutter emphasized the study of the more difficult higher organisms. This departmental commitment to the study of genes in eukaryotes, how they were regulated and expressed, generated much scientific excitement, and that in turn helped create a "big family atmosphere" within the department during what some have wistfully described as "the early days."

In addition to his administrative duties, Rutter maintained a large laboratory filled with a dozen or so postdocs. His major research emphasis was biological differentiation—how the cells of a developing organism, deriving from a single fertilized egg, ramify into more specialized cell types and tissues. And it was the pancreas, that curious dual-role organ, that was his "system" (biologists use the word "system" to connote a biological microcosm, a narrow area of focus, that they hope will illuminate much broader conceptual territory).

The advent of recombinant DNA technology provided Rutter's lab with a powerful approach to questions of differentiation. "When it became evident that you could consider cloning shortly after Herbert Boyer and Stanley Cohen had done those experiments," Rutter recalls, "then it was obvious that we should shift our program prominently, or even dominantly—and, as it turned out, almost solely—to the issues of isolating the genes as a prelude to finding out how they were expressed." The pancreas was home to an interesting collection of genes for such hormones as insulin, glucagon, and even somatostatin. "So the question was, which one of those sets of genes were we going to look at?" Rutter recalls. Insulin quickly became the leading candidate. "First of all," Rutter says, "it was a small gene and therefore in a sense more readily analyzable. Secondly, we knew a good deal about it. Thirdly, it was related to a disease and was of medical significance. . . ."

Rutter was a veteran in the field of pancreatic research, so when he convened a meeting with Goodman, Ullrich, and Chirgwin shortly after the Steiner seminar to iron out the insulin problem, the thrust of the discussion could not have surprised anyone. "At that meeting," Ullrich recalls, "Rutter, as the chairman of the department, decided more or less that my collaboration with Gerry Grodsky was not happening anymore. . . . He more or less forced us to collaborate with him." Grodsky confirms, "We got boxed out of it right away." Goodman did not protest the decision, Ullrich recalls, and neither did he. "I didn't really see what consequences that would have," Ullrich would say later, "only that there were more people involved in the work, but that the scientific problem would remain the same."

And so, around the beginning of 1976, the Rutter lab and the Goodman lab formed an unplanned alliance in quest of the insulin gene. Rutter denies the arrangement was awkward. Nonetheless, the Rutter-Goodman collaboration achieved a certain notoriety both within UCSF and outside the institution. One outside researcher familiar with the arrangement described it as a "marriage of distress." Another researcher in the insulin field recalls, "The fact that in Rutter's lab no one really was into recombinant DNA technology just made it difficult for Goodman and his people to handle."

Ullrich, getting his baptism in American science under unexpectedly close fire, found the situation "more or less a controlled competition within the same department." Other researchers within the department viewed the arrangement with dismay. In time, internecine competitions began to spring up between labs, between lab chiefs, between postdocs in the same lab, and between postdocs from different labs, as scientists all began to gravitate toward ambitious cloning projects. "It became a mad dash," one researcher remarked years later, "of odd bedfellows."

Axel Ullrich thrived—as healthy egos often do—in that most fertile and *infirma* of scientific soil, the place where conventional wisdom (or the lab director) insists something cannot be done while the individual scientist, or the enterprising cohort, insists it can. Howard Goodman, the lab chief, had a reputation for being intense, careful and cautious; "Dealing with people," one former postdoc recalls, "was not his forte." He insisted on absolute scientific correctness, and created what one researcher described as a "pressure-cooker lab." "I think in the beginning," says another former lab member, "Howard thought Axel was not a good enough scientist to invest all that time, money, and expense into his experiments."

Personable, self-confident, good-humored, and strong-willed, Ullrich was not easily discouraged. He was of medium height but broad-shouldered, with straight black hair falling across a broad forehead; and although he wore glasses and spoke with a very deliberate, Germanic kind of weightiness, the overall effect was one of physical presence: persistence, tenacity, endurance. He had the hands as well as the mind of a good scientist.

Ullrich was born in 1943 in German Silesia, in a town called Lauban, now part of Poland, and his childhood had been made itinerant by the geopolitical convulsions that reconfigured Eastern Europe after World War II. At the time, his father served in the German army; soon after, the family moved to another town in the Sudetenland, Kostany, which at war's end became part of Czechoslovakia. After the war, Josef Ullrich was a prisoner of war in northern Germany. "So my mother, my brother, and I were in Czechoslovakia after the border closed, after the end of the war," Ullrich recalls. "We had the possibility, the option, to leave, but then we couldn't come back anymore." The family finally relocated in early 1948 to a small community in the lower Rhine Valley called Rastatt.

There was no family background in science. Ullrich's father opened a grocery story; that small business financed the educations of Axel and his brother. Ullrich had an early interest in plants and animals, but was advised in school that you couldn't make a living as a biologist unless you became a teacher. "I didn't want to teach," he recalls. "I never liked teachers, and therefore I didn't want to be one." But he read about the "new" science of biochemistry, thought the prospects more appealing in that field, and decided to attend the University of Tübingen.

Tübingen was a most appropriate place for this career to get launched: it was there, in 1869, that Johann Friedrich Miescher discovered DNA. Miescher had isolated an acidic substance from the large nuclei of white blood cells, which he had collected from pus-filled surgical dressings; this so-called "nucleic acid" turned out to be DNA. After earning degrees at Tübingen and Heidelberg, Ullrich received fellowship money to do research in the United States. He was surprised by the competitiveness of the new environment. Surprised, but not unprepared.

Ullrich would routinely slip into the lab early in the morning, around two o'clock, when no one else was around, and sometimes work through until ten in the evening. "By the time we got into the lab, around nine-thirty or ten o'clock, he'd have half his day done," says Fran DeNoto, who was a technician in the Goodman lab at the time. He worked with great patience and perseverance, but he also prided himself on having "the killer instinct," always working toward the crucial solution. He savored his successes. DeNoto recalls showing up in the lab one morning after Ullrich had achieved some particularly encouraging results in an early-morning experiment; there was a bottle of champagne on ice, oysters already cracked open.

During most of 1976, unfortunately, such celebrations were few and far between. Although the collaboration between the Rutter and Goodman labs was less than a match made in heaven, the postdocs on both sides knew that success could be achieved only through scientific cooperation. Ullrich could not get messenger RNA without the expertise in pancreatic systems provided by John Chirgwin and Rutter's longtime research associate, the Swiss cell biologist Raymond Pictet. Rutter's group had no hope of isolating the insulin gene out of the welter of pancreatic messages without the cloning skills of someone like Ullrich. The result could not be confirmed without John Shine, a colleague of Ullrich's, who had mastered Maxam-Gilbert sequencing. So it fell to the postdocs, as it usually did, to do the actual problem-solving and get the data.

Two interrelated problems stood in the way of getting to the insulin gene. The researchers had to find a decent source of insulin messenger RNA and isolate it. The likeliest source was the pancreas of rats, but the rat pancreas, more than other possible sources, was aswim with a horrifically swift and destructive enzyme called ribonuclease (or RNase), and that complicated matters.

Ribonuclease chews up RNA. There is a lot of RNase in the rat pancreas because it's a digestive enzyme, and a powerful one. "Rats eat pretty nasty things," observes John Chirgwin, "so they have pretty tough digestive systems." When RNase meets up with RNA, the destruction of the RNA occurs within a matter of seconds. In the pancreas, RNase sticks to the digestive (or exocrine) part of the organ, and the insulin message sticks to the secretory (or endocrine) part of the organ, and ordinarily they are as separate as plumbing from wiring within the same house. But when the two tissues mingle, as almost inevitably they would during the preparation of messenger RNA in this kind of experiment, the effect would be as disastrous as a live wire hitting water. The enzyme utterly erased the RNA. "I would say the destruction is instantaneous," says Chirgwin.

How could they neutralize ribonuclease rapidly? Chirgwin, a scholarly and methodical British-born researcher, had spent nearly all of 1975 wrestling with the problem; it in fact blocked an entire avenue of research in the Rutter lab. If not overcome, it made the isolation of any kind of messenger RNA from the rat pancreas virtually impossible. In 1975 and 1976, nothing seemed to work, and John Chirgwin was getting nothing but "no" answers to his scientific questions, or as he puts it, "a year's worth of notebooks on how *not* to make RNA from pancreases."

Other people in the lab wondered just what Chirgwin was up to, including Bill Rutter. Rutter had the reputation both for being patient and for tolerating prickly personalities, members of his lab recall, as long as the scientist produced good results. Chirgwin's problem, as he himself admits, was that he wasn't producing any results, period. Finally, around the summer of 1975, Rutter gave Chirgwin what amounted to an ultimatum. As Chirgwin recalls it, Rutter in effect said, "Well, you've been here a year, and you're working on a project that no one thinks can be done. You have a couple more months, and if you get nowhere, you'd better change to a more sensible project or find somewhere else to go and work."

Chirgwin recalls replying simply, "Yeah, okay."

It was an indication of just how much pressure the postdocs were under. "Bill's attitude was, those people who take no risks made few gains," Chirgwin says now. "If you were willing to gamble, it was fine with him, but it was your career that you were gambling with, not his career. And there were a number of people who went through San Francisco who have sort of vanished entirely scientifically." That, of course, lay at the heart of a postdoc's paranoia: becoming a scientific nonperson, trapped in some gulag of insolubility.

In March 1976, Chirgwin achieved the first break in the struggle. After combing reports in the literature dating back to the 1950s, he had separately tried using two chemicals, guanidine and thiocyanate, because both were known to deactivate RNase. They did, but much too slowly—five or ten minutes simply was too long to wait. Then Chirgwin thought of pairing them together, using a compound known as guanidinium thiocyanate. As often happens in chemistry, where the total synergistic effect is greater than the sum of its parts, this compound knocked out all the RNase in a matter of seconds.

Chirgwin's discovery allowed the team to work with rat cells. The only problem was that they preferred not to. The insulin message was so rare and appeared in such small amounts that they would have to collect RNA molecules from huge numbers of rat pancreases, a singularly unpleasant task. To avoid that, Chirgwin, Pictet, and Ullrich poked through numerous animal pancreases in search of an alternative candidate.

Dog pancreases had lower amounts of ribonuclease, and might be a good bet. Both Chirgwin and Ullrich looked; the results were ambiguous. Was there likely to be more insulin message in a mature animal or in a developing fetus? At one point in 1976, Chirgwin traveled down to Houston, where, with Peter Lomedico of the University of Texas, he paid early-morning visits to slaughterhouses in order to collect the pancreases of fetal cows. Fetal cows, it was thought, might have less RNase and more insulin message. They did, but not enough to help the UCSF researchers.

The problem, as scientists like to say, was not trivial. Out of the Babel of RNA dispatches produced in every cell and tissue type, they had to pluck out the single message for insulin. The message was fragile, perishable, and in infinitesimally small amounts.

That is how far the UCSF group had gotten, which is to say not far at all, by the spring of 1976, when a new sense of urgency infected the insulin project. The presence of Wally Gilbert at the Eli Lilly meeting in May surprised everyone and made it clear that a formidable group of researchers at Harvard with an excellent track record was interested in the same project. Then came word from William Chick, within weeks, that the much-coveted rat insulinoma was not available in sufficient quantity to share with the San Francisco group.

It is difficult to convey to the nonscientist the impact of this news on the UCSF group. In terms of getting the right kind of RNA for their cloning experiment, it was like being told, at the outset of a race, that your competitor had the luxury of taking a tunnel through a mountain, while you were obliged to find a way over the top—by a route no one had ever successfully negotiated. "That set it up as kind of an East Coast–West Coast pennant race," Chirgwin remembers.

"Bill always thought he had the inside track to that insulinoma, and then was very shocked one day when he learned that it had gone to Harvard," recalls a colleague. "The UC story is that Gilbert snatched it out from in front of him." Regardless of what happened, Gilbert's sudden interest in insulin was viewed with some skepticism on the West Coast. As one of the UCSF postdocs put it, "We felt Wally Gilbert wanted to clone insulin just because it was glamorous." "Which was the same reason Howard wanted to do it, and why we wanted to do it," adds Ullrich.

There was nothing glamorous about trying to get good insulin RNA out of the pancreas, and the UC group was getting nowhere in its attempts. When scientists confront an impasse of this sort, an interesting dialectic develops. Many problems are soluble if one stoops to a "brute-force" approach. Brute force is neither subtle nor elegant. It is, rather, an all-out assault in terms of manpower, money, and time in order to barge through a difficult obstacle. Most biologists prefer to flash scientific art rather than flex technological muscle, advertise their ingenuity rather than their budgets. But sometimes they have no choice.

More and more, it looked as if the only way to get a decent amount of RNA was to go the brute-force route and isolate the islet cells from rats. No one was eager to do it. It meant killing perhaps five hundred rats and surgically removing their pancreases. It meant a lot of people working in production-line style. It meant a lot of money.

"The idea of making enough islets to do an RNA prep on," Chirgwin says, recalling the group's antipathy, "was just *appalling.*"

While Chirgwin pawed around for the insulin message in various beasts, Axel Ullrich also looked for the insulin message in his own menagerie of possibility. Normally the work in collaborations is split up. "But because of this initial awkward arrangement," says Ullrich, "I continued to do what I was doing, and John Chirgwin continued to do what he was doing." The search continued to yield nothing.

In Ullrich's case, the frustrations were compounded by pressures that had only marginally to do with the elusiveness of messenger RNA. He and the other German postdoc in Goodman's lab, Peter Seeburg, were locked in a generally amiable but occasionally tense competition of their own. Both were trying to clone important mammalian genes; both were stymied through most of 1976. Seeburg's target was a larger protein, growth hormone. Neither researcher was making much headway, which was not surprising, since both projects were state-of-the-art, on the very edge of possibility. But, as many workers in the lab point out, neither Seeburg nor Ullrich was getting much encouragement from the top, either. And they were not the types to suffer in personable silence. "Peter and I seek the conflict," Ullrich admits, "and seek to make things bluntly clear."

For Seeburg, the situation in California had turned especially sour. He had come to UCSF to work with Herb Boyer, but that arrangement didn't work out and he moved upstairs to the Goodman lab, which was the only other place in the university where recombinant DNA technology was sufficiently far along. When Seeburg expressed interest in going after the growth hormone gene, he says, Goodman responded with skepticism. The politics of this collaboration, too, became trickily competitive, because Seeburg had linked up with a growth hormone expert outside the department, a doctor in the medical center named John Baxter, and Goodman reportedly bristled at outside involvement on the project. Seeburg felt Goodman's discouragement so intensely that at one point he recalls discussing the possibility of throwing it all in and returning to Germany with his wife; as it was, he routinely returned to the ninth-floor lab after hours, when Goodman had gone home, to work on isolating the growth hormone RNA from rat cells with Joe Martial, a postdoc in Baxter's lab. During the day, he toiled halfheartedly at another project suggested by Goodman. As a fellow lab worker recalls, "Howard wanted basically to kick Peter out of the lab. He said, 'Well, it will never work.' "(Goodman declined several interview requests to explain his version of events.)

Ullrich experienced similar problems. Goodman, he says, "continually tried to stop me from doing these experiments. At that time, I had already invested more than half a year, and I said, 'No, I'm just going to go on.' He wanted me to stop, to not work on this insulin project anymore, because it was too complicated, it would never work, and all these things." As even Ullrich admits, Goodman was not alone in his skepticism. Some of Axel's friends, in and out of the lab, urged him to drop the insulin work. Given Ullrich's stubborn determination, the appeals merely poured gasoline on the fire of what was becoming an increasingly isolated and very personal one-man scientific crusade. "From a certain point on," Ullrich says, "I just felt that I wanted to know if I could do it, if it was possible."

Then, in September 1976, Goodman went to Japan on a long-planned sabbatical. The lab director's absence had a decided impact on both the mood and pace of the work, according to workers in the lab. "As soon as he left, things worked much easier," Ullrich says. "Everything was much more straightforward because we didn't have to worry about getting his okay to do certain experiments." It was roughly at this point, too, that the three prime movers of the insulin project—Ullrich, Chirgwin, and Pictet—reached a major point of decision.

Competition triggered what Ullrich remembers as a "pretty desperate" situation. Word had drifted back from the East Coast that the Gilbert group was making real progress, and the UC options were grim: they could bite the bullet and go for the RNA message by brute force, or they could abandon the project altogether. "We first decided on islets," Raymond Pictet explains, "when we heard that Gilbert had synthesized beautiful single-strand DNA." That was toward the end of the summer of 1976. Chirgwin and Pictet— mostly as emissaries, partly as petitioners—broke the news to Rutter.

As Chirgwin recalls, "Raymond and I sort of said, 'Well, this fiddling around isn't looking very productive, and if you really want to do this, it's just going to have to be a lot of people and a lot of money.' " They knowingly played on Rutter's competitive nature, and it seemed to work. "Bill thrives on competition," Chirgwin says. "When there was real competition, then he was willing to go in, twist arms, and get half a dozen people to come and provide the labor, and let us go and order a thousand dollars' worth of collagenase at a shot, and buy rats at six hundred dollars an order and stuff like that."

That is exactly what happened. The right arms were twisted, the right purse strings loosened, and the UC–San Francisco team, against considerable odds, went hunting for the rat insulin message in the elusive islets of Langerhans. The flurry of activity, on one level, would have amused the researchers back at Harvard. "It's always possible for people to work with that feeling of, you know, 'Somebody else is on my tail, I've got to do everything,' " Walter Gilbert would remark later. "It's only an illusion."

E i g h t
"Gel Crap!"
. . . .

In the blissful, unworldly self-absorption of graduate studies, Forrest Fuller spent his days and weeks in the Harvard Bio Labs struggling with his cloning experiments, unaware that a furious effort to clone the insulin gene was roaring on the other side of the country. "At this point," he recalls, "I had *no* idea what was going on on the West Coast. I was *completely* in the dark." Only Gilbert's lab, he thought, would be pursuing the project. "I envisioned taking two years," he says. "You know, something that I would do carefully over a period of two years. With globin, before I even got to insulin. . . ." Instead, Fuller began to rethink the whole project, making the painful discovery "that this was life in the fast lane and that this was going to happen a lot faster, and maybe with more people involved, than I had wanted. I was naive about what competition was all about, and what life in the fast lane [was like]—or, as we used to call it, Big Astounding Science."

Big Astounding Science. Self-mocking humor aside, science as it was practiced in the Gilbert lab was big, and different. Only a generation earlier, biology had proceeded at a far less breakneck pace. At the time DNA's structure was discovered and the genetic code cracked, the founding fathers of modern molecular biology typically exchanged ideas and insights by letter and the community at large customarily learned of new discoveries in the literature.

The molecular biologists of the 1970s traveled at an entirely different speed. Gilbert encouraged his group to use the telephone to talk out ideas, sound out the latest developments, track down the best enzymes; if setting an example was the best kind of teaching, everyone in the lab had noticed that the numbers on Gilbert's office phone had practically been worn off with use. Important new results often made the rounds by telephone, grapevine, and seminar presentations long before they appeared in the literature. As one biologist puts it, "You say something today, and someone else in Europe knows it tomorrow." One heard stories of scientists flying across continents or overseas and hand-delivering manuscripts to establish priority.

"There are two things that sound trivial, but turn out not to be, that Wally used to foster good science in the lab," recalls Debra Peattie. "We had no limits on what we could buy as far as lab supplies or equipment. . . . None of us was restricted in any way to what we could use to try an experiment, and I think that's very important, because I've seen the toll that having to worry constantly about money takes. It *does* have an effect on the ability to try novel things. And the other thing that Wally did, by example and by general encouragement, was to get us to use the telephone to, in essence, save time scientifically. And those two trivial things, as I say, aren't really trivial. I think that they're really fundamental in terms of the way in which science was done in the lab. Elemental, I would say."

Paradoxically, for all the urgency and speed with which projects were pursued in the lab, Gilbert had the reputation of maintaining a loose rein. He was not the kind of lab director who demands to see weekly progress reports; it was much more his style to drop in unexpectedly on a student, his visit announced moments in advance by the smell of cigar smoke, and simply ask, "What's up?" Gilbert himself didn't pretend scientists were insensitive to competition; he acknowledges what he calls "an underlying striving for renown in scientists." But what he attempted to teach, by example and by assignment, was the selection of significant problems to investigate; with his graduate students in particular, he recognized the crucial importance of setting up projects that were "reasonably difficult, but not absurdly difficult." And as his MIT colleague Phil Sharp once observed, "It's the choice of the problem that science is all about."

To competition, Gilbert says, he reacted "generally by ignoring it." His students generally did not feel any explicit pressure to win races and finish first. "Wally never goes around and says, 'We're going to get beaten. Hurry up.' But instead," recalls Karen Talmadge, "he would say once in a while something like 'Well, you know, you could put in some overnights on this.' After a time, he didn't need to say things like that. You always knew what was going to get you there faster. And one was driven equally by the desire to know the answer."

Which is not to say Gilbert, his protestations aside, totally ignored competition. "Wally did *not* like to be beaten," Talmadge recalls. "He didn't like it one bit."

The insulin project in a way typified one of Gilbert's most favored dictums, imparted so often to students that it became an unofficial lab motto: "It is as easy to solve an important problem as a trivial problem."

As Arg Efstratiadis and Forrest Fuller mounted their two-pronged attack on insulin, however, they discovered that the problems were neither trivial nor easy. Nor were they as far along as their California counterparts feared. Efstratiadis still concentrated on isolating the right insulin message in order to make a cDNA gene. Fuller tried to figure out how to plug this piece of DNA into bacteria and get it expressed. Neither made a lot of headway.

It was an oddly endearing alliance, Forrest Fuller of Salt Lake City and Argiris Efstratiadis of Lesbos. Fuller, a free-wheeling personality but exacting chemist, measured everything carefully, plotted elaborate controls, circled his intellectual prey with care and deliberation. Efstratiadis's scientific style, like his personal style, was confrontational and direct, with bolts of intuitive brilliance. He made for a difficult but desirable collaborator because he was enormously energetic, focused keenly on problems, and had a sensitive, almost artistic feel for the kind of adjustments required to make an experiment work.

"He was the kind of person," Fuller recalls, "who would do an experiment and if it didn't work right, he would change twenty things. And if it worked, he would never go back and find out which one of the twenty worked.

And he had almost these incantations he'd say over his cDNA." For someone with a strict chemistry background, some of these rituals were a bit hard to swallow; Fuller recollects "all these little 'mysticisms' about the way he did science. And originally you looked at him like, 'This is the master.' Now some of this, you knew, was a little bit off the wall and probably wasn't necessary, but it worked. And you didn't argue with something that worked."

It was Efstratiadis's almost innate sense of nucleic acid chemistry that impressed Lydia Villa-Komaroff, who worked in the same lab. "Arg just seemed to know when you needed to have larger volumes or smaller volumes, when you could push the enzymes, when you needed less enzymes," she says. "He could look at a gel and tell you why it had worked or not. He seemed to feel with the enzymes. *Often,* I might add, in the absence of knowledge. . . . But that didn't matter. I think that he does know that stuff at some level that's deeply buried but accessible to him. He had golden hands. His experiments worked the first time. Often. Not always, but very often."

Any illusions that the messenger RNA for insulin could be quickly and easily isolated from William Chick's rat tumors evaporated as the work progressed. To his chagrin, Efstratiadis had big problems just breaking the material down. The rat tumors were a tangle of connective tissue, much like the gristle in a tough steak. In order to extract RNA, the tumors had to be homogenized, or minced up. Efstratiadis, perhaps not the most patient of experimentalists, destroyed two homogenizers (the lab equivalent of a blender) trying to chop up the tissues. In desperation, he resorted to that most medieval of pulverizing technologies, the mortar and pestle. With "an awful lot of sand" thrown in, he managed to grind the tumors down.

Then Efstratiadis moved on to the biochemistry of trying to isolate the messenger RNA segments from the ground-up tumors. This is one of molecular biology's loveliest tricks, and it relies on the principle of "sticky ends" to clever effect. Every molecule of messenger RNA, regardless of the message, has a "tail" of at least one hundred adenines, a long, uninterrupted sequence of As. To pluck out the message by the tail, biologists embed long strands of Ts in a cottonlike filter and percolate the RNA through it. Because As and Ts seize on each other like sticky ends, the long strands of As and Ts cling to each other like burrs, and all the messenger RNA gets snagged in the filter.

Clever procedure, but with dispiriting results in this case. Spoiled a bit by his work on globin, where nearly all the messenger RNA in the blood cells codes for globin, Efstratiadis began to realize that the percentage in the tumor cells was less. Much less, certainly, than a motherlode.

Efstratiadis, working in his cubicle down on the first floor of the Bio Labs, kibitzing with the Gilbert group on a regular basis, spent the second half of 1976 simply trying to fish out the elusive insulin message. His initial aim was to isolate and purify the insulin message "to homogeneity"—that is, with very little contamination—before going on to the next step: converting the RNA into DNA. To do this, he had to run these molecules through an elaborate, almost Rube Goldberg assortment of gels.

If there is any word that constantly crops up in the conversation of molecular biologists, it is "gels." They are either running gels or checking gels or analyzing gels. The gel is an enormously versatile tool. Like an X-ray, it allows scientists to visualize otherwise inaccessible physical objects, in this case tiny molecules; like a ruler, it can measure the lengths of these molecules; and like a series of graduated sieves, it can collect molecular fragments of a certain size in one spot.

Gel electrophoresis is the proper name for it, and it is based on a simple but ingenious principle. Molecules of DNA and RNA, because of their slight electrical charge, can be pushed by an electrical current down through a thin vertical slab of dense gelatinous material. This material, which can be as soft as Jell-O or as stiff as plastic, is usually held together between two pieces of glass. Over a fixed period of time, smaller molecules travel faster down through the gel than larger molecules, so DNA or RNA fragments can thus be separated by size. Efstratiadis, for example, needed to find RNA molecules measuring about 450 bases in length.

Gels did not always run flawlessly, of course. You could spend weeks preparing RNA or DNA for a gel run, only to discover that impurities in a test tube produced unreliable, aberrant results. It was common knowledge, not to mention the source of much Bio Labs humor, that Arg Efstratiadis reacted with an angry biological expletive whenever something went wrong with one of these experiments. "Gel crap!" he would cry out. "A sheetload of gel crap!" Soon everyone else had adopted the same term.

As it turns out, 1976 was the year of gel crap. Efstratiadis collected RNA from the tumor cells and ran it through three different kinds of gels. The hope was that a predominant type of RNA—the insulin message—would emerge after trips through all three gels. Since the molecules were "labeled" with a radioactive isotope, usually phosphorus-32, their little molecular congregation would show up as a black band on a piece of film.

Such "three-dimensional" purification procedures involved great tedium, and so the researchers would kill time by thinking up songs like "The 50 Millicurie Blues." It went, Efstratiadis recalls, something like this:

> *If you want your message to be extremely hot,*
> *Use 50 millicuries of P³² and smoke a little pot.*
> *New England Nuclear's prices are rather cheap, you'll find,*
> *So save your money for the weed that often blows your mind.*
> *But if you run the experiment while your mind is blown,*
> *The chances are, my dear friend, you'll label your own. . . .*

Like slowly panning for gold dust, each of Efstratiadis's gel runs accumulated a bit more of the desired RNA. "In the end it gave pretty good results," he explains. "You could see a band coming out of the rest of the material." They knew it was the insulin message, but there was so little of it that it could not be converted into a gene. Efstratiadis kept trying, for the better part of a year.

Just as Efstratiadis struggled to find an insulin message hidden in all his gels, Forrest Fuller struggled to work out a scheme for cloning the gene in bacteria. The Harvard researchers were thinking ahead to the day when they could pop the gene like a cassette tape into bacteria and get the bugs to "play" it. In those early days of genetic cutting and pasting, Fuller—the Sorcerer's Apprentice—was among the first people in the Gilbert lab to dabble in such sophisticated recombinations. Just where were they going to put Efstratiadis's gene? And how? Those were the questions Fuller wrestled with for much of 1976.

He started by creating an expression plasmid—one of those ringlets of DNA that had been particularly engineered to efficiently churn out the transcript of a gene. In genetic engineering, the engine driving this process is the promoter—that little control region of DNA that precedes each gene. The promoter controls how rapidly and how accurately the information in the gene is transcribed into messenger RNA. Some promoters, souped up by mutation, are more powerful than others. What Forrest Fuller did was cut a so-called super-promoter out from a viral gene and transplant it into a plasmid provided by Herb Boyer's lab, pMB9. In theory, he could splice in any gene he wanted—an insulin gene or a globin gene or whatever happened to be handy—right next to this promoter. If everything was perfectly aligned, if there wasn't a single base-pair out of place, the bacteria would handle the rest. Spurred into action by this supercharged promoter, microbes would read the gene, copy it into RNA, and build the protein to order. Uncomfortable with what he considered the "self-flattery, narcissism" of christening new plasmids with one's own initials, Forrest Fuller bucked tradition. He called his creations "Operator/Promoter" plasmids—pOP (or "pop") plasmids, for short.

Next, he tried to figure out a way to splice a gene into those plasmids, and this led to a fatefully unsuccessful choice of strategy. As soon as Wally Gilbert had "rediscovered" that special flush-end glue for DNA (T4 ligase) in 1975, he and Fuller set out to prove that joining blunt pieces of DNA together would work as well as joining sticky ends. The technique, ironically, inspired the West Coast biologists to develop linkers, and as Herb Boyer later says, "The fact that Wally was pushing blunt-end ligation, or confirming it, gave us the confidence to go ahead." In the fall of 1975, Fuller tried cloning the globin gene by this method. It did not work. In the spring of 1976, just a week before Efstratiadis and Gilbert went off to Indianapolis for the insulin symposium, Fuller finally got the difficult procedure to work in a control experiment. Upon routine repetition, it failed. One day the technique worked, the next day it didn't. Fuller could get butt-end cloning to work with small pieces of DNA, but whenever he tried gene-sized pieces of globin, it didn't. Was it the enzymes? The recipes? The DNA? "I went through this bad spell where I couldn't clone any of this stuff," Fuller says, "and it was driving me nuts." In June, he again got the technique to work. But the lack of consistency obviously caused concern.

At this point, Fuller began to feel the competitive fires of the insulin race licking around his ankles. The heat became palpable at a meeting that summer at Cold Spring Harbor. It was not the Rutter-Goodman researchers but Keiichi Itakura who inquired about the progress of the Harvard team. "I felt a *real* sense, at that point, of competition . . ." Fuller remembers. "I got *really* defensive. I never handled that kind of competition really well. So I was sort of close-mouthed about it, but at the same time I tried to make it sound like we knew exactly what we were doing. Itakura was very nice, and he said, 'I think this would be a great thing to do if you could clone this.' " At the time,

Fuller did not realize that a second group in California, Itakura along with Herb Boyer's Genentech crew, had plans to clone insulin, too.

Around this time, Fuller also began to feel a bit embattled. Efstratiadis wanted him to use globin DNA for his cloning test runs, he recalls; Gilbert pushed him to go straight for insulin. "And when Wally pushed," says Fuller, "he *pushed.*"

Exiled to Long Island for the month of August, where James Watson's Cold Spring Harbor Laboratory offered the kind of P3 facility Harvard didn't have, Forrest Fuller kept waiting for Efstratiadis to produce the insulin gene, kept toying with globin. He didn't make any scientific headlines, but by the end of 1976 the Sorcerer's Apprentice began to read ominous implications into the insulin competition. In Fuller's view, Wally Gilbert had "been put on the line. He's got the tumor. He's got *the* insulinoma. Everything's in his court. And Wally's not going to be outdone by anybody, so his ego is out there on the line in all of this." That was not without ramification for Fuller, who began to realize that "I was on the line because Wally's ego was on the line."

At year's end, Fuller privately made an important resolution. Throughout the summer and fall, Efstratiadis had been teaching him how to isolate globin RNA, convert it into DNA, and in effect construct a full-length globin gene. But Fuller became increasingly convinced that Efstratiadis's insulin RNA, for all its trips through his "three-dimensional gels," was not sufficiently pure. "I've gotta make my own cDNA to insulin at this point," he began telling himself. At least initially, he would have to do this without Efstratiadis's knowledge.

Fuller reached another conclusion at the same time, however—less momentous, more reassuring. Despite the slow progress and frequent setbacks, the Harvard team still had the inside track. With the insulinoma, Fuller thought, they had "a big corner on the market" in terms of cloning. "You know, who was going to clone from these goddam isles of Langerhans?" he wondered. "It was impossible to get that message up to that kind of abundancy."

The fact that the Rutter-Goodman group was going after those "goddam isles" became apparent to a few discerning members of the scientific community in 1976 when there was a sudden run on a particular enzyme. "It became impossible," one researcher complains, perhaps with slight exaggeration, "to buy collagenase."

Collagenase is an enzyme used to break down connective tissue. What a blender does with blades, collagenase does chemically, reducing material with shape and fiber to a cellular soup. And somebody was buying up large amounts. The only possible explanation was that a tissue digest of gargantuan dimension was in the offing. "It turned out," recalls Donald Steiner, of the University of Chicago, "that Rutter bought all the collagenase that was available in the whole country to isolate islets in huge amounts for this project. In fact, there was a period where we weren't able to get too much collagenase. I wouldn't say it created a *severe* shortage, but there was a shortage for a while. And it seems that it was because of the amounts that were being shipped out to California."

Brute-force science left those kinds of footprints. The San Francisco team proposed no less daunting a task than to fish out the insulin message by isolating huge amounts of islet cells from the pancreases of rats. Beginning around the fall of 1976, after considerable preparation, half a dozen researchers—John Chirgwin, Axel Ullrich, Raymond Pictet, technician Walter Imagawa, and several volunteers—gathered in Rutter's ninth-floor lab to perform this mass pancreatectomy. They handled approximately two hundred rats in a week, and the work perforce took on airs (foul airs at that) of assembly-line surgery. For people used to working with their minds, it was the most unsavory kind of manual labor.

The term "brute force" is a bit of a misnomer here, however, in the sense that the actual surgery was tricky and delicate. Raymond Pictet organized the islet-isolation procedure on a mass scale, and he and Ullrich removed the tiny, diaphanous pancreas from each and every rat. These were passed on to John Chirgwin, who tossed them into the collagenase solution. Collagenase broke down all the connective tissue, but like most enzymes, it had to be handled with great care. Given too much time to work, it would chew up the islet cells as well. After several minutes, they ended up with a dirty soup, a grayish pancreatic purée, containing both exocrine and islet cells.

The rest of the isolation was fairly straightforward. The pancreatic cells were mixed into a special solution and spun in a centrifuge. The white islet cells would separate from other pancreatic cells because their density was different. After centrifuging, one could see a white band of cells in the test tube. Those were the islets. These were collected, installment by installment from two hundred rats, centrifuged again to form a small deposit, called a "pellet," at the base of the test tube, and quickly frozen. "At the end of the week," Chirgwin says, "we had this little pellet, maybe a couple hundred microliters of packed cells. Those were turned over to Axel, who then used the guanidine procedure that I'd worked out and made the RNA. It seems to me we got more than enough, and the secret, really, to that experiment was that what he got was *all* islets of Langerhans. It was as good as they could be gotten."

By the end of each session, dozens of rat carcasses had piled up in cardboard boxes in the hallway outside Rutter's lab. The air was sharp with rodent urine and feces. For their troubles, Pictet and Chirgwin and Ullrich and some volunteering colleagues had managed to produce about fifty milligrams of tissue. "That was very little, given the techniques available at that time," Ullrich says now. "It was just incredible. It's like a little speck of a pellet in a test tube. . . . It's two hundred rats and the work of three or four people for several days to get that." Ullrich processed the cells by the standard method to separate RNA from the rest of the "Gemisch," but with one important emendation. When he thawed out the frozen islets, he did it in a solution containing Chirgwin's anti-RNase agent, guanidinium thiocynanate. It totally knocked out any of that destructive enzyme lingering in the mix. Ullrich was left with just RNA—messenger RNA and the RNA that made up ribosomes.

"I got about half a milligram of *total* RNA out of this," he recalls. "Total RNA is the entire RNA content, which mostly consists of ribosomal RNA. The actual messenger RNA out of this total RNA was about thirty micrograms—thirty millionths of a gram. And that was not totally pure messenger RNA. I had no way even to test if this RNA was in good condition or not, because an analytical testing experiment would have used up all the RNA. I would have nothing left for the experiment."

Sequestered somewhere in that thirty millionths of a gram of material, Ullrich believed, was the message for insulin. The RNA was refrozen, theoretically safe from danger, but the minuscule amount of material did not breed calm. RNA is fragile and vulnerable to destruction; its chemical nemesis, RNase, is ubiquitous and indestructible. Molecules of RNase ride particles of dust and nestle in the whorls of fingers; it is so stable that when a smoker's fingers leave RNase on cigarette paper, the enzyme can be isolated, intact and completely active, later on in the ash. If any RNase found its way into a test tube containing the message—borne on a speck of dust, introduced by a wayward finger—it could erase all the messages, letter by letter, word by word.

Many researchers working with RNA had discovered, to their dismay, that preparations had become mysteriously and thoroughly bowdlerized by this enzyme. So, as Ullrich remembers the feeling, "having a few micrograms of this super-precious RNA, it makes you really nervous."

By the fall of 1976, the first and, by all accounts, messiest part of the project had been completed. The UCSF team performed several of these mass islet preparations during that period. At the expense of hundreds of rats, the UCSF postdocs had managed to get a good harvest of "goddamisles" out of the rat pancreases. It hadn't been elegant or pretty, but it worked. That was all that mattered. By brute force, they had managed to neutralize Harvard's advantage with the Chick insulinoma.

At this point, Axel Ullrich contemplated a return to scientific elegance. After the mass rat surgery, he took a vacation in Mexico. The cloning part of the experiment was up to him. Far from the pressurized "postdoc mill" of San Francisco, he pondered schemes for the next steps: turning the RNA into a gene, getting the gene into a cell.

N i n e

"Oh, God, What Are We Going to Do?"

. . . .

Just as Axel Ullrich began to think about a good plasmid for the next step in the cloning experiment, the crucial advantage of being at UC–San Francisco became perfectly clear to him. Five floors below the Goodman lab, at the very same university, in the very same building, in probably the most advanced plasmid design laboratory in the world, researchers in Herb Boyer's lab were putting the finishing touches on what would become their most famous—and, for one brief moment, infamous—creation. In other words, when Ullrich needed a state-of-the-art vehicle to smuggle his insulin DNA into cells for cloning, the solution lay exactly one elevator ride away.

During the early days of cloning, there were precious few good plasmids (or "vectors") available. A good plasmid needed to be small, had to replicate prodigiously once inside a cell (like a copying machine within a copying machine), and it needed certain features, known as markers, to tip off researchers when it had successfully carried a gene into a cell.

The workers in Boyer's lab had artfully engineered a new generation of plasmids; these were developed not for the exclusive use of the Boyer group, but for the scientific community at large, and as soon as their specifications became known, they were widely coveted by scientists around the world. The model that quickly won everyone's favor was a joint creation by "Paco" Bolivar and Ray Rodriguez of Boyer's lab: pBR322. This little ringlet of DNA became almost as much of a character in the race for insulin as some of the scientists.

In the grand tradition of postdoctoral defiance of authority, pBR322 was the quintessential after-hours creation. The idea, Bolivar recalls, came up one evening while the postdocs brainstormed over beers at the Chelsea Pub. Bolivar and Rodriguez proceeded to engineer this plasmid even though, as they recall, Boyer wasn't especially keen on the idea; he thought existing plasmids, like his own lab's pMB9, were good enough. So Bolivar and Rodriguez primarily worked the night shift to engineer this vector. On August 24, 1976, the two postdocs presented their *fait accompli* plasmid to the lab. Everyone, including Boyer, thought it was a great idea.

It had been engineered for convenience and safety in cloning, and its advantages were numerous. It was small, and therefore made cloning experiments easy to analyze. While bacteria duplicated every twenty minutes, this plasmid copied itself anywhere from fifty to four-hundred twenty times within the cell during those twenty minutes; in terms of copying a gene, it provided more bang per cell cycle. And, as Ullrich soon learned, its markers made cloning a breeze.

Finally, and perhaps most important from the public-safety angle, Bolivar and Rodriguez had produced a genetically neutered plasmid. Bacteria, although they reproduce merely by dividing into two identical cells, nonetheless indulge in a certain degree of genetic promiscuity and have the capacity to pass DNA back and forth between organisms. This capacity to swap DNA, known as conjugation, gave rise to public-health fears, because it raised the possibility that the laboratory strain of *E. coli*, if it escaped, might pass its genetically engineered genes to related strains of *E. coli* that normally inhabit the human gut. But pBR322 had a diminished ability to propagate outside the cell, its creators maintained, reducing the risk of transmission by many orders of magnitude.

It was, in short, a compact dream machine of a plasmid. There was just one problem. In November 1976, when Axel Ullrich began searching for an appropriate plasmid for his insulin gene, nobody was allowed to stick a mammalian gene into pBR322.

At that time, the power to approve the use of these molecules resided with the National Institutes of Health's Recombinant DNA Advisory Committee (RAC). RAC then had to pass it on to the NIH director for certification before the material could be used in experiments, as stipulated in the NIH guidelines. It was one of those areas where the rhythms of two different enterprises, science and the government bureaucracy, moved at radically different speeds. Molecular biologists often worked around the clock as they closed in on an experimental answer; the RAC committee held meetings four times a year. With the telephone, biologists routinely learned of, and applied, new ideas and techniques within twenty-four hours; guidelines and legislation arose out of months and sometimes years of turgid deliberation.

In a period of mounting public concern about the safety of recombinant DNA techniques, the NIH committee was not inclined to take precipitate action in approving the use of gene-splicing materials, even if they were ostensibly safer to use. In December 1976, Herb Boyer fired off a letter to the NIH committee, appealing for quick approval and certification of pBR322. The delay seemed inexplicable to Boyer: the laboratory of University of Washington researcher Stanley Falkow, a leading expert in bacterial genetics and drug resistance, had provided data indicating that these plasmids were much less likely to hop about than vectors currently in use. The NIH asked for more data. Until pBR322 received certification, Ullrich could not use it for cloning. Nor, for that matter, could his colleagues in the Goodman lab. Peter Seeburg and John Shine were making excellent progress on two other important mammalian hormones, growth hormone and placental lactogen factor.

While actual cloning remained in temporary abeyance, Ullrich worked up a lovely trick to find out if indeed the insulin message lay buried somewhere in those precious few drops of frozen liquid that had been isolated from the rat pancreases. It was a complicated procedure, but Ullrich's strategy—an idea, incidentally, that Seeburg claims to have developed independently—represented an exceedingly clever bit of biochemistry. He converted a small amount of his RNA into very "hot" radioactively tagged DNA, cut it with a particularly discerning enzyme named *Hze* III, and ran fragments down a gel in the hopes that a predominant molecule, ideally the insulin gene, would gather in the same place in the gel, where it would light up as a bright band and indicate that Ullrich was on the right track. That is exactly what happened.

"And that was probably the most exciting moment at that time for me,"

Ullrich recalls. "It looked beautiful—it came out later in our paper, this picture. It was clear that there were bands which you would only expect to see if there is one predominant species." Indeed, two distinct bands lit up on film. Measurements indicated one was about eighty base-pairs, the other 180. "We didn't know really if that was insulin. But you could still hope, because islets are known to make a lot of insulin. So if there was something predominant, it should have been insulin."

Exhilarated by this encouraging sign, proceeding now only half-blind, Ullrich went ahead and made double-stranded pieces of DNA out of the rat RNA. By January 1977, everything was ready. He had what promised to be an insulin gene. He had outfitted this gene with linkers, those man-made, sticky-ended adaptors. All he needed was a good place to insert the gene. The plasmid pBR322 was perfect. Unfortunately, it was still off-limits, hung up in the bureaucratic machinery. There was nothing to do but wait.

Having reached the penultimate moment of a fourteen-month odyssey, having labored against both technical and personal obstacles, Ullrich was not particularly enthusiastic about being held up by bureaucratic delays. He lay one step shy of proving a point everyone had said was impossible; at the same time, colleagues in the Goodman lab raced to prove the same point with different genes. Rumors continued to circulate about Gilbert's progress. "You would hear conversations in the hall almost every day about how Wally Gilbert's done this and done that and he's got a paper in press, and this and that," says former faculty member Brian McCarthy, whose lab was next to Goodman's. "And there was a real frenzy of people running around, trying to catch up." No one determined if the rumors were true or not. They merely elicited that Pavlovian response of scientists in competition: believe the worst and work even harder.

Herb Boyer, meanwhile, had gone off to attend the annual winter symposium in Miami, scheduled to begin January 9, 1977. The clamor for pBR322 grew only after Boyer described its virtues in a presentation at the symposium. Immediately following the symposium, on January 15–17, the NIH's Recombinant DNA Advisory Committee met in Miami, and Boyer took advantage of the timing to renew his plea for hasty certification of pBR322. But the administrative decision in Miami degenerated, according to the scientists' accounts, into a nightmarish scientific version of the parlor game Telephone. Somewhere along the line, the message from Florida got misstated or mistranslated—or, as some would later charge, conveniently misunderstood. The scientific ramifications were minimal, but the political ramifications were not.

At its Miami meeting, the NIH committee "approved" pBR322 for use—but not without qualification. The committee wanted to see more data from Stanley Falkow's laboratory proving low transmissibility of the plasmid before recommending certification by the NIH director. Certification could come *only* from the director, Donald Frederickson, and the vector could not be used in Ullrich's type of experiment until it was "certified."

The distinction between "approval" and "certification" was eminently clear to the committee that dreamed it up, and apparently was clear, too, to other researchers in the field ("*We* all knew very well what it meant," says one Harvard scientist). But as even RAC committee chief William Gartland admits, the terms could be confusing. To complicate matters, NIH rulings on recombinant DNA research at that time tended to be communicated by telephone. As William Rutter later pointed out, "There were never any written orders that had gone out."

Herb Boyer called up his lab from Miami in mid-January to pass on the news: pBR322 had been approved, but not certified. Ray Rodriguez, one of pBR's creators, took the call. In the roundelay that ensued, the message—as they say in biological circles—suffered degradation.

Rodriguez passed the word to Mary Betlach, also in Boyer's lab. Betlach called upstairs to the Goodman lab and told Axel Ullrich, who had been awaiting such a call. Ullrich heard "approved" and "not certified," according to his later recollection. "That was *exactly* the terminology," he says. "Of course, me, I was not really familiar with all these subtle differences in terminology, so I went ahead and used the vector. . . ." Ullrich had been in the United States about fifteen months and, according to a fellow lab worker, found English "very frustrating," so it is not difficult to imagine his confusion over these hair-splitting synonyms; a number of people interviewed both in and outside the Goodman lab concur that the mistake was probably accidental. But some of the people working in the UCSF labs took a dimmer view of the proceedings, casting a cloud over the experiment from which it never entirely emerged. Says a former UCSF postdoc, "I don't think there was any innocence in the whole building at that time. . . . I think we all sort of knew that it was better to ask for forgiveness than to ask for permission."

In any event, Ullrich strode into Bill Rutter's office on January 17 or 18 and announced that pBR322 had been approved. Rutter recalls saying, "Great! Well, let's go ahead and use it, huh?"

116
. . . . . .

"Okay," replied Ullrich. "I'll just go ahead and do it." That same evening, Ullrich began a history-making experiment. In so doing, he also broke the NIH guidelines.

There is nothing further removed from the hue and cry surrounding genetic recombination than the act itself. When molecules of DNA mix in a gene-splicing experiment, when they find their opposite sticky-ended number and silently rejoin, it is a momentous biological event that takes place completely out of view. There is nothing to see: all the ingredients are colorless, like water. There is nothing to hear, save the nervous clicking of hand-operated pipettes. Enzymes and plasmids and DNA and bugs commingle in the milky plastic well of Eppendorf tubes, little test tubes about the size of a ballpoint pen cap. In experiments like the one Ullrich was about to perform, the reactions often take place in a volume of fluid about the size of a human teardrop.

This particular experiment—slipping a mammalian "gene" like rat insulin into a simple organism like bacteria—had to be performed in a P3 facility, which UCSF had recently outfitted on the tenth floor. Ullrich had the pBR322 ready, and he had the hoped-for insulin genes with linkers prepared as well, and although the operation of putting them together was certainly more complicated than making a cocktail, it did have elements of mixing ingredients together in the right sequence and in the right amount to get the desired result.

Like many other complicated biochemical procedures, cloning was—and still is—exceedingly difficult to perform with consistency. In a standard cloning procedure, there are several dozen steps to take. Optimum conditions had to be worked out for each one of those steps. The concentrations of the enzymes had to be right. The temperature at which reactions occur had to be right; if a colleague had inadvertently left an enzyme sitting out on the bench too long, it could go bad, as inevitably would the experiment, too. The amount of time in which ingredients were permitted to interact had to be right. The best concentration and pH of the reaction buffers (buffers are the solutions in which enzymatic reactions optimally occur) had to be figured out. The bacteria could be finicky, especially the weakened lab strains required for this type of experiment. Under the best of circumstances, cloning in those days was an operation with very low efficiency. A 1 percent success rate in slipping a plasmid into bacteria generated euphoria. When you were dealing with

117
. . . . . .

billions and billions of bacteria, of course, ninety-nine failures out of a hundred, or even 999,999 out of a million, become statistically tolerable. All it takes is one cell to start a colony.

During the last two weeks in January, benefiting from a clever trick developed by John Shine that prevented plasmids from closing on themselves without the inserted DNA, Ullrich slipped his presumed insulin gene into the plasmid and then moved on to perform the procedure known as transformation. If the plasmid is a smuggler's bible, then transformation is the operation that gets it aboard the ship. Transformation is not a cheery experience for bacteria. In theory, you barrage the cell wall of *E. coli* with caustic chemicals, which perforate the bacterium's outer ramparts, leaving pores through which plasmid DNA can squeeze into the cell. In practice, the cells take a beating. The trauma is so great that most don't survive the experience, which is another reason cloning has such a high attrition rate. But out of the slough of dead cells a few stout-hearted bugs will survive, and since at the bacterial level, life itself is the sole meritocracy, these survivors overcome the disruption and begin multiplying the way bacteria are supposed to. The whole mix is then spread on plates.

It is in this crucial phase that the sophisticated markers of pBR322 contributed revolutionary ease to cloning. By adding penicillin to the nutrients, Ullrich killed any bugs which didn't have pBR322, because this plasmid contained a gene with resistance to penicillin. Only bugs containing pBR322 could survive that first elimination round. Next, Ullrich took some of the survivors and tried to grow them in the presence of another antibiotic, tetracycline. And in fact he hoped to find some that didn't. This is because Ullrich had inserted his insulin DNA smack dab in the middle of pBR322's tetracycline-resistance gene, severely handicapping its ability to resist the antibiotic. Would any of these survivors have inserts? Would any of the inserts be rat DNA? Within twenty-four to forty-eight hours, Ullrich would begin to have some answers.

At the time, Ullrich lived in an apartment on Parnassus Avenue, a two-minute walk from the entrance to Moffitt Hospital and the UCSF biochemistry labs across the street. When he returned to look over the plates, the results delighted him. Five colonies had grown overnight, five little white pustules of bacteria, each about the size of a pinhead. Three of them turned out to be "false positives," giving a positive result without in fact containing the desired insert. But two clones seemed to have legitimate inserts. "The big question was," he says, "was one of them insulin or not?"

The only way of finding out was by DNA sequencing. John Shine, the Australian postdoc, was the resident sequencing expert, having immersed himself in the technique since the time Wally Gilbert gave a seminar at UCSF. Until the gene was sequenced, they had no way of knowing exactly what they had. The work proceeded slowly: sequencing was arduous and messy work.

No one had ever glimpsed the actual genetic structure of a hormone before. Precious few had even seen a mammalian gene of any kind. What they soon hoped to behold was nature's unique script for insulin—the letters and words for the protein itself, the stage directions, the whole package. The answer lay in those two little pustules of cells. According to tradition, Ullrich christened these clones with his own initials: AU-1 and AU-2.

Back at Harvard, the researchers in Gilbert's lab had no inkling that the California researchers had progressed so far. Nonetheless, the rivalry between the Harvard and UC–San Francisco students and postdocs began to be a matter of record in the only publication of record that mattered, the *Midnight Hustler*. The second issue of the *Hustler* wasted no time in making this clear. "GILBERT HUSTLERS OUTMUSCLE BOYER CARTEL IN DUAL MEET" read the front-page headline. "Coast Crew Crumbles as Gilbert's Gapes" added the subhead.

By any criteria, the *Midnight Hustler* was not an ordinary newspaper. It made its debut in the spring of 1977, and it wasn't long before it achieved a certain underground notoriety in the Harvard Bio Labs and even on the West Coast. In its pages readers could find everything from inside jokes about molecular biology to dour observations about credit-mongering among colleagues. It was recombinant DNA's first satire broadsheet, a cross between a lab manual and *Lampoon*. Each issue was Xeroxed, circulation was about fifty, and it was the implicit policy of the editors—Allan Maxam, Karen Talmadge, Debra Peattie, Lorraine Johnsrud, and Philip Farabaugh—that no cow was sacred.

And so it was that in the May 1977 issue, under a photograph of "Our Founder"—a picture of Walter Gilbert in open-necked shirt, squinting through his dark-rimmed glasses—readers discovered a breezily written imagi-

119
. . . . . .

nary account of a biological track meet between the Boyer and Gilbert laboratories. Spiced though it is by jargon and technical references, it is generally comprehensible:

The long awaited battle of the biochemical titans took place over the weekend in Cambridge. An overflow crowd of two Harvard administrators cheered the event from the glass pipes overhead. The first contest Friday night featured DNA sequencing and, as expected, the Gilbert methylators swept the field. Sequencing captain Alan [sic] Maxam executed a perfect strand separation, the first in league history. The Boyer forces were severely hit early in the round by the Gilbert lead-off sequencer, Greg Sutcliffe, who produced the sequence of the tetracycline promoter of pBR322 in three hours.

Boyer's hopes of recouping in the second round Saturday were shattered by unexpected Gilbert depth in the insulin contest. Peter (Sevag) LoMedico [sic], newly acquired from Texas for cash and an undisclosed amount of chloroform, pushed the Gilbert hustlers over the top, aided by the butt-end ligating of Forrest Fuller. Before the second round Fuller was quoted as saying "Nobody knows more about butt-ends than I do!" and the performance of the Gilbert insulin team proves it.

The third round Sunday was highlighted by brilliant performances. Who can forget Jeffrey Miller successfully streaking an eight-sectored plate for single colonies in 28 seconds, or Rich Tizard's world record gel pouring feat—two sequencing gels, one agarose gel, one short 8% acrylamide gel, and one pH 7.9 imidazole gel in the astounding time of 7 minutes? What about Gilbert Himself burying Boyer in the climactic paperpushing contest with the four foot square blueprints for the new biochemistry building?

As his team headed for the lab safety showers after the meet, Gilbert was modest. "Boyer has a solid team. We were clearly outclassed by Mary Betlach in the vector contest, and Bad News Heyneker made a strong showing with his cloned synthetic operator."

What will be the next challenge for the Gilbert international array of talent? Gilbert refused to discount rumors that he was

grooming his lab bosses and postdoctoral individuals for a run
against the Khorana stable. . . .

Gilbert Himself was particularly fond of the article, one graduate student
recalls, because "he really did see it all as sports. There wasn't anything
personal in it; he just loved the fun of the competition." The satire did not
so much reflect the rivalry over a single project as the comradely competition
between two groups, East and West, on the cutting edge of molecular biology.
Indeed, the Harvard researchers still mistakenly believed the Chick insulinoma gave them an advantage in the insulin race.

Even though only two or three people in the group were actually involved
in the insulin work, the Gilbert lab as a whole rallied around such projects.
A kind of hometeam partisanship developed, particularly when a project was
patently competitive. As Debra Peattie recalls, it functioned as a "spectator
sport," and the regular group meetings in the Tea Room to discuss research
kept the lab group up to date on particular projects—who else was doing it,
how they were going about it. "That was part of the *esprit de corps,*" Peattie
says, "being involved even from a distance."

The progress reports from the insulin corps were not terribly encouraging.
By the spring of 1977, Argiris Efstratiadis was still trying to isolate good insulin
RNA from the troublesome insulinoma, and Forrest Fuller was still trying to
make "butt-end" ligations work for cloning. But as often happened in a large
laboratory such as Gilbert's, there were several unrelated projects underway
which would fortuitously dovetail some months later with the insulin work.
One of them, pursued by a graduate student named Stephanie Broome,
involved designing a highly sensitive assay. It would be capable of detecting
as little as ten picograms—ten *trillionths* of a gram—of a protein, any protein,
in bacterial cells. This would provide an extremely discriminating way to test
for the presence of a scarce protein—something like rat or human insulin—
being manufactured in a bacterial cell. Graduate student Greg Sutcliffe, meanwhile, had embarked on the most ambitious Maxam-Gilbert sequencing
project to date: determining the exact sequence of the penicillin-resistance
gene in pBR322.

Sutcliffe first considered tackling the sequencing project on February 8,
1977. By coincidence, the Cambridge City Council had, on the previous
evening, finally rescinded its moratorium on recombinant DNA research in

the town. A citizens' panel known as the Cambridge Experimentation Review Board, after conducting exhaustive hearings, concluded that recombinant DNA experiments could resume in Cambridge without endangering the health of citizens. Many experiments were off-limits to Harvard, however. The university still didn't have a P3 lab.

The UC–San Francisco researchers already had their P3 lab, but they were beginning to feel the same community heat that had made things uncomfortable for the Cambridge biologists. On the morning of February 4, 1977, the *San Francisco Chronicle* carried a front-page story by its respected science writer David Perlman under the headline "Tough Rules on Creating New Forms of Life." For anyone engaged in recombinant DNA work in the state of California, the news was troubling; for the Rutter-Goodman insulin team, the date would have special, sobering significance.

"State Health Department officials," the story began, "have drafted a tough bill to regulate the controversial research field of gene-splicing and to control industry experiments applying the new technology, the *Chronicle* learned yesterday." The legislation, as outlined in the story, proposed restrictions more rigorous than the NIH guidelines, aimed at both academic institutions and industry. Laboratories would be subject to state inspection; reports would have to be filed; all researchers engaged in recombinant DNA work would have to register with the state. The penalties for violators were sobering: $500 fines and forced closure of laboratories. The proposal had been drafted by Marc Lappé of the California Health Department's Office of Health Law and Values, and the *Chronicle* story quoted state health officials as saying that California governor Jerry Brown would "push for its passage."

This was disturbing stuff to any would-be cloner, but it had special resonance in San Francisco. On that very same day, February 4, Herb Boyer and Brian McCarthy convened their lab groups to discuss safety procedures for the P3 laboratory on the tenth floor. Howard Goodman's research group attended as well. In the course of the meeting, Boyer briefed the assembled scientists on the regulatory status of pBR322, according to a later account of the meeting, and he specifically mentioned that it had not yet been certified by the NIH for experiments like cloning insulin. This could hardly have escaped the attention of Ullrich and Shine: according to the record, they both attended the February 4 meeting, and by that time Ullrich says he had already obtained clones using pBR322. Clones, it was now clear, that were in violation of the NIH guidelines. Nor was that a secret. Ullrich maintains (and other lab members confirm) that he gleefully spread the word as soon as he obtained this successful result. "Everybody congratulated me, and everyone was excited," Ullrich recalls. Yet neither Ullrich nor Shine said anything on February 4 about the premature use of the vector. Why? At the time, Ullrich was still running preliminary tests on the clones. "Obviously, I didn't know if I had succeeded, so there was really no reason to talk about it," he says. In any event, it made sense to wait and see. No one considered it a big deal. "I never really felt funny about it . . ." he says. "I really believed it was okay."

It formally graduated to a big deal later that month, however. Howard Goodman had returned, prematurely, from his sabbatical in Japan on February 10 and had been informed of the latest insulin results. If Ullrich expected vindication at this point for his perseverance, however, Goodman's reaction fell somewhat short of the desired affect. "He had looked at my data and didn't believe any of it," Ullrich would recall later with a laugh. "And I also told him at that time that I used pBR322 and found out later that it was not really approved." Goodman, according to Ullrich, did not seem perturbed by the infraction, just unconvinced by the data. Soon after, Goodman and Ullrich traveled to Park City, Utah, at the end of February to attend an ICN-UCLA scientific meeting. It was during this trip that the use of pBR322 assumed crisis proportions, and thrust the UCSF scientists into an unusual dilemma where scientific achievement and social responsibility tugged in opposite directions.

The bad news came from one of the conference participants, William Gartland, who headed the NIH's Recombinant DNA Advisory Committee. At a March 1 session of the Utah meeting, he briefed the gathered scientists on the NIH regulations and explicitly made the point that pBR322 still had not been certified by the NIH director and thus could not be used for the cloning of mammalian genes. "That's when I sort of first really knew officially that we had it in the wrong plasmid," says Ullrich. "But I think it was known before that already. . . . It was never really clear." According to officials at UCSF, this was when Goodman first became aware of the violation.

This public reiteration of pBR322's status was unambiguous and only made the good news from San Francisco difficult to exult over. While Goodman and Ullrich were in Utah, John Shine had continued to sequence the

DNA from Ullrich's clones, checking to see if the inserted DNA indeed coded for insulin. Some partial sequences were coming out on the gels. They looked good.

"We called during the meeting," Ullrich says, "and it was actually insulin. One of the two—the largest of the clones John Shine looked at—was insulin. So we were super-excited, but of course immediately there was this concern. What do we do? Because it's in the wrong plasmid.

"I'll never forget that moment. Howard and I were sitting there thinking, 'Oh God, what are we going to do?' "

The timing could not have been worse. At the end of that same week, even as they wondered aloud about what to do, distinguished scientists were convening in Washington for a highly publicized National Academy of Sciences forum on recombinant DNA. Critics were expected in strong numbers, press coverage would be heavy, there inevitably would be renewed calls to end recombinant DNA work altogether. In fact—irony of ironies—the very triumph Ullrich and his colleagues had just achieved, the cloning of insulin, was to be the subject of a debate several days hence at the Washington forum, a debate whose premise was that such an accomplishment was months if not years away.

The Rutter-Goodman collaboration was the first to clone the insulin gene, which had been invoked as one of the heralded targets of the gene-splicing technology. And yet the plasmid they used had not been federally certified. Obviously they couldn't breathe a word of it. "You couldn't really be happy about it," Ullrich remembers. "Oh, it was terrible."

T e n

"Tell Me About
Insulin"

Almost to the day when Axel Ullrich's clones unequivocally became bacteria *non grata,* one of recombinant DNA's fiercest critics stepped up in front of a hostile audience of scientists in Washington, D.C., and vowed, "You ain't seen nothin' yet." His name was Jeremy Rifkin, and his passions boiled over like biology's latter-day version of John Brown. "The press here, the critics, think that this is a question of the public interest groups versus the scientists," he said. "Wait until the Protestants, the Jews and the Catholics, the Methodists, the Presbyterians and the Baptists all over America start to realize the long-range implications of what you gentlemen are doing here tonight." It had the sound of a popular uprising. Pandemonium ensued.

The spring of 1977 perhaps saw the highwater mark of public clamor over recombinant DNA research. The protests took many forms: hearings before the California state assembly, public forums in university towns where the research was contemplated or actively pursued, concerns about the use of gene-splicing in biological warfare, and the increasingly widespread conviction in Washington that some form of federal legislation would be necessary to oversee both industrial and academic work.

No event of the period so dramatically captured the stormy public passions swirling around the issue as the three-day forum sponsored by the National Academy of Sciences in Washington, D.C., which began on March 7, 1977. Cambridge mayor Alfred Vellucci crashed an afternoon press conference, declaring he hadn't been invited to the forum and demanding to know who was going to control this research. That same theme, reflecting somewhat less personal affront, dominated much of the initial discussion: why hadn't the public been invited to play a role from the very beginning of the debate, in 1974 and 1975, in deciding how or if recombinant DNA work should proceed? Why hadn't religious figures and philosophers been invited to ponder the ethical and moral impact along with scientists?

To that end, a group calling itself the Coalition for Responsible Genetic Research demanded an immediate cessation to recombinant DNA experiments. George Wald, his fellow Nobel laureate Macfarlane Burnett, Ruth Hubbard, and Lewis Mumford were among the sponsors, and it was associated with the environmental group Friends of the Earth. The coalition called for a thorough discussion of the social, political, and moral, as well as scientific, implications before work resumed. As if to underscore the particular delicacy of the historical moment, when the genie seemed even then to be squeezing out the throat of the bottle, Hubbard told a reporter, "This is the time to clamp down on it, not in a year's time when it will have become widespread."

The tone of the National Academy affair was established in its very first moments. A spontaneous demonstration broke out. Protesters sang "We Shall Not Be Cloned." A banner quoting Adolf Hitler—"We will create the perfect race"—was unfurled and tauntingly waved in the faces of the scientists until one biologist, in a fit of pique, ripped it apart. Then Jeremy Rifkin, head of a group called the Peoples Business Commission and an outspoken opponent of genetic engineering, took the podium during the evening's welcoming ceremony.

Rifkin, a former lobbyist, had made genetic engineering his cause, and he brought to it a moralistic fervor noticeably absent from other aspects of the debate. To some, his was the loudest, if not lone, voice addressing the long-term ethical issues involved with the research; to others, these moral concerns masked a more general philosophy that attacked science and technology. He minced no words in his introduction. The forum, he declared, was "rigged," and such meetings were "legitimizing functions for the national press."

To get a flavor of both the substantive arguments and exaggerated, hyperbolic rhetoric of the debate, consider Rifkin's opening comments on the "central issue" involved in recombinant DNA research:

> My friends, the real issue is not whether the laboratory conditions are safe or unsafe, although obviously there is a problem with potential viruses and bacteria getting out of the laboratory and endangering the health and well-being of millions of people. But that is not the central issue. We could have legislation passed this spring by Congress for safety regulations, and it still would not detract from the central issue we are facing.
>
> The real issue here is the most important one that human-kind has ever had to grapple with. You know it, and I know it. With the discovery of recombinant DNA scientists have unlocked the mystery of life itself. It is now only a matter of time—five years, fifteen years, twenty-five years, thirty years—until the biologists, some of whom are in this room, will be able literally, through recombinant DNA research, to create new plants, new strains of animals, and even genetically alter the human being[s] on this earth.
>
> Some scientists in attendance will say this is sensationalism, this is emotionalism. Well, it is because this technology is sensational, and because it hits right to the basic emotional core of life itself. For three generations of Americans weaned on Huxley's *Brave New World* the long-range implications of experimentation in this field are ominous. And the precedents are being set right now with the scientists in attendance here.
>
> You can't hide from the fact any more convincingly than the physicists were able to hide from the knowledge that they had when they split the atom. They knew what that could lead to and what it would lead to. Biologists doing DNA research also know what this technology could lead to and what it will inevitably lead to unless there is public reaction.

Most scientists would find it amusing, even ten years after the fact, to know they had unlocked the mystery of life. Nature divests her innermost secrets with a parsimony that has kept biologists busy since Aristotle. But that was about the only amusing aspect of Rifkin's harangue. Could bugs crawl out of the lab and cause epidemics? There was no hard evidence to suggest so, but Rifkin probably realized the mere suggestion would send shivers down the public spine. Would human beings be altered, even cloned, by genetic engineers? No one except writers of science fiction had expressed any interest in cloning humans, and biologists didn't even believe it could be done. Had the biologists, like the physicists with their split atoms, unwittingly unleashed destructive forces with all the problematic social and moral fallout of a nuclear bomb? The same litany of charges, by its very repetition, left a residue of credibility. For the scientists, the most troubling thing of all was that people—lay citizens, reporters, senators, congressmen—were paying attention.

In what must surely be one of the more amusing ironies of the era, the cloning of human insulin was hotly debated on Tuesday, March 8, as a distant theoretical possibility in Washington even as the first crucial step had been accomplished in California.

None of the groups actually working on insulin took time off to argue the case (although Herb Boyer attended the forum and made no secret of Genentech's intentions to manufacture insulin). But Irving S. Johnson of Eli Lilly & Co. (one of eight pharmaceutical companies which helped sponsor the forum) was on hand to argue the benefits of genetic engineering, and Ruth Hubbard was there to renew her arguments about potential risks. It began as a genteel exchange of views. By the discussion period the exchange had assumed chillier tones very much consistent with the forum as a whole.

Compared to some of the other claims made for recombinant DNA insulin, Irving Johnson's presentation seemed a model of restraint. There were no glistening promises and no exaggerated statements. "There is no shortage of insulin at the present time," Johnson declared. "There was no shortage of energy a few years ago. Our concern is whether there will be a time when a shortage of insulin may exist." He went on to present what may have been the most detailed and even-handed public analysis of the insulin picture during the entire period when recombinant DNA insulin was debated.

Johnson estimated the world diabetic population at sixty million, thirty-five million of whom were in less-developed countries. The U.S. population was conservatively estimated at five million; only 3.5 million of them were diagnosed cases, and of them more than a third—1,250,000 in all—required insulin as part of their treatment. Johnson also noted that the American diabetic population was increasing at about a 4 to 5 percent rate.

At the same time, the number of units of insulin retrieved from beef and swine pancreases had dropped sharply between 1970 and 1975. Theoretically, the maximum amount of insulin available from pancreatic sources was more than enough for diabetics, Johnson reported, but in practice, collecting pig and cow pancreases was "unpredictable and uncontrollable by man." Drought conditions, for example, caused fewer head of cattle to be raised in some parts of the world; elsewhere, bad growing conditions reduced grain yields, which meant that livestock normally raised in feedlots had to be graze-fed, and that simple dietary change resulted in lower amounts of insulin in the pancreases of the animals.

To these aberrant developments were added the traditional problems of pancreas collection. The pancreas from a $400 steer commanded all of 75 cents on the market, and a pig pancreas went for about a quarter; some of those pancreases were snapped up as a food delicacy while others were simply too contaminated to be salvaged. The picture that emerged from the succession of facts recited by Johnson was of a pharmaceutical supply completely beholden to other, more powerful economies. That was rarely mentioned in public as one of the attractions of recombinant DNA insulin; to Lilly, Johnson would admit later, control of the source material was "one of the driving forces in our own mind."

Johnson's point was simple. Although there was no imminent shortage of insulin, the long-term possibility of one existed; in tortuously benign and unalarmist language, he had "presented data suggesting the prudence and logic of at least entertaining the possibility that additional sources of insulin should be seriously considered." Johnson believed the technical problems in genetic engineering, though considerable, could be overcome. "We can ill afford," he said in summary, "to wait until an absolute shortage develops before initiating studies to develop these alternatives. I believe we have a good deal of time before that possibility happens, but we should use that time well."

Ruth Hubbard had expected Lilly to "put forward a very strong case for why it was necessary to produce insulin in this particular way. And when it came right down to it, of course, they didn't do anything of the sort." So Hubbard began her talk by commenting that she and Johnson would make "very agreeable opponents."

Hubbard instead directed her remarks to the philosophical biases inherent in deciding what is a "potential benefit" of a new technology, "the hazards of looking for technological fixes as solutions to complicated diseases of metabolic control and other complicated problems." Recombinant DNA, she implied, symbolized the quick fix. Genetic engineering could create a new product, but it could not cure diabetes.

Like Johnson, Hubbard spoke of alternatives—but alternatives in medical practice and attitudes, not in technologies. One of the ironies of insulin therapy, she noted, was that while a diabetic might spend 10 cents a day on insulin, the cost of the paraphernalia—the disposable syringe—was, at 12 cents, actually slightly greater. So why not encourage sterilization and reusable syringes to reduce the economic cost rather than talk about cheaper insulin?—a point, significantly, which Irving Johnson did not address. To those listening closely, this was one of the few public hints that the economics of making recombinant DNA insulin might be lousy.

Hubbard went on to suggest that since social and environmental factors were implicated in the disease, perhaps those influences should be investigated to understand diabetes better. "There are lots of questions that we have to answer in order to lick diabetes," she concluded, "and diabetes is a major health problem. But what we don't need right now is a new, potentially hazardous technology for producing insulin that will profit only the people who are producing it. And given the history of drug therapy in relation to other diseases, we know that if we produce more insulin, more insulin will be used, whether diabetics need it or not."

It was in the discussion portion of the program that a real disagreement, unsettling in its scientific acrimony, broke out. It captured, in microcosm, everything that went wrong in the recombinant DNA debate.

Stanley Cohen, of Cohen-Boyer fame, rose to contest Hubbard on two points of her talk. In a chilly exchange of civilities, he challenged the suggestion that many diabetics could be treated with diet and did not need insulin. Then he took exception to a more technical point made by Hubbard. It involved the presence of certain enzymes, called proteases, in *E. coli* which could snip away portions of insulin's molecular packaging in a genetically engineered organism, leaving an active molecule.

"Perhaps you could comment on the second point, your statement that it is likely that *E. coli* would have the enzymes to chop off—"

At this point, MIT biologist Jonathan King jumped into the fray. "No," he intervened, "that there *might* be enzymes."

"I believe that *E. coli* has intracellular proteases," said Hubbard. "Are we agreed on that?"

"We are talking here," said Cohen, "about some very specific post-translational processing of a complex molecule, not just some random protease activity."

It is likely that the vast majority of readers are unable to follow the gist of this brief exchange, yet therein lies one of the crucial paradoxes of the entire debate. Hubbard's position was the more alarmist, Cohen's the more commonsensical—*if* you understood the science. For anyone unversed in the world of proteases and proteins, it was easier to hear the alarm than the common sense. Cohen was suggesting that those cellular interactions *were* very specific; and, it now appears, *E. coli* bacteria do *not* readily convert proinsulin into active insulin. And if those assertions were true, then the hypothetical fear of an epidemic of insulin-spewing bugs was not realistic, either. But who in the public could discern these things?

The argument was symbolic on another level as well. Here was Stanley Cohen, one of the godfathers of recombinant DNA, sharply disagreeing with Ruth Hubbard, who, along with her Nobel-laureate husband, George Wald, was among the more venerable presences in Harvard's famed Bio Labs. If these people couldn't agree, if these serious scientists conversed in such clipped tones, what was the lay public to make of the dispute? Who spoke the truth?

It was Jonathan King, the Science for the People member, who concluded the exchange. As Cohen aggressively challenged the scientific argument against recombinant DNA, King invoked political and philosophical objections. "We are told that we could make insulin in *E. coli.* It is recognized that there are many barriers. There are extraordinary barriers. We need P4 facilities; we need ingenious molecular biologists. But," he continued with evident sarcasm, "we can overcome those barriers, and therefore we must."

Why couldn't more cows and pigs be raised? King asked. Why not synthesize insulin chemically? Why did recombinant DNA provide the only solution? "These are not scientific imperatives," he said, "these are social choices, absolutely and completely, and there should be no confusion about it whatsoever."

Irving Johnson, in that soothing Midwestern twang of his, replied that some of King's statements struck him as "somewhat irrational." He noted that the chemical synthesis of insulin required more than two hundred steps and produced so little material that the cost of the final product would be prohibitive. And he added a line that indicated that Lilly personnel as well as everyone else were in the dark as to what was going on on the ninth floor of the Medical Center at UCSF. "I did not say that we could make insulin with *E. coli,*" Johnson insisted. "I don't know that. In fact, the people at Harvard probably have a better concept of that than I do, because they are much closer in terms of inserting the gene for insulin than we are."

If there was any voice of alarm at all in the insulin discussion, it came not from Johnson but from a high-ranking NIH official named DeWitt Stetten. Stetten recalled that supplies of insulin declined to dangerously low levels immediately following World War II and added, "By all odds I think we are perhaps in a more perilous position with respect to insulin supply than Dr. Johnson's remarks led me to believe. . . ." This attitude may have conditioned Stetten's reaction when a very hot political potato, postmarked in San Francisco, was dumped in his lap within a fortnight.

Not everyone was happy with the outcome of the National Academy forum. Jonathan King observed that recombinant DNA was offered "as the best of all possible worlds" while those who dared criticize it represented "a lunatic fringe who will be swept away under rational progress." For all the information that had been presented to the public, the divisions in the scientific community were just as publicly apparent, and the well-established rhetorical antipodes—paeans to a roseate recombinant future on the one hand, howls of apocalyptic foreboding on the other—were expressed with a disquieting disregard for any middle ground. The fact of the matter was that during years of debate, everyone from Lewis Thomas to Jeremy Rifkin had weighed in with their versions of the "real issue." Yet that single real issue was as slippery as a greased pig and its two-bit pancreas.

In periods of great ambiguity and uncertainty, however, apocalyptic visions seem to carry disproportionate weight. Erwin Chargaff, professor emeritus at Columbia University and a respected biochemist whose eloquence has the sweep and economy of a sword in the act of beheading, left his audience with a particularly dreary assessment of the entire debate. "The juggernaut of scientific majority opinion," he observed, "is much too strong for a few individuals to have any effect. I can only hope that the names of the many workers of the first rank who have assured us that nothing can happen will be remembered," he said, issuing his opinion in the locution of a curse, "when something does happen."

Irving Johnson made several other comments of note to the gathering. "It appears inevitable," Johnson told the forum, "that there will be legislation in this area." Indeed, in March 1977, two bills had already been introduced in Congress; the Department of Health, Education and Welfare—overseers of the NIH and its budget—was preparing its own request for legislation; and within several months of the NAS forum, two more regulatory bills were on the docket. Congressional hearings had been scheduled by Representative Paul Rogers for later that March. State legislatures in New York and California, meanwhile, had prepared legislation even more stringent than the NIH guidelines.

Johnson was hardly alone in predicting federal legislation. Anyone reading contemporary accounts of the controversy could have reached no other conclusion. "Whether scientists want it or not . . ." *Nature* reported, "legislation is definitely coming." *Science News* noted that National Academy president Philip Handler "said most of the scientists who attended the Academy's recent forum on recombinant DNA research grudgingly concluded that federal legislation is inevitable and perhaps even desirable, partly to 'terminate the feckless debate which has offered outlets for anti-intellectualism and opportunity for political misbehavior while making dreadful inroads on the energies of the most productive scientists.' " Feelings, needless to say, were running high on both sides.

The specter of legislation may have been more ominous to many academics than it was to Eli Lilly. "I would like to suggest to you," Johnson told the scientists assembled at the NAS forum, "that [legislation] is something that we are going to be able to live with a lot more comfortably than you. In industry we are used to having people stand over our shoulder and suggest what we do, and I don't think you are going to like it."

Johnson was right. While his admonition was still fresh in the air, the Rutter-Goodman researchers back in San Francisco struggled to figure out a way to handle the pBR322 problem. The idea that outsiders might suggest what they do stirred bristling indignation. One postdoc summed up his feelings this way: "When your life is built around science, you put an awful lot of yourself into it and you have a rather low tolerance for bureaucratic assholes wasting your time and making a fuss over what are basically petty legalities, at least from your point of view as a scientist at the bench." That sentiment was by no means atypical. And attitudes like that among those "at the bench" posed problems for the lab chiefs. "The administrators could not condone any kind of lack of adherence to the federal guidelines," says one former UCSF postdoc. "But the postdocs were being completely honest and candid in saying that they were a joke. They were."

It fell to William Rutter to handle the problem. Since that day in mid-January when Axel Ullrich came into his office to say that pBR322 had been "approved," Rutter and his group had depended primarily on progress reports from the Goodman group, which had the expertise in molecular biology. This technical gap may have produced a gap in communication. When Rutter finally learned that the guidelines had been violated, it came as a complete shock. He was at a meeting in Houston at the time, but rushed back to San Francisco as soon as he heard the news.

"I couldn't believe it, on the one hand," Rutter says now. "But on the other hand, there wasn't at that time any pejorative or punitive aspects of having done anything wrong, even if it [hadn't] been approved. I mean, the attitude was, 'Okay, well, let's *do* do this thing according to agreement.' But everybody was so convinced that pBR322 was a better vector system to use that approval was a matter of formality. That was also the attitude of everybody else who had been associated with that program. It seemed so obvious. I'd say it's not like having borrowed the countess's bracelet or something like that and suddenly discovering it was a serious misdemeanor. On the other hand, it was clearly a serious matter."

Goodman and Ullrich, meanwhile, returned to San Francisco from Utah. According to university accounts, Goodman ordered work on the insulin experiment to stop. According to Ullrich, another one of the first things Goodman did when he got back to San Francisco was to call up Eli Lilly & Co. to try "to get a deal." Lilly officials confirm that they were contacted by Goodman at about this time. Meanwhile, there were anguished meetings, and a flurry of phone calls to officials at the NIH in Bethesda, Maryland, to confirm pBR322's status. First the UCSF scientists tried to handle the situation quietly, diplomatically. Rutter was on the telephone with people at the NIH, asking when pBR322 was to be certified and what caused the delay. Ullrich, too, called up the NIH.

To hear colleagues and even rivals tell it, no grave scientific sin had been committed, nor even any unspoken etiquettes of competition breached. As a member of Gilbert's lab later put it, "I don't think they should have done it any differently; I don't think I would have." Others shared the sentiment of Lydia Villa-Komaroff, another molecular biologist at Harvard, who felt that "politically, it was stupid." Scientifically, however, no one could seriously fault the work. It was just that the problem sprang up at a time when magazines and newspapers were full of headlines like "Total Ban Sought on Genetic Engineering."

Within the UCSF biochemistry department, however, not everyone regarded the incident as an innocent oversight, and the dismay in some ways reflected the deepening factionalism that recombinant DNA fever brought to the department. "The idea to skirt the rules was to beat Wally Gilbert. That's really the crux of the issue," says one faculty member. "Everybody knew at the time that pBR322 was the best vector by far. So it wasn't science; it was just playing to win."

The political significance of the violation may have seemed like a minor technicality to the scientists involved, but the potential political damage was not lost on the NIH. The harangues of the NAS forum still echoed in Washington, and the institutes at that time faced a difficult and contentious battle with Congress on just how rigorous the NIH could, or should, be about enforcing guidelines in an area of research it also funded. Hearings, in fact, were scheduled to begin March 15 in the House of Representatives on federal legislation. Was this the right time to disclose the fact that scientists had violated, albeit accidentally, the very guidelines they had voluntarily sub-scribed to? Would the importance of the scientific success be obscured by an emotional response to a harmless violation and adversely influence the out-come of federal legislation?

Beginning on March 2 and for about the next seventeen days, those questions hung like low, heavy clouds over the Goodman and Rutter laboratories. Recollections about this period, about conversations and group discussions, tend to be circumspect, vague, and incomplete. The UCSF group was led to believe that pBR322 would be certified within days, although Ullrich, who also made calls to the NIH, concedes this might have been "wishful thinking" on the part of the researchers. "Everybody kept saying, 'Well, pBR322 is going to be approved daily,' " Rutter says, disgust in his voice. *"Daily!"* As Ullrich and others suggest, the UCSF group hoped that the long-rumored certification would come through; the awkward situation, bureaucratically created, might in turn bureaucratically disappear. It did not.

Then, in mid-March, Rutter made "informal" contacts with a high NIH official about UCSF's use of the uncertified vector. This official, Rutter recalls, "said, 'Oh God, let's not do this at this time because we're having this big problem at NIH with respect to control. There'll be a strong reaction about this.' And he was careful to say, 'Was there any danger? Is there any danger? Did you do it with P3?' [I said,] 'Yes, of course.' Then the issue was simply how to handle it tactically, and we decided of course not to use the pBR322 until it was approved."

Documents from the NIH indicate that Rutter's call went to the deputy director of science, DeWitt Stetten—the same official who had spoken about the insulin supply problem with such concern at the National Academy forum. Stetten spoke to Rutter between March 16 and 19. He recommended that the clones be destroyed.

What had been a signal scientific achievement one week had become a slithery political football the next. In Washington, where small cracks in the institutional facade of federal agencies provide fingerholds for a thousand critics, the pBR322 incident might assume repercussions far beyond the semantic confusion between "approval" and "certification." Critics could exploit it as an example of the NIH's inability to enforce its regulations over a controversial new area of research, which could have given that much more impetus to legislative sanctions. And that, just as clearly, was something neither the scientists nor the NIH wanted. So, although the premature use of pBR322 was known within the laboratories at UCSF and at fairly high levels of the NIH, not a word of it reached the public's attention. As even the scientists admit, that was not an accident.

The sense of paranoia enveloping the UCSF group amounted to a psychological state of siege. This feeling is captured in a revealing statement made by Rutter several months later, as he recalled the sensation of finding his research group in the midst of the recombinant DNA controversy. "The issue," he said, "appeared to coalesce various concerns regarding research on atomic energy, biological warfare; the specter of a biological holocaust was portrayed. This issue was used by some, including scientists, as a vehicle for their own political or philosophical motives. The press, among others, sometimes fanned the flames of controversy. Cloning molecules was linked to cloning humans. Imaginary and sometimes sensational scenarios were depicted. This combination of science and fiction made good copy. Consequently, the value of scientific exploration has been questioned. Scientists have sometimes been depicted as sinister rather than useful members of society. Further, and in an unprecedented fashion, repressive and punitive legislation was considered. We felt," Rutter said, "that directly informing the NIH would inevitably lead to public disclosure and debate about this incident and that would exacerbate the whole situation."

What to do with the clones? In a series of despair-filled meetings, the principal members of the research team—Rutter, Goodman, Ullrich, Chirgwin, Pictet, and Shine—discussed their options. The politically expedient course of action would be to destroy the clones, even though they were known to contain the insulin sequence. Any decision to destroy Ullrich's clones, though, would complicate matters, and morale, in the Goodman lab, for Ullrich's lab mate and friendly rival Peter Seeburg had by that time, apparently unbeknownst to UCSF officials, also used pBR322 prior to certification. And Seeburg had achieved another major breakthrough with the uncertified plasmid: rat growth hormone clones.

Destruction of landmark experimental evidence is not, to say the least, a routine occurrence. Not surprisingly, the drastic step of destroying the clones was not universally embraced. John Chirgwin realized that the insulin genes in Ullrich's clones could be snipped out with restriction enzymes and salvaged before the clones were destroyed, with no one the wiser; he also realized that if he could think up this ploy, the others could, too. So he went to Rutter with a highly unusual request. He asked that his name be removed from any paper growing out of the work if the group chose "the dirty way out." Chirgwin didn't necessarily suspect that this was happening, but as he recalls, "No one got up and made a virtuous speech saying, 'Under no circumstances should this be allowed to happen!' "

Other members of the group insist, however, that it did not happen that way. "We did everything right to the best of our knowledge at the time," asserts John Shine. "But to argue at that point may well have been detrimental in the long run. For the whole area of recombinant DNA. Which we didn't want at all, you know. Our futures were involved in it. And we all felt that it had so much potential that to perhaps mess it up for a bureaucratic sort of mistake like that would be ridiculous. So we had no choice, really, in the end."

Slightly more than two weeks after the problem arose, the researchers claim, they reluctantly decided that Axel's bugs should be put to sleep. "It was just a tremendously difficult time period," Rutter recalls, "because I think not everybody agreed with the decision." On March 19, Ullrich says, the test tubes containing his precious insulin clones were emptied into a bucket, then covered with hydrochloric acid. In just a few moments, eighteen months of work dissolved in acid in the bottom of the pail.

As Ullrich described this sequence of events, slouched in a chair in a windowless office, voice low and a correlative sag in his shoulders, he had no difficulty making it sound like the scientific nightmare of a lifetime.

"It was this crazy situation where you, as scientists, know better than the administrators, and it's just a legal thing," Ullrich said. "Not even a legal thing. Just a procedural problem. We *knew* pBR322 was safer than pMB9. I knew that I did not do anything knowing better. The only mistake I had made at that time was that I had naively just taken somebody else's word . . . and on that basis, without waiting for some written approval, clearly that was . . ." The words trailed off.

"This rush," he added quietly, "was also caused by this competition that we knew was on the other side of the country."

Once more into the breach.

"It didn't seem like an impossible feat to do it again," Rutter recalls. "We had RNA. We had *good* RNA. We had lots of it. It wasn't as if you could do it once out of a thousand times at that point, because I think there'd been a lot of work done on both sides to make sure the methods worked. There was a good degree of confidence, although you never can tell. It was the first time it'd been done."

In fact, Ullrich says there was very little RNA left, but he had held about half of his precious insulin DNA in reserve, and this was readied for the second effort. It was the middle of March, and Ullrich and his colleagues were in the same boat as at the beginning of the year: their insulin cDNA was all dressed up with no place to go. Finally, on April 18, NIH director Donald Frederickson certified another plasmid, pMB9, for use. It wasn't quite as versatile or convenient as pBR322, but it would do in a pinch. In a sequence of events executed so rapidly and flawlessly that it raised eyebrows anew in some quarters, Ullrich recommended the cloning experiment on April 23. By late April, the team knew they had at least four colonies which contained inserts. Ullrich renamed these new clones AU-1, AU-2, AU-3, and AU-4. Did any of them contain insulin?

Ullrich harvested some of the plasmid DNA, cut out the cloned rat DNA with restriction enzymes, and prepared the DNA for sequencing. The final step, the confirming step, was to determine the exact A-T-C-G sequence of the inserts and see if they corresponded to the known protein sequence of rat insulin.

John Shine, master sequencer, shepherded the experiment over this final threshold, although his services were much in demand. Peter Seeburg recalls having to wait, and none too happily, with his growth hormone clones while Shine sequenced the insulin clones first. "To be honest," Seeburg says, "I was pissed that growth hormone sat on the shelf while the insulin was being sequenced, but there were so many more diabetics than growth-hormone-deficient people, so it made sense." Shine, meanwhile, ran Ullrich's DNA on his gels and left the photographic record of the DNA's nucleotide sequence in the darkroom overnight to develop.

The following morning, Shine walked into the little room on the ninth floor that was used as a darkroom to take a look. "We were hopeful, but far from sure," Shine recalls. "We developed the X-ray film and Axel was in this little room there with me. . . . And whilst it was still sort of dripping wet, we hauled [out] the film and looked at it."

It wasn't the best sequencing job ever run, Shine concedes. But in the first picture they saw a long sequences of Ts. That was a good sign; it was what would show up in a piece of DNA copied from messenger RNA, which had a long complementary tail of As. They began to read in from the other end of the gel. Ullrich was standing there with the amino acid sequence of insulin. Shine knew the genetic code by heart, of course, so he started reading the three-letter words in the sequence and saying the amino acids out loud. "We could see phenylalanine, valine, lysine or something—whatever it was," Shine recalls. "And Axel said, 'Hey, that's it! That fits here! What's the next one? Is it such and such?' 'Yes, it is.' And so we knew then—in the space, really, of less than a minute after we'd actually taken the film out—that we actually had an insulin clone."

Out of the roughly 2.5 billion base-pairs of DNA in rat chromosomes, out of that dense tangle of information, Ullrich and his colleagues had reached in and, for the first time, plucked out the gene for a medically significant hormone, copied it, and made it available for extensive analysis. "That was one of the best, most important moments in my life," says Ullrich. "That determined a lot." His pAU clones had indeed turned to gold.

As soon as Shine completed the sequencing, the group was able to piece together fragments of the insulin sequence, and an intimate view of the gene began to emerge. It measured 354 nucleotides in length. One could clearly delineate the three segments of the insulin molecule—the B, C, and A chains—that formed proinsulin, which matured into insulin when the C chain was snipped away. The data provided a partial but incomplete look at the signal sequence, the "pre" part of preproinsulin, and suggested mechanisms by which the various chains were clipped.

In both scientific and personal terms, it was a great triumph for the Rutter-Goodman collaboration. Despite not having the Chick insulinoma, the group had managed to get enough insulin message to manufacture a cDNA "gene." Despite the pBR322 fiasco, Ullrich and company had still breezed to the finish line ahead of the pack, with the first insulin clones. In a larger sense, the work went far beyond winning the first part of the insulin race. The suite of techniques they put together while cloning insulin instantly became a how-to manual for molecular biologists all over the world.

There was, however, little time for celebration. The group worked twenty hours a day to prepare a paper as soon as possible (it was still believed that the Gilbert group was hot on their tails). There was some additional paperwork involved in this case, because the decision had been made to apply for a patent as well. By May 12, less than three weeks after the cloning experiment had commenced, a polished version of the paper announcing the cloning of a rat insulin gene was in the hands of the editors at *Science* magazine.

This first phase of the insulin race had a rather unusual coda. In May 1977, Wally Gilbert was back in the Bay Area for a series of seminars. On a Friday afternoon toward the end of the month, he found himself sitting with a group of about a dozen graduate students from the University of California in the Japanese Tea Garden in Golden Gate Park. They were drinking tea, eating fortune cookies, and discussing recombinant DNA research. Herb Heyneker of the Boyer lab was there as well. At one point, Gilbert left the table. There was a single cookie remaining, and "Bad News" Heyneker removed the enclosed fortune and inserted a hastily scribbled replacement.

Gilbert returned to the table and, after considerable coaxing, was finally persuaded to take the last cookie. "He opened it up," recalls Heyneker, "and was reading it, read it again, and he started laughing so hard . . ."

Written on the piece of paper was: "YOU WILL CLONE THE INSULIN GENE SHORTLY."

Like a lot of fortunes, it flattered the ego but veered considerably from reality. Even as the group sat in the park, the information service at UCSF was making arrangements for a news conference to announce the Rutter-Goodman achievement.

Since Gilbert probably already had inklings that the UCSF team had cloned the rat gene, he proved to be an unusually good sport about getting beaten. But Gilbert had one more encounter with the UCSF researchers before the day was through, and that one was considerably more awkward.

Heyneker, a great admirer of Gilbert, invited the Harvard scientist back to the nearby laboratories on Parnassus Avenue, which overlooked Golden Gate Park. On Friday afternoons, people in Howard Goodman's laboratory customarily gathered in the ninth-floor hallway for an end-of-week party. It was an obvious destination. On that particular Friday, one of the lab members was even absorbed by the latest issue of the *Midnight Hustler*, a copy of which was occasionally sent to UCSF by the Harvard scientists. Axel Ullrich and Goodman were standing together in the hallway when who should round the corner with Herb Heyneker but Wally Gilbert, the rival whose progress on insulin—real or rumored—had tormented their lives for months. Gilbert walked directly up to Goodman and, from one man of few words to another, said, "Tell me about insulin."

Goodman said nothing. Caught off guard by Gilbert's sudden appearance, he just stood there. "Howard was a terrible communicator in the best of times," John Shine recalls, "and Wally just put him on the spot." "It was really awkward," Ullrich recalls, "because nobody really said anything. Nobody wanted to confirm it." Ullrich and Shine made futile attempts at small talk. A look of concern flickered across Gilbert's face. He could later dismiss the strange interlude as "just a general side of the competition," but undoubtedly Goodman's agitated reaction registered, in its own way, as silent confirmation of the UCSF success. Goodman stared back, saying nothing, looking nervous and discomfited. Finally he "just took off," as one observer recalls, without even responding to Gilbert's appeal to tell him about insulin.

Goodman had the reputation for being exceedingly reticent. In this case it was as if there was too much to say.

Around one in the afternoon on May 23, with full-throated ease, Howard Goodman and William Rutter and other members of the UCSF rat insulin team met the press for the first time. Because insulin continued to be a magic word, the response was almost overwhelming.

There can be no doubt that there was considerable political advantage to publicizing the work. Although the actual paper had not yet been published, it was due to appear within weeks in *Science.* The speed with which it was rushed into print (with some unusually significant typographical errors) suggested that there were political overtones to the editorial handling of the work at the magazine—that the experiment was an important stepping-stone to gaining public acceptance of recombinant DNA research and should be publicized without delay. When Rutter and Goodman reported to the assembled press on May 23 that they had succeeded in cloning the rat insulin gene, it sent the right kind of ripples through the rest of the country.

Accounts of the work appeared on the front pages of the *New York Times, Washington Post,* and *Los Angeles Times* the next day. The *San Francisco Chronicle* said the work "offers proof that the much-debated field of combining genes from unrelated organisms can in fact provide crucial insights into the ways by which chemistry governs heredity." The British magazine *New Scientist* observed, "The California team's experiment will stand as a landmark in genetic engineering research." The scientists were asked if the work followed the NIH guidelines; no one volunteered an account of the pBR322 interlude. Given the controversy that would erupt within months, it would have been simple to acknowledge the mistake, perhaps—as Chirgwin suggests—in a footnote to the paper. Instead, as the *San Francisco Examiner* reported, "The UC group emphasized that the gene-isolation work was carried out according to the safety guidelines put forth last year by the National Institutes of Health."

One of the more optimistic comments to emerge from the press conference was Rutter's assessment of the next step. Rat insulin could be *expressed*—that is, produced by bacteria—within six months, he contended, and expression of human insulin was only a year or two farther down the road. Inserting the human gene into bacterial cells was P4 work, requiring the equivalent of a biological warfare laboratory; Rutter said that if necessary, his team would go to the Army's Fort Detrick facility. The next stages of the work were laid out.

For all its superficial success, the press conference created very real problems at UCSF. "There was a feeling of unreality about the whole thing," recalls Chirgwin; other workers in the Rutter group were, in his opinion, doing equal or better work, and the burst of publicity "made me feel a little awkward." He was not alone. No other phenomenon has come to so sharply define tensions between scientists, particularly that intimate enmity between lab directors and the postdoctoral fellows who work with them, as the press conference. Public recognition of scientific work inevitably involves simplification, and with it, an untidy compression of credit; the process becomes positively slapdash, however, in the setting of the press conference, where reporters represent a broad spectrum of expertise and ignorance, where requests to explain the medical implications of basic research often elicit oversimplified or exaggerated claims, and where journalists, by habit and convention, reflexively cite the head of the participating laboratory for an overview of the work, the summational comment, the key quote. The contributions of younger scientists are invariably compressed, simplified, or sometimes completely ignored when chronicled for popular consumption.

That was the rude lesson on May 23 for the San Francisco postdocs. Axel Ullrich, first author on the paper, was by consensus the key mover in the insulin project. His role, along with the rest of the postdocs, was noted in the lower paragraphs of the ensuing stories—if noted at all. Meanwhile, Goodman, who had been on sabbatical during most of the critical work and, in the view of others in the lab, had hardly expressed enthusiastic support for the project, flew in from out of town for the press conference.

The attention heaped on Rutter and Goodman caused great consternation among the postdocs, and indeed throughout the whole department. "Of course we were not very happy about that," admits one, expressing a widely shared sentiment. "Especially Goodman, who really had *nothing* to do with it. And he took all the credit. It was just incredible." Another member of the insulin team observed that what he unhappily refers to as "the-boss-did-it" syndrome began with the May 23 press conference. Even the lab directors, he says, believed their press clippings, and were swept up in the unexpected fame that public attention conferred. These appear not to be simply postdoctoral plaints; as Brian McCarthy puts it, "They did the work when Howard was gone, and after he had discouraged them from doing it. When he got back, he rushed in and took credit. . . . With respect to Bill, he hadn't given much advice because he wasn't in that area [of research]. . . . They have a legitimate grievance."

The person who stood to lose the most credit was of course the person who had done most of the actual cloning work: Ullrich. "At that time," Ullrich concedes, "I was really . . . I was clearly not happy about it, but I thought, 'That's the way it works around here.' But then this continued and became even worse later on." The ill will that developed had a decidedly sour effect on the subsequent collaboration between the Rutter and Goodman laboratories. Indeed, bad feelings began to perfuse the whole department. The rush to clone medically important genes had a "boomtown" feel to it, as one postdoc put it, and some researchers began to complain that increased secrecy reflected attempts at personal financial gain on the part of the researchers. "After the initial successes," Chirgwin recalls, "people were sniffing Nobel Prizes and multimillion-dollar stock options further down the road. The lab chiefs could sort of see that they were going to be in the National Academy, and at least had a shot at Stockholm further down the line. That was when the enticements of power and fame, as a major corrupting force, were just appearing." Rutter concedes that "we hadn't planned on a lot of publicity, and I don't think we handled it particularly well," but insists that the lab chiefs deserve credit for setting up the programs in the first place.

The family atmosphere that had once marked relations was supplanted by factionalism, suspicion, and a certain degree of secrecy. "In 1977 following the cloning of the rat insulin DNA," wrote Keith Yamamoto, a faculty member at UCSF, "a fellow faculty member confided that the ensuing patent application and press coverage seemed to be affecting the motivation of some of his post-doctoral colleagues, and that the potential for financial gain was an apparent consideration in the planning of their experiments." "Most people who have a historical perspective will tell you that there were really two phases to the department," says Brian McCarthy. "Before recombinant DNA and after recombinant DNA." The watershed event, McCarthy says, was probably the rat insulin experiment.

The San Francisco press conference also had the effect of whipping up public expectations for insulin. Again, as in the parlor game Telephone, information suffered incremental but significant changes as it was passed along from experts to the public. Possible benefits began to sound like certain benefits. The possibility of cheaper therapeutics became the certainty of cheaper therapeutics.

At the May 23 press conference, William Rutter was quoted as saying there were a hundred million diabetics in the world; two months earlier, Irving Johnson of Eli Lilly had put the figure at sixty million, emphasizing that, since diabetes was an unreported disease in many parts of the world, the numbers were not firm. Rutter indicated that twenty million worldwide needed daily doses of insulin to survive; Lilly's figures suggested less than half that number. The UCSF News Services/Publications office, justifiably renowned in science journalism circles for the quality and professionalism of its operation, nonetheless included this passage in its press release: "Although slaughterhouses provide beef and pig pancreases as insulin sources, a shortage of the hormone *is developing* as the number of diabetics increases worldwide. Moreover, some diabetics develop an allergic reaction to beef and pig insulin, with deleterious effect. *Manufacturing human insulin in bacteria would solve these problems* and in addition might make insulin available at a lower cost than it is now, they [the UCSF scientists] declared" (italics added).

The distinctions are subtle but important. There was no evidence of an imminent insulin shortage, as even Eli Lilly acknowledged. There was no proof that human insulin would be nonallergenic. And there was good reason to doubt that a genetically engineered human insulin would be cheaper than animal insulins. Eli Lilly, for one, never claimed that it might be cheaper at the National Academy forum, for a very good reason. "I really didn't think it would be," says Johnson, "particularly initially." A cheaper, better insulin was certainly the aim, and a highly commendable one, but nobody—especially scientists, whose reputations tower or teeter upon a foundation of defensible facts—should have substituted probability, much less certainty, for possibility. But certainty—cheaper, better *human* insulin—was the message that began to trickle down to the public. Everyone, in a sense, was an accomplice to this conspiracy of hopefulness. Who didn't want to hear about wonderful new ways to treat disease?

When the rat insulin paper finally appeared in the June 17 issue of *Science,* everyone in the scientific community could appreciate the resourcefulness of the San Francisco team. The researchers were swift to establish the historic context of their work; the first citation, ten words into the paper, referred to Frederick Banting and Charles Best. Exactly fifty-five years after those University of Toronto researchers first isolated insulin, science had progressed to the point where a letter-by-letter, word-by-word rendition of the gene could be spelled out and examined. It was a prodigious endorsement of recombinant DNA, but it also built upon half a century of outstanding research on insulin. The UCSF work merely added the latest chapter to the unfinished biography of this much-studied molecule.

The triumph of the San Francisco group sparked understandable disillusionment in other quarters. Back at Harvard, the news was greeted with both shock and disbelief—shock that the UCSF team could have cloned the gene so quickly, disbelief that the work could have proceeded so flawlessly in so short a period of time. "We became all very depressed, okay?" recalls Arg Efstratiadis. "We were scooped. And not feeling happy about it." Forrest Fuller recalls feeling "jealousy, and anger at the system in terms of being held back. Some of that anger got displaced at the California group. Simultaneously, there was this feeling of comradery—*somebody* was able to succeed and to beat the system. They had done something the hard way, and they had done some beautiful work." Perhaps the most succinct reaction appeared in a front-page article in the May 25 edition of the *Harvard Crimson*. The headline said it all: "DNA Results Irk Harvard Scientists."

A certain emotional plasticity is required of scientists, however, because many, perhaps most, experiments do not work, and for every winner of a race, there are sometimes a dozen losers. Argiris Efstratiadis, as much a philosopher as a scientist in the midst of his disappointment, remembered an old-world story about the time Christopher Columbus returned from his first voyage to the new world. On the day of Columbus's return, a huge crowd gathered at the dock, and a young boy, unable to see over the crowd of onlookers, was picked up by a man and hoisted above his shoulders to take in the scene. As Columbus's boat docked to triumphant cheers, the little boy looked discouraged, and then asked the man plaintively, "Is there anything left to discover?" The boy's name, Efstratiadis likes to point out, was Ferdinand Magellan.

For the young Magellans of molecular biology, Efstratiadis knew, "there's always something left to discover." After a disastrous spring and summer in 1977, his old-world fable would come true.

146
· · · · · ·

PART II

E l e v e n

"In the Real World, You Don't Really Believe These Things . . ."

. . . .

The Rutter-Goodman group's finest hour was soon followed by Genentech's worst. "In research," says Herb Heyneker, displaying that endearing foreign penchant for latching on to our clichés, "you often go through deep valleys to reach the highest peaks." For Herb Boyer's lab, the deep valley came in June 1977. It was, Boyer admits, "a very difficult moment."

The success of the Rutter-Goodman team in cloning the rat insulin gene reiterated—as if the point needed reiteration—that other scientists were perfectly capable of cloning medically important genes. Then a minor controversy arose with publication of the June issue of *Smithsonian* magazine, which contained an article by science writer Janet Hopson detailing her three-month stint as a technician in the Boyer lab. It was a generally favorable account of the scientists and their work, but Hopson nonetheless reported that UCSF's P3 laboratory had been temporarily closed because of "messy conditions" and also observed that "among the young graduate students and postdoctorates it seemed almost chic not to know the NIH rules." With legislative initiatives creeping forward on both national and state fronts, here was further evidence, recombinant DNA's critics could claim, that legislation was necessary. Boyer fired off a letter to *Smithsonian* refuting Hopson's allegations, but the damage had been done; as the magazine noted in a rejoinder, a member of Boyer's lab had read and concurred with the conclusions of the article before it appeared in the magazine.

The pressures on Boyer were particularly onerous at this time, according to former members of the lab. The fact that Genentech was sponsoring the research at UCSF of both Herb Heyneker and Paco Bolivar became the subject of heated debate in the corridors of the medical center and in sessions of the academic senate. Would the university be reasonably compensated for work undertaken on its premises? Would Genentech's scientists share information with colleagues as freely as colleagues did with them? Charges of conflict of interest were aired openly; suspicions of greed, of using knowledge obtained by publicly funded research for private gain, were whispered about as well. Boyer argues that he and Bob Swanson took great pains to negotiate a fair and equitable relationship with UCSF, but not everyone shares that view. A current faculty member who was present at the time says, "The university just got ripped off, financially and just in the way things were set up." Such feelings, in Heyneker's view, were "ninety percent sour grapes."

Those who did not react with indignation to the incursion of business into academia looked on with the smug assurance that Boyer's venture was doomed from the outset. Hardly anyone but Boyer and Swanson and a handful of colleagues—Itakura, Riggs, Heyneker, Bolivar, and Scheller—believed Genentech would amount to anything. And even they had their moments of doubt. The somatostatin project moved slowly along, and then suddenly, inexplicably, ground to a halt in June.

The plan, very simply, was to construct a human gene and induce bacteria to make the protein. Keiichi Itakura had begun working on the gene in the summer of 1976, even before officially transferring from Cal Tech to the City of Hope. He was given workspace on the main floor of the medical hospital building at City of Hope, a short distance away from a wing for patients. "Workspace" is a deliberately neutral choice of word: it looked less like a laboratory than a simple hallway. Within six months, however, Itakura had turned this long, narrow corridor into the most advanced synthetic DNA facility in the world; with its profusion of foul-smelling reagents, it soon acquired a certain malodorous infamy at the City of Hope.

Itakura, cheerful and industrious, installed fume hoods, huge metallic cabinets that sucked away the noxious chemical fumes (a special filter system was installed, and Art Riggs insists a major effort was made to ensure the safety of both lab workers and nearby patients). He filled the lab with benches, equipment, and supplies. He recruited and trained technicians. Then he began to assemble the raw materials for his assault on the somatostatin gene. It was the type of work that requires extraordinary patience; building DNA, even by Itakura's "rapid" method, was a slow and laborious process.

Synthetic DNA chemistry, like the laboratory to which Itakura was relegated, was a narrow, odd-shaped area of research. The goal—converting inert chemicals into molecules capable of self-replication—related to nothing less than the origins of life on the planet from primeval commingling chemistries. From a more mechanistic, atomic point of view, it represented a series of complicated reactions in which chemical girders and struts were added and subtracted and purified. Bit by chemical bit, atomic group by atomic group, a skimpy skeletal array of atoms would grow into a functional scrip of DNA.

As in any kind of writing, you started out with letters and built them into words. The "letters" in this case, the As and Ts and Cs and Gs, were isolated from (of all places) the sperm of salmon, and could be purchased from chemical supply companies. A single gram of deoxycytosine, or "C," for example, cost about $9.80; a pound of Gs ran about $2,800. It took days of work to convert these simple letters, chemically joined to a sugar group and called nucleosides, into nucleotides. Nucleotides were letters with DNA's complete set of atomic appendages, including the phosphate group that joined with the next sugar to form the helical backbone. Once you had converted nucleosides into nucleotides, you in effect had prefabricated units, like building blocks that fit together. Whenever you wanted to add a strut between a sugar and a phosphate, for example, you had to block off (or "protect") the other components to make sure they didn't participate in the reaction. It was like plugging the holes in a Tinkertoy joint so that only a single hole remained open and available for a chemical reaction.

"Essentially it's protection, deprotection, protection, deprotection," Itakura explains with a laugh. "The reaction itself is very simple. It's finished in two or three hours. But after each reaction you have to purify, and purification takes more time than the reaction itself. . . . The reaction never works one hundred percent, and also there are side reactions, undesirable reactions, that contaminate it [by] about five percent or ten percent. So you have to purify each step. Otherwise the impurities accumulate after several steps and you can't use it."

The object was to string three letters together to form what was called a "trimer," which corresponded to one of the three-letter genetic words that coded for an amino acid. These trimers, like words, were then strung together to form short genetic phrases. Since somatostatin was spelled out by fourteen amino acids, Itakura faced the immense task of stringing together forty-two nucleotide assemblies for each DNA strand, eighty-four in all. Then he had to figure out a way of putting them all together to form a stable double helix. The pioneer in the field, H. G. Khorana, had developed the method of constructing overlapping segments and using the zip-lock quality of base-pairing to assemble the gene. That was the approach the City of Hope team—Itakura, with initial help from Richard Scheller—took.

Not surprisingly, the work did not proceed smoothly at first. They tried synthesizing a piece of somatostatin DNA in the summer of 1976; that didn't work. Then they tried cloning some synthetic DNA into a plasmid—a plasmid, incidentally, developed by Forrest Fuller and donated by Gilbert's lab. That didn't work either. Then mistakes began to crop up in the synthetic DNA. It was wrong by a single base-pair, but wrong nonetheless. "Maybe some kind of side reactions accumulated," Itakura admits, "and gave a wrong base." In any event, he was forced to go back and try again.

This was exactly the kind of problem that skeptics might have predicted, and in this newly accountable business environment, delays had to be explained. A meeting was held in the office of Thomas Perkins, who was bankrolling the venture; one participant describes the atmosphere as "tense" when the scientists described the setbacks in DNA chemistry. The businessmen demanded assurances that somatostatin could be made; the scientists, of course, said, "of course." Itakura was encouraged to try again.

Itakura, in fact, tried two more times. Finally, in the spring of 1977, Itakura shipped his DNA—packed in ice, delivered by air courier—to the Boyer lab in San Francisco. Herb Heyneker and John Shine cloned and sequenced it again, and this time the synthetic DNA was letter-perfect. They had a bona fide gene. But did the gene work? Would it make somatostatin?

No one had posed that question of bacteria before. Other scientists had asked bacteria—unsuccessfully—to make a mammalian protein, but no one had put in an order for a human hormone, not to mention a synthetically derived human hormone, and so the researchers didn't quite know what to expect as they moved on to the expression stage of the experiment. Riggs and Itakura were optimistic, though. In their NIH grant application, they had predicted a harvest of ninety grams of somatostatin per three hundred liters of genetically engineered bacteria.

This stage of the experiment introduced a whole new magnitude of problem. It was not merely enough to plug a gene at random into a plasmid and, in effect, let 'er rip. Itakura's gene needed to be placed *just so,* right next to the genetic control unit called the promoter, which allowed the sequence to be spotted, read, and copied by the bacteria. The gene also had to be "in frame"; in other words, the enzymes had to see CAT ATE RAT and not ATA TER ATC. If the gene was not aligned almost perfectly, right down to a single base-pair, it would be as if Itakura had written out the sequence in invisible ink.

It fell to Riggs and Boyer to devise a strategy to make the cloning experiment work, and the scientific decisions they made here, in 1976, still have felicitous business ramifications for Genentech many years later. They figured out a way to hoodwink bacteria into making a human hormone, and they did it by exploiting the wealth of knowledge that had accumulated about the *lac* gene. They would lash Itakura's human gene to the bacterial gene for beta-galactosidase (or beta-gal). Riggs in particular deserves credit for what one colleague describes as a "kind of long-range vision of things." Like many biologists, Riggs realized that if you fused two genes together (such as somatostatin and beta-gal), the protein that emerged from this hybrid gene would be fused, too, and unsalvageable. So, benefiting from the versatility of synthetic DNA, Riggs suggested that Itakura use a short DNA link between the human gene and the bacterial gene. This link, a mere three bases in length, coded for the amino acid methionine, and methionine acted like a dissolving stitch because the researchers could chemically make it dissolve later on. In short, they could unfuse the protein and free the somatostatin molecule. This may sound like a minor touch, but it in fact overcame a problem that stymied other researchers for several years—several very crucial early years in the development of Genentech. It also added an element of safety, the researchers claimed, because they knew somatostatin could not be biologically active while the molecule's front end was fused to another protein.

With enzymatic scissors and glues, Herb Heyneker and Paco Bolivar reduced the ruse Boyer and Riggs had planned to straightforward engineering. First, they grafted the *lac* promoter, along with the initial bit of the long beta-gal gene, into their plasmid. Then they spliced in Itakura's human gene right next to it (or "downstream"), so that any enzyme that started to read beta-gal would soon be reading somatostatin. At this point, the great logistical advantage of using the synthetic gene became apparent, for pBR322 *still* had not been certified for use by the NIH in this type of experiment. In fact, to insert genuine human DNA into pBR322 would, according to the NIH guidelines, have to be performed in a "maximum security" P4 laboratory.

But a synthetic gene was not genuine human DNA. It was similar to, but not identical to, the human gene, so the Genentech researchers legally side-stepped any regulatory constraints on their work. "Weird but true," Heyneker confirms. "It was basically an omission. People did not realize at that time the power of synthetic DNA." Boyer considered it a "throwaway" advantage of synthetic DNA. "It's sort of like the games people play with the IRS," he says. "Once you have a set of rules, you find a way around it. I felt it was just a way of dealing with the regulations." There was a brief flap at UCSF over the use of pBR322 for this experiment, but the reaction in no way rivaled the storm brewing over the Rutter-Goodman insulin experiment.

Herb Heyneker and Paco Bolivar cloned the gene in late spring of 1977. Colonies appeared on plates right away, indicating that the cloners had successfully inserted somatostatin DNA into the bacteria. But how could you tell if the bacteria were *making* somatostatin? The drill was to grow up potfuls of each colony and crack the bugs open; this internal liquid could then be tested for specific proteins. If there were any somatostatin molecules floating around in the soup, radioactive antibodies "programmed" for somatostatin would bump into them and hang on in a biochemical bearhug. When the sample later passed through a machine called a gamma counter, "pings" of radioactive decay would register, indicating the presence of the protein. Art Riggs and his technical assistant, Louise Shively, prepared the test. One of the reasons Riggs had argued so strongly for somatostatin instead of insulin was that this assay was sensitive enough to pick up a few molecules per cell.

On the morning of June 16, 1977, the fledgling corporate hopes of Genentech balanced on the kind of numbers that came out of the gamma counter. Swanson had made a special trip down to the City of Hope to watch; Riggs, Itakura, and Shively joined Swanson on this momentous occasion in the young company's history. The stakes, for everyone, were considerable.

Riggs and Boyer, of course, had established academic reputations and would survive any setback to go on to other projects; but all the controversy swirling around Genentech at UCSF certainly did little to diminish the investment of Boyer's pride in the outcome, and Riggs could hardly forget that the NIH had declined to fund this project. Itakura, meanwhile, was trying to establish himself as the premier synthetic DNA chemist. "Lack of success," as Riggs puts it, "would *not* have helped his career."

"The person who was most concerned," Riggs says, "was the person that had the least detailed knowledge of the chemistry and recombinant DNA work that was going on, and had gambled the most. Had, in fact, staked his career on success. And that was Swanson." Bob Swanson had quit his job at the venture capital firm of Kleiner & Perkins and gambled everything on this protein with a funny name. He had grown anxious with the setbacks and delays over the previous year. He would drop by Boyer's lab every day to see how the work was progressing, and he made frequent trips to the City of Hope. Especially in light of the Rutter-Goodman triumph of the previous month, the company urgently needed to rack up a success. Riggs's expectations were a bit more modest. "I was quite prepared," he says, "to do the whole thing, get a few molecules per cell, and declare it a fantastic scientific success." Riggs considered the chances good. "But," he recalls, "one of the exciting—the *most* exciting—aspects of it was that we had no guarantee of success. We might not have been able to pull it off. If there had been something in our know-ledge that was wrong, or we were missing something to actually get these foreign proteins produced in bacteria, if the genetic tables were somehow wrong . . ."

And so they all stood around nervously in the narrow equipment room on the second floor, hovering, awaiting the results. The gamma counter was a creaky old warhorse of a machine; lights would scoot up and down its antiquated instrument panel as the numbers came spitting out on adding-machine paper. The numbers started to come out, and the scientists could tell immediately that they just weren't right. "Actually it turned out to be less than two molecules per cell," Riggs recalls. "The initial results were nothing."

The result plunged Swanson into shock. "He was at the end of his rope, in more ways than one," says one participant. "He was close to running out of money for the project."

"You see your career and the investment and the energy go flashing by," Swanson recalled later; he made it sound like a brush with death, the entrepreneur seeing his economic life flash before his eyes. "You say, 'Oh my God!' Then it was a matter of sitting down and saying it should work, why wasn't it, and then figuring out a solution. It was a *very* scary time." When word of the fiasco reached the Boyer lab, it capped a miserable month of bad news. The mood turned noticeably grim. Lab workers recall that the atmosphere became "very heavy"; the golden touch that had seemed to anoint all the experiments during the two previous years suddenly turned leaden. "Herb Boyer went through, and Genentech went through, a really down period," John Shine recalls, "when it looked like they were going to be beaten everywhere because they'd chosen the synthetic route, which wasn't the best one."

Art Riggs, whistling in a graveyard of bad data, tried a few more test runs and tinkered with the results for several more days. Nothing changed the implication of the initial entry made in his lab notebook on June 16, 1977. They are the two most routine and painful words in a scientist's vocabulary: "Negative results."

Meanwhile, the longest summer in Forrest Fuller's life, and easily the most depressing, began in May 1977 when news of the UCSF triumph in cloning rat insulin reached the East Coast. There had been warning signs ahead of time, the rumors that often precede the announcement of major work, but thinking the California group was way behind, Fuller hadn't put much stock in the news.

"I had heard rumors, but I didn't want to believe them, and I think that the first time I heard it was when it came out in the paper," Fuller says. "There were all these guys sitting out there in California to clone it, and I was pretty shattered. And Wally was, too. He said, 'Well, we have expression. We can still get expression first, maybe.' I was *so* caught up. Every day I was coming into the lab, and I was working ninety-hour weeks, doing the same damn experiment over and over and over, trying to make it work. And it was just a small technical glitch. . . ."

Argiris Efstratiadis tended to wear his emotions, including disappointment, on his sleeve, but at the same time he made light of the entire episode in an inspired parody that appeared in the *Midnight Hustler*. "California scientists Howard Goodman and William Rutter," it began, "announced today that they synthesized and cloned the soul gene. This extraordinary breakthrough will certainly revolutionize religion." The article, full of biological esoterica and wordplay, went on to explain how the scientists had isolated a soul tumor that was "overproducing soul mRNA" in a devoutly religious young man who had been forced to watch pornographic movies, drink vodka, and spend a night at a motel with two MIT coeds. To this achievement, wrote special *Hustler* correspondent Platerito di Alma, "Wally Gilbert said 'Pfui.' This last statement gave grounds to the *Harvard Crimson* to speculate that the Gilbert group was working on the same project and was scooped. However, Harvard sources wishing to remain unidentified said, 'No, no, we are trying to do just the opposite.' This obscure statement puzzled the Bio Labs population for a whole day, until it was disclosed that Forrest Fuller is dating Linda Blair [referring to Blair's role in *The Exorcist*]."

Wally Gilbert did not exactly say "Phooey," but in a sense he didn't need to: as a well-known scientist who had publicly identified himself with the project—indeed, who had testified at public hearings about his intention to clone insulin—getting "beaten out" by the West Coast group, as he later put it, clearly represented a setback.

The Harvard work continued to go badly. Arg Efstratiadis was going through large amounts of the Chick insulinoma. In the midst of one disappointing attempt at enrichment after another, the awful truth was becoming apparent: this vaunted source of rat insulin message was, in fact, not as terrific as presumed. As one researcher later put it, it "doesn't make beans worth of insulin." They knew the insulin message was there, however. They knew this because of tests run by Peter Lomedico, the insulin researcher from the University of Texas, who had settled into the Gilbert lab for his postdoctoral fellowship in the spring of 1977. Almost as soon as he joined the Gilbert group, Lomedico was performing these tests, known as cell-free translations, for Efstratiadis.

There was a problem, however. Lomedico was running these very same tests for Forrest Fuller, too. Fuller had by that point taken the not unprecedented step of quietly competing with his own collaborator. He wanted to make his own insulin cDNA, without Efstratiadis's knowledge. His main preoccupation remained the cloning problem, though, and all the obstacles he encountered, all the missteps he took, were like one archetypal nightmare that compressed all the bad experiences of all the pioneer cloners.

First, he still couldn't get "butt-end" ligations to work when he tried to get gene-sized pieces of DNA into plasmids. Week after week, he ran control experiments with globin cDNA, trying to prove that you could, in effect, cement these blunt-ended pieces of DNA together from two different species without having to use sticky ends. Transformation—smuggling the recombined DNA into the bacteria via the plasmids—proved to be a major headache for Fuller, too. Unlike the California group, the Harvard team was having fits with a new laboratory strain of *E. coli* known as Chi 1776. This strange, crippled little bacterium—denounced by some lab workers as a "wimp"— owed its creation to the recombinant DNA controversy.

Pressure had mounted to design an infirm lab strain of *E. coli*, a class of bacteria so genetically debilitated in so many functions that it would be unable to venture beyond the lip of a laboratory dish. A microbiologist at the University of Alabama named Roy Curtiss III set about to design this ill-starred creature after the Asilomar meeting of February 1975, and his bicentennial gift to the research community a year later was the enfeebled, gimpy bug he had dubbed Chi 1776. Gelded of essential enzymes, Curtiss's bacteria could not build their own cell walls or synthesize their own DNA; they were made especially sensitive to detergent, sunlight, bile, and chemical pollutants—the type of environmental insults a bacteria on the lam might encounter. Deprived of these enzymes, the bugs could live only at the pleasure of researchers, who supplied the crucial missing chemicals to the growth media.

The result was, indeed, a lame bug. So lame, in fact, that it reproduced only sluggishly and seemed especially ill-equipped to survive transformation experiments. These problems would have to be overcome if the Harvard team hoped to express rat insulin—to get the bacteria to actually make it. Only then, too, could they move on to Gilbert's long-range goal: making human insulin.

These were ambitious, highly visible goals for that period. In light of the slow progress, they began to seem too visible to Fuller. He was, in his own words, "one lowly little graduate student"; all the other researchers working on insulin were either senior scientists or postdocs, trying to establish or maintain reputations; egos were caught up in the bid for priority, the race to be first; all the techniques were so new that everyone was improvising technical solutions to unforeseen problems as they went along; and there was the sense that things were happening so fast that to hesitate, to fine-tune, to linger over details was an unaffordable luxury. That, at least, was how things looked to Fuller. He began to perceive it as a major point of conflict between himself and Wally Gilbert.

There is a self-explanatory phrase biologists use to make this kind of stylistic distinction. They speak of doing projects "fast and dirty." The philoso-phy is implicitly conveyed in the following rebuke, made by one of the molecular biologists working on insulin, about a colleague (and not about Fuller): "He had this tendency to get into these details, losing the real perspec-tive, the real goal of these activities. He certainly was pretty good, but he would go on and on and do all kinds of strange experiments without really—he didn't have the killer instinct, as we say. You know, he didn't go for it. And work toward the experiment that needs to be done." The other side of the coin, of course, is that the methodical approach often keeps "fast and dirty" work from becoming so dirty that it is meaningless. Both approaches are necessary. Science has its yin and yang, its imperatives of haste and patience, each holding the other—with a certain amount of tension—to the fecund middle ground.

In the view of other members of the lab, Gilbert more and more began to see Forrest Fuller as a scientist who didn't know how to work toward the crucial experiment; Fuller increasingly perceived Gilbert as a scientist who was interested only in results, leaving the details to others. In the pressure of the insulin race, according to members of the Gilbert group, both attitudes cal-cified into what became an unfortunate personality clash. Their working relationship, after all, was hardly that of peers; Fuller was still a novice. There was no more of the Sorcerer and his Apprentice, no more big blue Ws streaked on plates. If anything, this particular Sorcerer's Apprentice was learning that what had escaped from the laboratory was his control of the project. A sense of disillusionment set in. With it came a stubborn and in some ways fateful withdrawal.

In order to escape Gilbert's hectoring, Fuller took to working on his experiments at night, spending as little of the day as possible in the Bio Labs. He felt under enormous pressure to have his blunt-end cloning procedures in working order for the insulin project, for Gilbert had arranged to use the laboratory of Susumu Tonegawa in Basel, Switzerland, that summer to at-tempt the insulin cloning (Harvard still didn't have its P3 facility). Fuller remembers, "Wally keeps asking me, 'Are you ready yet?' And I keep saying, 'Another four weeks . . .' or 'Another two weeks. . . .' In truth, Fuller's experiments had not achieved any consistency. Sometimes the blunt-end cloning technique appeared to work, sometimes it didn't.

All of Fuller's bad luck boiled to a head in June 1977 as he made final preparations for the trip to Basel. Efstratiadis had slaved months to prepare one microgram of rat insulin cDNA—one *millionth* of a gram—and this he gave to Fuller for further workup prior to the trip. To this minuscule amount, Fuller planned to attach linkers. This would give the rat DNA a set of couplers on either end. It would be this rat insulin "gene"—a complementary DNA with linkers on either end—that he would plug into bacteria while he was in Basel.

At one stage in these preparations, the material had to be placed in a vacuum desiccator, an instrument that dries materials under a vacuum. The DNA rested in an Eppendorf tube—an inch-long vessel. Ordinarily, the tube would have a cap of Parafilm (a plastic similar to Saran Wrap) with holes poked in it, but Fuller admits he had neglected this step, and when another Harvard researcher went to retrieve material from the same drying chamber, both the vacuum and Fuller's hopes suffered calamitous rupture.

"He popped open the drying chamber too quickly," Fuller recalls, "and the air came in and threw out two-thirds of this cDNA into this alkali. It was on the bottom as a drying agent, okay? So this is how things go: here was this *gold* sitting there, and I hadn't covered it." Horrified, Fuller realized that more than half of the precious genetic payload for their experiment in Basel had vanished in a gust of man-made wind. "It was all I could do to keep myself calm," he recalls.

Questions raced through his mind. "What am I doing? What have I just done? It's my fault, you know. I can't tell Wally. Wally can't know this." Gilbert, Fuller admits now, never learned about the mishap. Fuller later regarded the incident as "probably the worst day of my life at that point, I think. Not losing out to these guys in California. It was losing that two-thirds of the material."

As it was, relations were tense. They were due to leave for Basel in a matter of days. Now some 0.6 micrograms of Efstratiadis's DNA was gone. Quietly, without telling anyone, Forrest Fuller put linkers on what remained of the DNA and then girded himself for what promised to be a difficult month in Switzerland.

The full extent of the difficulties between Gilbert and Fuller, or at least Fuller's anxieties, was symbolized by a scene at Logan Airport in Boston at the beginning of July when the two scientists checked in for their flight to Switzerland. When the Swissair clerk asked about seating assignments, Fuller recalls, Gilbert requested nonsmoking. Fuller almost reflexively asked for smoking. He was known to smoke a cigar now and then, but he makes it clear that wasn't why he asked for a seat in the smoking section. He just needed to escape the Sorcerer's shadow.

Fuller had about six weeks to produce results in Basel. He would rise at eleven in the morning to go to the lab and return at three the next morning. One time the hotel cook played country-and-western music while he ate breakfast; another time Gilbert came into the lab after dinner and sang opera arias while Fuller spliced genes. He must have made a strange and exotic sight as he walked back to the hotel alone in the wee hours of those summer nights: stocky, with a determined stride, shoulder-length hair, cowboy boots scraping the pavement. He was not the kind of person who showed outward signs of uncertainty or doubt.

But again, the bacteria wouldn't cooperate. Fuller tried cloning by blunt-end ligation, and that didn't work. He tried cloning by a procedure known as tailing, and that didn't work either. Fuller, by his own admission, "scientifically freaked out." "I tried it every way when I was there. I tried everything that they knew in Susumu's lab," he says. His laboratory notebooks from the period record a dreary litany of failure. "Negative. Negative. Negative . . ." they read, interrupted by unscientific but heartfelt annotations like "Aagh!" and "Shit!" By the second week in August, Fuller had resorted to one final attempt. This last round of experiments produced very few clones: one colony in one experiment, two in another. Four, two, zero, one—the numbers of colonies were distressingly low. After five weeks in Switzerland, Fuller was coming back nearly empty-handed. He had exactly ten candidates which might contain an insulin insert. Out of time in Susumu Tonegawa's laboratory, he planned to analyze the material back in Cambridge. "I held my breath," he recalls, "and went home."

Nothing came easy to Fuller during this trip, though, not even coming home. He had dutifully written "Bacteria" on his customs declaration form. "The guy asked me what they were and I went through this whole song and dance, scientifically explaining what it is, and I said, 'Do you want to see them?' And he looks at me and he says, 'Not particularly. But you wrote something down here, and I have to write something down now.' So he wrote something down and let me go."

That was the only light moment those bugs were to provide. Forrest Fuller spent the next three weeks in Cambridge analyzing the colonies. It appeared they all contained some sort of DNA insert. Every attempt to establish that it was insulin, though, proved negative. The Swiss gambit had failed.

That same summer, about the time Fuller was returning from Europe, the future of Genentech took a dramatic turn for the better, and it hardly stretches matters to assert that this reversal in fortunes was directly related to an obscure scientific fact: the amino acid called methionine happens, by chance, not to be present in either somatostatin or human insulin. That seemingly insignificant bit of biochemical trivia in large part allowed Genentech to burst out ahead of the biotechnology pack.

What had gone wrong in the June disaster? The City of Hope and Genentech team had no definitive explanation, but they had their suspicions. The fused precursor of somatostatin, in tiny amounts, was probably being made in the bacterial cells, but the bacteria almost immediately recognized the small molecule as foreign and chewed it up. "It had two strikes against it," Riggs says. "It was small *and* it was foreign."

After discussion, the group reverted to what Riggs called "Plan B"—one of the backups scientists usually have in mind when Plan A fails. If somatostatin was being destroyed because it was a foreign protein, as feared, perhaps they could camouflage the hormone's foreignness by patching Itakura's gene into a spot farther along the bacterial gene they were using. In other words, they would bury the somatostatin gene at the end of the larger beta-gal gene, "The obvious and only convenient thing we could do was put it back at the end of beta-galactosidase," Riggs recalls with a nervous laugh. "You might say we did it in desperation." An enthusiastic Boyer conveyed the strategy to Herb Heyneker. "Let's *bury* it," he said. "Bury it. Bury it behind beta-gal."

Heyneker received some help on this concept from colleagues at UCSF, a collegial exchange of data that later led to a messy dispute. Beta-galactosidase is a battleship of a gene, more than three thousand base-pairs in length; somatostatin, at only forty-two bases, was by comparison a piece of molecular fluff. By chance, there happened to be a splicing site (recognized by *Eco* RI) near the very end of the beta-gal gene, and researchers in the UCSF laboratory of the late Gordon Tomkins were able to tell Heyneker that a gene grafted on at this site would be "in frame"—that is, read properly, with no displaced letters. Was this crucial information? One researcher from that lab, who asked not to be named, says it was; Heyneker says it wasn't. "They felt they were giving us invaluable information," says Heyneker. Riggs says he and Boyer knew the location of the site by May 1977, *before* the initial failure, and Heyneker argues that with synthetic DNA, the Genentech group could easily remedy any out-of-frame problems. "I was grateful for the information, but I did not consider it invaluable. I did consider it helpful."

It was then a fairly straightforward sticky-ended exercise to attach Itakura's somatostatin gene onto the back end of the beta-gal gene. Enzymes scanning this strange hybrid gene would recognize the familiar protein beta-galactosidase, and would simply transcribe the entire text. The end result was, in Heyneker's words, "a garbage protein"—a large lactose-digesting protein with a tail of human hormone wagging at the end.

So they had figured out a way to get this hybrid gene inserted into a plasmid, and they had also figured out a way to get somatostatin unfused and out again. Sticking in that methionine was like putting a ready-to-tear perforation between the disposable beta-gal and the desired somatostatin; the fused proteins could be separated not by ripping them apart by hand, of course, but by the addition of a very discriminating chemical called cyanogen bromide. It would find every link of methionine in the amino acid chains and cut right through it. It was the test-tube equivalent of threshing grain—the small somatostatin molecule could be cut away from the beta-gal like kernels of wheat from their enveloping chaff. To be sure, Genentech's scientists—and investors, for that matter—led rather charmed lives. Somatostatin does not contain the amino acid methionine. If it did, somatostatin itself would get chopped up in the reaction.

It took about two months for Herb Heyneker and Francisco Bolivar, working in Boyer's lab at UCSF, to reconstruct these fused genes and insert them correctly into their customized plasmids. They performed an elegant bit of cutting and pasting—perhaps the most elegant and involved that had been done to date. Every day, Bolivar recalls, Bob Swanson stopped by to check up on the progress. By August, they finally had created eleven colonies—a cloner's dozen. Now the task was to test each one to see if there were any somatostatin molecules riding piggyback on beta-gal.

The transformed bacterial colonies were induced to churn out this white elephant of a molecule, neither enzymatic fish nor hormonal fowl. The hybrid protein, biologically inactive, just built up inside the cells. It remained for Riggs down at the City of Hope to rupture the cells, collect the debris, and separate out the material he wanted.

By August 15, 1977, they were ready for another run through the gamma counter. It was late in the afternoon. No expectant entrepreneurs or onlookers gathered this time around the machine. Just Riggs and Itakura. The first numbers rolled out. "This one might have something," Riggs thought. Four out of the eleven looked very promising indeed. By the time the assay was over, they had a pretty good idea that they were the first scientists in the world to express a human protein in bacteria. More remarkable still, that "human" gene had been built from scratch from common over-the-counter chemicals.

How do scientists react to signal achievements like this? Cautiously. "In the real world," says Riggs, "you don't *really* believe these things until they're repeated. So we smiled and shook hands and said, 'Looks like we got it.' And then talked just briefly about what we could do to confirm it."

Just like that, they packed it in. Another day at the office. Riggs had promised to take his son to see the Los Angeles Dodgers play that evening, so after the historic data were logged in his notebook, without fanfare or celebration, having duped bacteria into producing a hormone normally made in the human brain, he picked up his son and went out to the ballgame.

In the latter days of August, there were people to be called and celebrations to be held and papers to write, and, yes, a patent to prepare. Boyer was delighted, of course, by the news. Between the setbacks on somatostatin, the controversy about Genentech at UCSF, and the negative press, this, finally, was *good* news. Yet Paco Bolivar recalls that Boyer, unlike Swanson and the others, was notably calm, almost diffident about this spectacular development, and Bolivar believes this was because Boyer was the most confident of—and therefore the least surprised by—success. "I was feeling that maybe it was not going to work and I was scared about that," Bolivar admits now. "Heyneker was more confident. And I think Boyer was the most confident of all." Boyer's phlegmatic reaction may just have been another way of saying, "of course." Swanson arranged a dinner at an Italian restaurant in Los Angeles for all the scientists. "Immediately after that celebration," recalls Roberto Crea, a young Italian who joined Itakura's group toward the end of the somatostatin project, "I remember we had another dinner with Boyer, Swanson, Itakura, myself, and Adam Kraszewski [another DNA chemist in Itakura's lab]." Insulin came up. "I remember," says Crea, "somebody at the table said, 'You're crazy. You're crazy because it's going to take us two years to do that.' "

"You really think you can make it?" Swanson wanted to know.

Crea replied, "We can make it in six months. With this library of codons we are building, it's going to be much faster."

"If you can make it in six months," replied Swanson, "you go ahead."

Human insulin, of course, had always been Swanson's intended encore to somatostatin. Perhaps Crea, who had joined the team only at the beginning of May, did not realize that insulin had been the next target all along. Nonetheless, Robert Swanson surely did not mind hearing so optimistic an assessment of its feasibility. Somatostatin in a sense was scientific show-and-tell; human insulin was a money-maker. Doing human insulin, and doing it first, might very well establish Genentech's preeminence in the emerging field of biotechnology.

T w e l v e

"Recognition
Is Our Bread"

Under the best of circumstances, scientists view journalists with wariness. When Nicholas Wade, a reporter for *Science* magazine, began to place a series of disquieting phone calls to high officials at the NIH as well as to members of the Goodman and Rutter laboratories at UC–San Francisco around the beginning of September, that wariness veered toward paranoia. Wade had been tipped off about the pBR322 episode; now he was seeking clarification. There were obvious questions to ask. Was it true an uncertified vector had been used? Why hadn't the NIH been informed? Why hadn't the researchers said anything when the news was announced?

Wade seemed to have been privy to the most intimate conversations between people in the UCSF labs. He knew, for example, about the conversation in which John Chirgwin told William Rutter he wanted his name removed from the paper if there was any improper handling of the problematic clones. "He must have had a ten-page dossier on all the gossip in the department," recalls one UCSF scientist, "and he knew who wasn't on good terms with whom at the postdoc level between labs. He knew more about it than anybody *there* did. . . . It was quite remarkable."

As the phone calls continued, suspicions began to grow in San Francisco that there was a "Deep Throat" in the UCSF biochemistry department, a disgruntled source feeding information to *Science.* Some people even felt that the source, whoever it might be, was particularly intent on embarrassing Howard Goodman. Rutter began to telephone *Science*—not Wade, but editor Philip Abelson—in an attempt to determine what the article would say, what the extent of the damage might be.

All the paranoia of the summer seemed thoroughly justified when the *Science* article came out. Under the title "Recombinant DNA: NIH Rules Broken in Insulin Gene Project," Nicholas Wade's article appeared in the September 30 issue of the journal. It painted a less than flattering portrait of the UCSF collaboration. "The secrecy and suspicion surrounding the insulin gene experiment, together with perhaps a touch of resentment, helped to fan rumors within the department alleging that the NIH rules had been broken and even that the experiment published in *Science* might not have been performed as described," Wade wrote. The article went on to mention the ill feelings generated by the press coverage and conflicts created by potential commercialization of the research. UCSF officials, Wade reported, blamed most of the problems on poor communications by the NIH.

The article obliquely revived one very touchy issue in the biochemistry department. Wade had alluded to the notion that the research report in *Science* might not in fact have reflected the way the experiment had been conducted. Ever since the rat insulin gene was cloned back in April and May, there had been suspicions—no evidence, just suspicions—that the researchers had not destroyed the original renegade batch of clones as described. Perhaps a few of the clones were saved, or perhaps the insulin-containing inserts had been cut out and saved prior to destruction. Were that the case, the insulin-cloning team would knowingly have introduced incorrect information into the scientific literature, at least in part for competitive reasons. "Many people around the department didn't believe they were destroyed," recalls Brian McCarthy, a member of the faculty at the time, "for a very simple reason. If they'd gone through all that work and created something which, albeit against the rules, was what they wanted, and there was no danger, why make work for yourself? Why destroy it and then go back and do it over again? People, especially scientists, don't do crazy things like that." McCarthy's point, he wishes to emphasize, is based on common sense, not on any evidence or knowledge that the clones weren't destroyed. Axel Ullrich, in rejecting this version of events, suggests that it reflected the heightened emotions and factionalism running through the department at the time.

With publication of the Wade article, brush fires sprang up around the insulin breakthrough, and once again the fires had to be put out. The University of California at San Francisco issued a statement through William O. Reinhardt, acting dean of the school of medicine; while conceding the infraction, Reinhardt stressed that the premature use of pBR322 was "entirely safe." The response emphasized that there had been no danger to public safety. What is interesting, however, is that at no point did anyone in an official capacity, either a university official or one of the scientists involved, address the issue of hushing up the original incident when recombinant DNA, for better or worse, was a matter of great public interest. Only safety, not ethics or scientific credibility, was at stake here.

The *Science* article, and the subsequent flurry of press inquiries it inspired, did not produce the kind of publicity universities seek to cultivate. Yet the unauthorized use of pBR322 had been rumored in molecular biological circles that summer, usually provoking no stronger reaction than amusement; there was little sentiment among scientific peers that the group had done anything questionable in terms of safety or even competitive zeal. Indeed, if the sin had been considered truly egregious, it is likely that the fuss would have been louder and occurred sooner. Several members of the insulin team, including Rutter and Shine, scoff at the suggestion that the work was conducted in secrecy. Axel Ullrich, however, makes the following point.

"You know, secrecy you find in science," he says. "Everywhere. Wherever you work on a competitive project. That's not only secrecy caused by some illegal actions, but it's mainly because you're competing with other groups and there is a lot of recognition, and recognition is our bread. That's what increases our market value. It gives us jobs and so on. So that's nothing new. It's always been like that. Just read *The Double Helix*. Same thing. And," he adds matter-of-factly, "it always will be."

Even an innocent mistake could be shaped by political opponents into something more dangerous to the entire recombinant DNA field. Senator Adlai Stevenson III of Illinois, one of the NIH's harshest and most persistent critics, wasted no time in turning it into a political football. The Science, Technology, and Space Subcommittee, which he chaired, had scheduled hearings on recombinant DNA research; Stevenson, it was no secret, wanted federal legislation. The *Science* article provided ammunition to the critics, who maintained that the NIH was incapable of enforcing its own rules. Not long after, invitations from the committee went out to William Rutter and Herb Boyer.

It is rare enough that biologists are called to testify before Congress; it is rarer still to hear them describe and defend, against considerable political sarcasm, a single experiment. Not only the insulin experiment, but in a sense the mores of contemporary biology came under blistering scrutiny on the morning of November 8, when William Rutter and Herb Boyer seated themselves at a table in front of Stevenson's subcommittee. All of science drew a deep breath as they began to testify.

In his opening statement, Rutter recounted the sequence of events of the pBR322 episode, acknowledged its inappropriate use, and insisted that "there was no question of a coverup." His comments steadfastly reflected the university's position, which in effect blamed the National Institutes of Health (indeed, both Stevenson and NIH Director Donald Fredrickson conceded that the guidelines were "very confusing" on the issue of certification). Communications with the NIH were informal to the point of confusion, Rutter suggested, and the NIH decision-making process was excessively bureaucratic. In terms of safety, he added, the uncertified pBR322 was "ten thousand to a million-fold better" than other vectors which had already received the NIH's blessing. The implication was clear: despite proof of pBR's superior safety as a plasmid, six months elapsed between pBR's approval at the committee level and its certification by the director. That begged an underlying question— which served the public interest better, use of an uncertified but safer material or, as Rutter would later put it, NIH "boondoggling"? In making these points, Rutter implicitly betrayed a scientific notion of time quite alien to his interlocutors. Six months was a modest interval in a massive government agency, almost an eternity to scientists in a race.

Why hadn't the NIH been formally notified? Rutter raised the issue himself, but at no point in the public testimony did he mention his contacts with NIH official DeWitt Stetten; only when subsequent documentation of the incident was assembled did it become clear that someone at the NIH knew of the incident as early as March. Rutter conceded that "the argument for not informing NIH was . . . related to the inflamed social and political climate that

169
· · · · · ·

existed with respect to recombinant DNA technology at that time." And he used a curiously revealing word in a scientific context—"supremacy"—in analyzing that decision. "The benefits of the experiments were obvious," he told the subcommittee. "Reporting work would have even indirectly secured the supremacy of our cloning experiments with insulin. In retrospect, this might have been the best course of action."

Two strong messages emerged from Rutter's comments. First, feelings of persecution ran rampant among the scientists (indeed, Rutter would later refer to the Stevenson hearings as "inquisitional" and a "witch hunt"). Second, the public simply could not be trusted with discussion of this sensitive, emotional issue. Those may seem like extreme positions, but Rutter merely enunciated in public what the vast majority of scientists felt privately: the guidelines, created in an atmosphere of public hysteria, were ridiculous, unwarranted, meddlesome. Nonetheless, one of the last things he said, apropos of the lessons implicit in the pBR322 flap, had a ring of truth to it. "Since time and respect are among our most highly valued possessions," he told the subcommittee, "the combined loss of these is a powerful sanction indeed."

Unfortunately for Rutter and UCSF, more than just respect and honor were at stake. Pending an NIH investigation, federal grants—according to correspondence between UCSF and the NIH—were subject to "conditions and restrictions" if UCSF officials failed to explain the incident in adequate detail. In such cases, according to William Gartland of the NIH's RAC committee, usually individual researchers, and not the institutions themselves, faced funding sanctions.

Next came Herb Boyer. Since he was sitting next to Rutter before the Senate panel, the senators naturally presumed that he was involved in the pBR322 episode—the plasmid, after all, had come from his laboratory. Thus Boyer spent a fair amount of time attempting to explain that he had nothing at all to do with the Rutter-Goodman rat insulin experiment.

Whatever goodwill Rutter and Boyer earned with their statements to the committee, it began to erode as they tried to explain the "gory details" (Boyer's terminology) of the pBR322 fiasco. Stevenson, every bit the prosecuting attorney, launched into a sharp and unremitting cross-examination. The two UCSF scientists got off to a bad start when Boyer tried to reconstruct how "approval" and "certification" got mixed up. Elizabeth Kutter, a member of the NIH's Recombinant DNA Advisory Committee, had informally told him, he explained, of the decision of the NIH group in Miami to grant tentative approval to the plasmid.

"The exact wording," Boyer recalled, "was that the plasmids were approved, were given tentative approval pending receipt of [additional] information from Dr. Falkow."

"Dr. Rutter, is that what you heard?" asked Stevenson.

"That nuance had not reached me," Rutter replied. "As far as I know from Dr. Kutter, she did not mention that aspect."

"Are you suggesting that that is a nuance?" Stevenson shot back.

"That was an unfortunate term to use," Rutter agreed.

"Tentatively approved subject to the receipt of additional data. Is that a nuance?" Stevenson continued. "That is accurate. That is what in fact had happened."

Stevenson probed further; the biologists searched their foggy memories for unrecollectable details. At one point, a frustrated Boyer said, "You must remember that this is—trying to recall all of this information in retrospect. I didn't write it down exactly."

"Isn't that exactly the point?" replied an exasperated Stevenson. "Would reasonable men, let alone scientists, have proceeded without some confirmation, without something in writing, without something that wouldn't put you in this preposterous position today of trying to reconstruct all of this? You say you don't want legislation. If there is legislation, you gentlemen would be the authors of it. I cannot, for the life of me, understand how reasonable men could have relied on rumors."

And so it went. All the social opprobrium heaped upon scientists, upon all the supposedly arrogant molecular alchemists of DNA, seemed to come crashing down on Rutter and Boyer. It is difficult to read through the Senate testimony without concluding that the two UCSF researchers waged their best arguments in scientific, not public, forums. Both came unprepared to discuss the pBR322 episode in detail, however, creating what Boyer describes as a feeling of being "set up." Still, several aspects of their account of the pBR322 episode were so painfully clumsy, and their utter distrust of the public so transparently evident, that Adlai Stevenson and Senator Harrison Schmitt had almost no choice but to come down hard on the two biologists. Schmitt, more like a lawyer than a former astronaut, zeroed in on one of the key discrepancies: Axel Ullrich and John Shine had attended the February 4 lab meeting in which Herb Boyer reported that pBR322 was not certified and could not be used. How then, Schmitt wanted to know, could an entire month pass before news of the lack of certification reached Rutter or Goodman? It was a good question. No one had a good answer. "If you deserve any criticism, and I think you or your lab certainly deserve some criticism, it is for not carrying on more responsible scientific research," Schmitt said. Rutter was forced to agree.

Similarly, Stevenson's sharp questioning about the operation of UCSF's special P3 laboratory revealed an embarrassing absence of control and an even more damaging administrative attempt to cover up the damage. Log-books had been backdated ("retroactive entry" was the term used), details were sketchy and incomplete. In fairness to the UCSF researchers, many other university laboratories involved in recombinant DNA work during that period, perhaps the vast majority, would have been hard-pressed to provide any better documentation or explanation to as implacable an inquisitor as Stevenson. It was a period when scientists had no previous experience with this type of administrative control, bridled against outside intervention, and in any event tended not—even in the best of times—to observe such rudi-mentary clerical work as signing and dating their own lab notebooks, to say nothing of logbooks.

Yet the infractions, if not serious in and of themselves, suggested to the public that scientists had arrogated to themselves the responsibility for de-ciding which federal standards deserved to be observed, which regulations could be quietly flaunted. It was on this symbolic level that the UCSF viola-tions, however innocent in origin, seemed so grave. The retroactive dating of the logbook, the inexplicable gaps in the pBR322 episode, the deliberate decision to withhold information from the NIH and the public—all came under heavy fire from the Stevenson committee. In explaining why he was "very disturbed" by the incident, Harrison Schmitt chided Rutter and Boyer by saying, "I think the rationale that some knowledge in the hand[s] of the public would be dangerous or counterproductive is not really an acceptable rationale, even though I will admit that sometimes a little information can cause great misunderstandings. We find that out in front of this committee quite frequently. It also is generally inconsistent," he added, "with those people who argue against the idea that there is some knowledge science should not seek."

The Senate performance earned mixed reviews back in San Francisco. A number of people on the ninth floor credited Bill Rutter with going to Washington and taking responsibility for an infraction that had, after all, taken place in someone else's laboratory. Howard Goodman did not fare so well. If Goodman had flown in from out of town to share credit for the rat insulin experiment at the press conference, people wondered, why hadn't he appeared in Washington to share responsibility for the infraction that had occurred in his own laboratory? A rather less generous interpretation of events came from some of the postdocs. According to a researcher in Rutter's group, "In the lab, the feeling was that Bill didn't want to acknowledge that he wasn't in charge of the experiment because then he couldn't claim credit."

Some researchers took time out to read an account of the testimony, and those familiar with the history of the insulin race as well as with Rutter's reputation were amused by one brief exchange. Harrison Schmitt asked if the UCSF lab was under any kind of competitive pressure.

"We knew about the competition," Rutter replied, "but I think it is fair to say that it did not dominate our demeanor."

The case of Herb Boyer is a little more complicated. Members of his lab felt he'd been dragged into something about which he had little knowledge and virtually no responsibility. What undoubtedly made it a particularly unpleasant experience was that since the inception of Genentech and the tensions it had stirred in the biochemistry department, Boyer's decision to go commercial had come in for criticism by faculty members, including Rutter and Goodman. Now, scientifically estranged from the two and their celebrated insulin success while his own group struggled with the synthetic DNA approach to insulin, Boyer somehow had been associated with the controversial aspect of their work without, of course, enjoying any of the credit associated with its achievement. Having been a lightning rod for all the criticism surrounding the commercialization of biology, Boyer suffered from this last indignity, according to others. "He caught a lot of flack, and he was hurt by that," recalls one former lab member. "It caused him to withdraw, and that hurt him scientifically."

Boyer's withdrawal, as it turns out, coincided with Genentech's emergence as an entrepreneurial outfit that could deliver on its promises. On December 1, 1977, in the Cordovan Room at the Biltmore Hotel in Los

Angeles, there occurred the curious phenomenon of reporters and science writers showing up en masse for what in fact was month-old news. The Genentech-sponsored researchers in Boyer's lab at UCSF and Art Riggs's lab at the City of Hope jointly revealed for the first time—officially, at least—that their bacteria had been programmed to make the human hormone somatostatin. More even than the rat insulin work, this achievement brought the promise of recombinant DNA work closer to the threshold of therapeutic application.

But no development in the field could be without its incumbent controversy, of course. Somatostatin was no exception.

Almost from the moment that Art Riggs recorded his positive results on August 15, there was pressure from three different fronts to publicize the work—not only from Bob Swanson, who understandably saw a splashy press conference as a means of pushing Genentech's name into the field of investor vision, but also from UCSF and the City of Hope. "Both institutes need support both from NIH and from private donations," Riggs explains, "and publicity is a value to them also. So there was pressure to publicize, really, from three different sources, from three different directions. Our position was 'Wait till it's accepted.' " Riggs was old-school when it came to this point, probably more than any of the other scientists involved on the project. Disclosure, he argued, should not precede acceptance of the work by a refereed journal (a journal, that is, where scientific peers scrutinize papers prior to publication to ensure novelty and importance). "In effect, I drew the line," he recalls. "I said, 'Nothing can be done until the paper's been accepted. And then whatever our public relations people want to do with it after that, I'll go along with it.' " Rigg's view prevailed.

As a precautionary step, the researchers contacted the editor of *Science*, Philip Abelson, to sound out his interest in the work. "He was interested," Riggs recalls, "and said that, in effect, if reviewers liked it, it would be published quickly." With this comforting sign of expeditious treatment, the scientists then submitted the manuscript to *Science*. It, too, became a political football. This time, however, it was the scientific community that called the plays.

On November 2, 1977, *Science* received the landmark paper on the somatostatin experiment. That same day, the results leaked out (if testimony before a heavily attended Senate subcommittee hearing can be said to have been "leaked"). Appearing before the Stevenson subcommittee on November 2, Stanford University molecular biologist Paul Berg and National Academy of Sciences president Philip Handler both disclosed in public what Art Riggs and Keiichi Itakura had privately celebrated in August: human somatostatin had been cloned and expressed. Berg called the experiment "ingenious and elegant"; Handler referred to it as "a scientific triumph of the first order." David Perlman, the science writer of the *San Francisco Chronicle*, expressed dismay. He explained why in a letter to *Science*.

"A double standard of scientific announcement seems to be operating here," he wrote. "The 'orderly processes' of refereeing and publication remain in force for journalists and the public. But when the political process is operating in Congress—in this case, apparently, the specter of political regulation for a new field of science—then the rules of science go by the board, and the public learns of a new scientific triumph via a congressional hearing rather than through the pages of *Science* or the annual meetings of the American Society of Biological Chemists." Perlman argued that political debates on legislation were inappropriate forums for scientific announcements, and concluded his letter, "The propriety of Dr. Handler's testimony, however politically useful, should, I believe, be widely discussed." It wasn't.

A month later, on December 1, Genentech and the two research institutes reiterated the news at a well-attended press conference. Genetic engineering generated tremendous publicity; much of it—despite the critics—was favorable. Stories heralding the work appeared in newspapers from coast to coast. News reports about the experiment suggested that recombinant DNA could deliver on its promises, and that in turn had a desirable political impact.

Although City of Hope and UCSF ran the show, the press conference was as commercially useful to Genentech as the Senate testimony had been politically useful to the field of recombinant DNA. Even though Art Riggs conceded to a newspaper reporter that it was unlikely somatostatin would ever be produced commercially, a few astute observers immediately sensed the import of Genentech's involvement in the project. Stories soon appeared in *Business Week* and on the financial pages of the *New York Times* about the "tiny San Francisco company" that had scored a "biomedical research coup." This was the kind of publicity money couldn't buy—here was an industrial company achieving scientific milestones. The company was already capitalized to the tune of about $1 million, and Bob Swanson was sure to find investors more willing to finance the next project. Which was, as all the papers reported, human insulin.

For the biologists involved, however, somatostatin was a watershed *scientific*, not entrepreneurial, event. "We did not take the natural gene for somatostatin, so we did not use a natural gene," Riggs points out. "We didn't copy nature. We sat down and we designed our own gene—from beginning to end, you might say. We designed it first on paper and then we made it. So in my mind, when it really worked, we had the final decisive proof that Watson and Crick were right, that the genetic code was right, that all the people who worked on the protein synthesis machinery were right, and that, my God, this chemistry that we've been using . . . well, let's say all these clever and esoteric chemical techniques that are being used to figure out what's going on inside the cell—you sit back and say, 'My God, that was all right! Science really works!' "

But for Forrest Fuller, science did *not* always work. By the fall of 1977, the morale of the Harvard insulin team had drooped considerably. Fuller tended to withdraw even further from the collaboration. He decided to go all out to make his own insulin gene. Frozen rat tumors had piled up, big as grapes, in the lab freezer. There was plenty of raw material. Making rat cDNA, he knew, directly competed with Arg Efstratiadis. Mindful of the possible consequences, he plunged ahead.

Fuller also devoted a good portion of October and November to more control experiments. He felt he was making progress on getting the cloning system to work, but the beginning of the end came when he gave a group seminar one afternoon in the third-floor Tea Room. This large room, overlooking the Bio Labs courtyard, was so named for the daily four-o'clock gathering for tea and cookies; student presentations of research in progress, however, took place in a considerably less genteel atmosphere. On that particular day, Fuller spoke only about his globin experiments. He made no reference at all to insulin, and that omission, he feels, was his fatal mistake.

"Wally called me into his office and said, 'What have you been doing on insulin?' I said, 'Essentially nothing. I've been working on a system to do globin.' " In retrospect, Fuller regards this as the final straw to Gilbert.

"It wasn't a blowup, actually. He explained to me how I was a lousy scientist and I should probably drop out of science and probably do something else. I think his intention was to make me mad and say, 'No, I'm going to go back in there and do that insulin thing and make it work!' But by this point, I had really had it. After I got back from Switzerland, our relationship was really bad. All through 1977 our relationship was bad, but now it had fully deteriorated. So I just sort of said, 'Well, I can understand that you feel as though I'm slowing down the insulin project. And if you feel that way, I don't want to be in the way any longer, and I will be happy to remove myself from the project.' I think he just got livid. I said I'd like to work on the globin project. He wasn't going to let me. I had to basically plead with him a little bit to give me two months to work on globin construction."

Before going on vacation in December, Forrest Fuller made a last attempt to get globin expressed in his beloved "pop" plasmids. He created a number of clones and diligently searched through them for signs that globin was being made. Again, no signal. It was the last blow. "So here was the final—really the final—drop on my whole plan of what I had wanted to do when I came to Harvard, or to Wally's lab, which was to get expression of a eukaryotic gene . . ." Fuller says.

The disintegrating rapport between Gilbert and his self-styled apprentice was of course closely monitored by the lab group. The process was painful to watch, for Fuller was well liked; it was as if an animal were dying a slow death in the midst of a close-knit herd. "He ended up being on a project that was very competitive, and he was still learning the trade," says one lab member, "so it was just an unfortunate situation." Adds another: "We were aware of what was going on to some degree. And it *was* painful. It was painful because we all could empathize directly with Forrest. We were all in the same general situation, being graduate students. And then once he was asked to leave the lab, it was painful at a different level because it brought up a cloud of doubt. You know, can this happen to us, too?"

The stubborn part of Fuller maintained its bristle to the bitter end. "My biggest problem with it," Fuller would say later, "was I had sort of a laid-back attitude, having come from the West. And of course Easterners do not like laid-back attitudes. It wasn't that I wasn't intense about my work, or that I wasn't working hard. It's just that I absolutely refused to run around like a crazed rat and give the appearance of being on the edge of a nervous breakdown, which is what was desired. Other than that, it didn't matter what you did. It just only mattered that you looked like you weren't relaxing. And that you accomplished something. If you accomplished something that was visible, tangible, it didn't matter how hard you had to work to do that. That wasn't important."

The bottom-line problem for Fuller, unfortunately, was that after four years as a graduate student in the lab, he was still working on the controls and had yet to produce those tangible results. On a less visible or less competitive project, such failures might have served as valuable learning experiences, the usual dreary prelude to more successful work. But insulin was a high-stakes venture. Gilbert had clearly reached the limit of his patience, the reservoir of which was infamously finite. Fuller knew his days in the Gilbert lab were numbered.

When Wally Gilbert mentions how he "had to kick him out of the laboratory," there is not the slightest tremor of remorse in his voice. The verdict was delivered after Fuller returned to Cambridge from his Christmas break, in January 1978. Fuller had a month or so to find another Harvard lab (he moved to the lab of Helga Doty and earned his Ph.D. without further trauma). In its unsentimental and democratic way, the *Midnight Hustler* acknowledged the incident. It bid fond adieu to Fuller with a simple parting shot. The newspaper reproduced one of the Bio Labs requisition forms, ostensibly filled in by Fuller and addressed to "Genentech" in "Frisco, CA." One "insulin cDNA clone" read the order. The price: $100,000. At the bottom, they used the rubber stamp of Walter Gilbert's signature to approve the purchase.

Gilbert, of course, was not immune to the treatment. Fuller's banishment from the lab was an alarming development for many of the graduate students, and there had been a spontaneous, "highly charged" meeting in the Tea Room when the lab group questioned Gilbert on his handling of the affair. "The tensions were there," recalls one graduate student, but Gilbert's explanations seemed to allay certain fears and, "By the end of the meeting, things had calmed down completely and Wally was in control, and it was as though *he* had called the meeting, which is not originally how it happened." Efstratiadis thought Fuller had been kicked out "for no significant reason," and claims to have told Gilbert so. In an article wryly titled "Gilbert Gives Gabfest," the *Midnight Hustler*'s anonymous correspondent began an account of the lab leader's self-defense thusly: "Guru Walter Gilbert last month revealed his philosophy on the leadership of a research group at the frontiers of science. Gathered around, eyes wide, ears open, his students and colleagues listened attentively as he wove a tapestry of individual guidance on a background of desperate world medical need. . . ." By now, some members of the lab viewed the insulin project rather sardonically. For the cover of that particular issue of the *Midnight Hustler,* someone dug up a photograph of a suitably bedraggled rock group named Klone; superimposed on two of the musicians were the faces of Gilbert and Efstratiadis. "The Bio Labs Midnight Hustler," read the headline, "in conjunction with N. Sulin Bigbucks presents: Klone." The Gilbert lab, clearly, was no less susceptible to frictive emotions and conflicts than any other lab. The *Midnight Hustler,* however, provided students and postdocs with a device to "talk back" to the lab chief, and that proved to be a remarkably effective way to vent steam.

By the fall of 1977, Arg Efstratiadis had reservations about pursuing the insulin work any further. He was infamous for juggling a dozen different projects at once—"all of them," as he liked to point out, "brilliant." But Wally Gilbert wanted to push ahead on insulin. Efstratiadis agreed. With Fuller's departure, the experimental plan changed, too—blunt-end ligation no longer seemed the way to go.

They needed a new cloner, however, and it was Efstratiadis's idea to recruit his friend and colleague Lydia Villa-Komaroff. She sat back to back with him in the small wooden carrels of Fotis Kafatos's first-floor laboratory; they had worked, literally and figuratively, within shouting distance of each other for several years. When Efstratiadis first broached the subject, Villa-Komaroff had good reason to listen.

She had been working with Efstratiadis on one of his less successful ventures, a series of experiments investigating the formation of the eggshell of silkworms. The fact that the work hadn't gone well, during a year at Cold Spring Harbor and another six months or so back at Harvard, influenced Villa-Komaroff's response. As she recalls, Efstratiadis made the insulin work sound like a brief project. "Look," he told her, "we gotta have somebody do this cloning. It's a *real short job.* All you have to do is make a clone for us. It won't take any time at all."

"Okay," replied Villa-Komaroff. But her response was conditional. "I'm going to do all the controls *my* way. If I'm going to do this, we're not going to do it quick and dirty. We'll do all the controls."

Eftratiadis said, "Sure."

When Efstratiadis said "real short," Villa-Komaroff figured he meant perhaps two months. Little did she realize that it would ultimately involve nearly a year of intense labor, a trip to England, gas masks and rubber raincoats, the "dance of collaboration," patent attorneys, and, in the end, one of the classic papers of the early cloning days, leading to a frenzied final-stretch run for human insulin.

Thirteen

"Unsnowbound at Last"

. . . .

It was during the Mexican Revolution, or so goes the family legend, that a young man named Encarnación Villa was trying to flee the country when his northbound train was overtaken by revolutionary troops. The soldiers ordered everyone off the train and offered them a choice: they could donate something valuable to the revolutionary cause, fight with the rebels, or be shot. Encarnación Villa told the rebels he had nothing of value to give and refused to abandon his family. He was about to entertain the third option when none other than the revolutionary leader himself, Pancho Villa, appeared on the scene. Upon learning Encarnación's last name, the rebel general ordered his release with the exhortation "Have many sons with that name!" Encarnación Villa obliged by having a son named John, who in turn had a daughter named Lydia, who more than half a century later would find herself in the midst of a different, more invisible kind of revolution.

The development of recombinant DNA, like the first golden age of molecular biology, was largely a male-driven revolution. Female scientists were rare enough on the cutting edge to qualify as curiosities; among the twenty-six speakers listed on the program of Lilly's insulin symposium in 1976, for example, only one was a woman. This says more about the old-boy network of molecular biology than about the intellectual aptitude of women for the work. A few, nonetheless, managed to muscle their names onto important ground-breaking papers. The granddaughter of Encarnación Villa was one of them.

Lydia Villa-Komaroff was a logical choice for the insulin job, for she had been among the first molecular biologists to attempt cloning; like the other early pilgrims of both genders, she had learned by trial and error, mostly error, what worked and what didn't.

If it was generally unusual for a woman to play a key role in such a competitive project, it was not unusual at the Gilbert lab. Gilbert enjoyed a particularly favorable reputation among his female graduate students. "He has a sense of fair play as far as your sex is concerned," says Debra Peattie. "Another leader of a group may not have selected Lydia, even though she had the expertise, because she was female. As far as I'm aware, that sort of way of thinking just isn't part of Wally's grain." "I guess his expectations weren't any different for the men and the women. He just sort of assumed," Villa-Komaroff believes, "that if you were serious about science, you'd go about it in a certain way. And that was not even a question in his mind, which is very very rare."

Villa-Komaroff was a good choice, too, for reasons that—technically, at least—had nothing to do with her scientific acumen. She got along with fellow workers very well. Amid all the prickly temperaments, she managed to keep everyone talking—to her, if not to each other; and although that is a talent never acknowledged in the scientific literature, it reflects the kind of intangible social lubrication that can prevent a high-powered scientific endeavor from overheating and blowing up. Small, athletic, with the straight dark hair and deep brown eyes of her Mexican-Indian heritage, Villa-Komaroff was extremely personable and self-aware, almost bubbly with enthusiasm. That she was not easily cowed by reputation was clear to a large crowd of molecular biologists one summer when, during one of the "traditional" food fights at Cold Spring Harbor, she fearlessly emptied a bottle of white wine onto James Watson's head. "*Half* a bottle," she says, insisting on scientific accuracy.

Her working relationship with Efstratiadis typified her collaborative skills. "Some people found him impossible to work with; I enjoyed it," she recalls.

"There was a time, just before I went to Cold Spring Harbor, when we were doing more yelling than getting anything done. It got to the point where it was no longer productive. But that wore off very quickly. And when we came back, it was a very synergistic interaction. Because Arg, even though he was opinionated . . . he did have this remarkable intuitive sense of how to make an experiment work. And in fact he was, beneath that exterior, very caring about what was going on, too. So once you understood that, it was . . . it was just a matter of yelling back."

Perhaps growing up as the eldest of six children gave her basic training in that sort of give and take. Although the name Villa-Komaroff vaguely suggests some princess from a long-vanquished house of Eastern European royalty, she grew up in Sante Fe, New Mexico. Her father was a schoolteacher and musician and, it would later turn out, a diabetic; her mother had been interested in botany, which was as close as anyone in the immediate family got to biology. Like many other would-be scientists, Villa had an early fascination with nature, and remembers thumbing through old turn-of-the-century natural history books belonging to her grandmother. She particularly recalls being impressed by an uncle who worked as a chemist at Sandia Laboratories in Albuquerque. "He showed me a paper once, I remember, when I was in the third or fourth grade. . . . It was very impressive, with the label and a very complicated title with a scientific name in it of some sort. I was admiring it and he said, 'Well, that's easier to write than an English paper.' And that impressed me a *lot.*"

By age nine, she decided she wanted to be a scientist, and the National Science Foundation ratified her choice with a minority scholarship during high school to work in a lab in Texas during one summer. When she reported to the University of Washington in the fall of 1965, she had pretty much decided she wanted to be a chemist. "That," she says, "didn't last very long. An adviser told me that, really, women didn't belong in chemistry." So she switched to biology. Then it was history of science. And then biology again. During an academic career described by its perpetrator as "very checkered," Villa-Komaroff finally gravitated toward developmental biology.

When in 1967 her college boyfriend, a medical student named Tony Komaroff, went off to Washington, D.C., for his internship, she finished her junior year at the University of Washington and followed him to the East Coast. With some difficulty, she gained admission—but only as a junior—to

Goucher College, a small private women's school in Baltimore, as a biology major. Two years later, Tony Komaroff planned to move to Boston to begin his residency; by then they had married, so Lydia looked for a graduate school program in Boston, too. By then, however, she had begun to exploit the "old-girl network" as nimbly as her male counterparts. Loretta Leive, for whom Villa-Komaroff worked as a summertime lab technician at the NIH, had graduated from MIT and urged her young technician to apply there. In 1970, to her considerable surprise, Villa-Komaroff was accepted into the graduate program at the Massachusetts Institute of Technology. Even the acceptance letter reflected the high-powered environment she was about to enter. It had been signed by Salvador Luria, graduate officer of the biology department, who just a year earlier had won a Nobel Prize in physiology/medicine. Intellectually as well as geographically, it was about as far as a small-town girl could get from New Mexico.

At MIT, Villa-Komaroff went to work with David Baltimore, one of the brightest young investigators in molecular biology. He had already achieved recognition as codiscoverer (with Howard Temin of the University of Wisconsin) of reverse transcriptase, that nearly alchemical enzyme that allows researchers to copy RNA back into DNA, and Villa-Komaroff recalls, "We all *knew* that we were in the lab of someone who would win the prize someday." Doing a joint project for Baltimore and Harvey Lodish, Villa-Komaroff spent four years studying the virus that causes polio. On the basis of her graduate work, she received a Helen Hay Whitney Postdoctoral Fellowship in 1975 (the same year that Baltimore won his Nobel).

From MIT, Villa-Komaroff moved to the Harvard laboratory of Fotis Kafatos for her postdoc. Kafatos's lab had focused its developmental studies on an insect system, the silkworm. Developmental biology had been of passionate interest to Villa-Komaroff ever since her undergraduate years in Seattle. More important, in her little first-floor study carrel, she could literally eavesdrop on the revolution in recombinant DNA work going on, for Argiris Efstratiadis was wrestling with the globin problem at that time.

She immediately realized—as Efstratiadis did—that this technique could provide an extremely detailed and revelatory view of how organisms developed. With recombinant DNA, they could perhaps isolate and study the genes that kick in on a precise time schedule to form the silkworm's eggshell. When the regulatory climate grew chill in Cambridge, in fact, Villa-Komaroff was already down at Cold Spring Harbor, at the laboratory run by James Watson on Long Island, working with Tom Maniatis. Like Forrest Fuller, who was also a Cold Spring refugee at the time, they struggled valiantly but vainly to get cloning to work. Grim words recur in Villa-Komaroff's conversation when she describes that long scientific dry spell, which stretched from the spring of 1976 to February 1977: miserable, terrible, very bad, *not happy.*

Failure is an indispensable part of the education of scientists, and it is not an elective course, especially in high-caliber labs where the most difficult problems are attacked. "You go in with a high sense of expectation," Villa-Komaroff says. "Things are happening all the time. If whatever you are working on *isn't* happening, then it is a very difficult situation. And with people who have very large labs, like Wally or David Baltimore, and have a lot of interests like they do, then a student or a postdoc who's having difficulties doesn't get the kind of support that maybe would be beneficial. On the other hand, people *know* that when they get there. And it's a self-selected group. Nobody who needs a lot of babying or mothering or reassurance is going to go into one of those labs, because they know it's not going to be there. You have to be able to know that a lot of it has to come from yourself."

The year at Cold Spring Harbor was Villa-Komaroff's first real taste—more of a force feeding, actually—of scientific failure. Unlike Forrest Fuller, she was lucky that it occurred while she was working on a relatively low-profile project. Silkworm eggshells were seldom bruited about in congressional testimony as a benefit of gene-splicing. She worked on cloning her eggshell genes. Nothing clicked. A year passed. She'd been doing nothing, it seemed, but "listening to the geese on Long Island." Then Arg Efstratiadis approached her to do insulin. She'd recently turned thirty and needed a successful project. She said yes.

To judge by all the precautions taken, the insulin project qualified not only as Big Astounding Science, but also as a Clear and Present Danger. Thanks in part to Villa-Komaroff's prior affiliation with MIT, Walter Gilbert had gained access to MIT's P3 laboratory for the work. In the open and informal environment of molecular biology, the P3 lab was a troubling anomaly: the door was locked, and a biosafety committee decided who could use the keys.

Located in MIT's lyrically named Building E17 on Ames Street, the lab reflected what might be called the interior design style of High Caution. Room 540 stood at the very end of a long corridor on the fifth floor, part of MIT's Cancer Research Center. A sign to the right of the door read "P3 Authorized Personnel Only" and a bright Day-Glo-orange Biohazard symbol warned that potentially dangerous experiments were underway within. A little porthole in the door looked into an antechamber to the lab. Filled with disposable pastel-colored garments, this became Lydia Villa-Komaroff's biological changing room for many months.

Each time before entering, Villa-Komaroff would slip on blue plastic shoe covers, milky translucent gloves, and a yellow gown. The lab itself was small, perhaps eight by twelve feet, with two benches, a refrigerator, an incubator, two centrifuges, one hood, and the official P3 radio, usually tuned to rock music, occasionally to classical. The room had negative-flow ventilation, meaning that the air pressure inside the lab was lower than in the hallway, so outside air always flowed in; any wayward airborne bug would have to buck a stiff breeze to escape. Nothing left the lab alive, save the researchers. Every scrap of refuse produced in the lab was put in bags; the bags were decontaminated and put in cans; the cans were put in autoclaves; and after the refuse had been pressure-cooked at high temperature, *then* the garbage was tossed out. All this precaution made the work appear dangerous, probably much more dangerous than it actually was. In part because of public concern and in part to anticipate any unforeseen problems, gene splicers grudgingly observed these safety extremes *as if* the work were hazardous. Most scientists considered potential problems highly unlikely, however. Villa-Komaroff considered the weakened lab strain of bacteria, Chi 1776, to be a "wimp," and they also knew that 1776 could not even establish a colony in a "nude" mouse—a mouse without an immune system or intestinal flora (that is, normal intestinal microorganisms like *E. coli*) of its own. "So we weren't really worried about that," she says.

There were also elaborate medical precautions. Although Harvard *still* didn't have its P3 facility in operation, it did have its biosafety committee, which had to approve faculty applications to do any experiments involving recombinant DNA. The uncertainties about the insulin work prompted a rather involved medical drill. For this relatively straightforward cloning experiment, doctors did an EKG and an EEG, drew blood serum to be stored, did a urinalysis, and performed a complete physical examination on each member of the Harvard team. Any illness had to be reported to medical authorities. It was up to the researchers, though, to define "illness."

Villa-Komaroff recalls telling herself, " 'Well, hell, I'm not going to report when I get a cold. If it has strange symptoms or if it's unusual or if it's *really* bad, then I'll report.' That never happened. The other thing we were supposed to do was if we were taking antibiotics, we weren't supposed to do the bacterial work. And that makes sense, actually. Because if you're taking antibiotics, that means that your own flora are reduced and so you're suscepti-ble to colonization by bugs which *are* drug-resistant." And pBR322 carried resistance to both tetracycline and penicillin.

One early fear was that even the enfeebled 1776 *E. coli* bacteria carrying as payload the rat insulin gene, were they to escape laboratory glassware, might colonize the gut of a researcher, start churning out insulin, and send a person into insulin shock, for an excess of insulin in the blood creates a dangerous condition of low blood sugar known as hypoglycemia. Jonathan King, the MIT biologist, had further suggested that the fusion of bacterial and mammalian proteins, if exposed to the immune system, could provoke a bizarre autoimmune reaction. Antibodies might develop to intercept not only *E. coli* proteins, but perhaps also the attached insulin molecule. The individual, according to this theory, could in effect become allergic to his or her insulin.

Villa-Komaroff, like the other Harvard researchers, viewed these sce-narios with, to put it mildly, extreme skepticism. For her graduate school thesis, she had worked with a live, though weakened, strain of polio virus. "I was much more comfortable working with this recombinant DNA stuff than I would have been working in, say, a microbiology lab, where they're taking cultures from people who are *sick.*"

Villa-Komaroff began working on the insulin project full-time in January 1978. True to her word, her very first experiments were controls designed simply to improve the efficiency of the cloning procedure. True to the Gilbert style, she also started to work the telephones. Out of it came a beautiful and novel strategy for cloning the rat insulin gene. Like most experimental plans, this was not decided at a strategy session or set down in writing. In fact, Villa-Komaroff did not exchange a single word directly with Wally Gilbert about the insulin work until months later.

First off, Efstratiadis and Villa-Komaroff decided to junk the idea of cloning by blunt-end ligation, to which Forrest Fuller had dedicated two and a half years. They would instead use a different technique known as "tailing." They also entered the experiment armed with the genetic "road map" of the penicillinase (or amp) gene in pBR322, which had just been sequenced by Greg Sutcliffe of Gilbert's lab. What did this tell them?

The Harvard researchers realized they could splice their insulin gene into a special location within the plasmid pBR322 known as the *Pst* site. This site was special for two reasons. First, it allowed them to easily snip out the many copies of the gene after cloning—a convenience often difficult to obtain. Second, this insertion site interrupted one of the two genes in pBR322, the one that coded for penicillinase. Bacteria make penicillinase inside the cell and then ship it to the cell membrane, where it intercepts penicillin before the antibiotic—poison to the microbe—can get inside. If the Harvard researchers spliced the gene for insulin *within* this larger penicillinase gene, perhaps the bacteria would not only make the hormone but *ship* it.

This was a teasing intellectual question. Perhaps the most remarkable point of all about it was Wally Gilbert's prescience. In applying to use the MIT lab back in June 1977, Gilbert had proposed just such a strategy, and even predicted the possibility that insulin would be secreted by the bacteria— before Greg Sutcliffe had sequenced the gene, before Villa-Komaroff joined the team, before the new cloning strategy had even been adopted.

Through most of January 1978, Villa-Komaroff and Efstratiadis did test runs of the experiment at Harvard. Villa-Komaroff practiced putting tails on her plasmid at different temperatures, with different concentrations of enzymes, for differing lengths of time—the grunt work of science. Conceptually if not physically, it resembled pinning the tail on a donkey, because in a land of colorless liquids in teardrop amounts, the work was certainly blind. "There are not a lot of visual cues," says Villa-Komaroff.

Efstratiadis knew that Richard Firtel, a scientist on the West Coast, had been tinkering with the so-called G-C tailing technique, so he urged Villa-Komaroff to give him a call. Like many such scientific conversations, it resembled nothing so much as the exchange of recipes. Within a week, Villa-Komaroff had prepared her own batch of the restriction enzyme *Pst* from scratch, fiddled with G-C tailing, and then run a test. The plasmid went into wimpy 1776, and this time the dry run worked.

By the end of January, everything was ready. Villa-Komaroff put a long CCCCCCCCC tail on the rat insulin gene provided by Efstratiadis, and a complementary set of GGGGGGGG tails went on the plasmid the next day.

On February 2 the two segments of DNA, one bacterial and the other rodent, with their sticky-ended G-C tails, were mixed together. The result: five millionths of a gram of pBR322 with foreign DNA, the insulin DNA, tucked inside the penicillinase gene. The big question, of course, was whether this DNA could be cloned.

As it turned out, they had to wait longer than anticipated for an answer. Just as Villa-Komaroff was about to find out, just as she was set to manipulate nature by slipping these recombinant plasmids into Chi 1776, Mother Nature struck back at the Harvard team with soft vengeance. Just when there was pressure to do the work as quickly as possible, when there had been a period of disquieting silence from the West Coast, when no one at Harvard knew where the Rutter-Goodman group stood exactly in their efforts to express rat insulin—just when Villa-Komaroff had everything ready, it started to snow in Boston.

And it snowed. It snowed all day Monday, February 6, and through the night and into the next day. By the time it stopped, two days later, Boston was digging out from a twenty-five-inch snowfall. Local headlines proclaimed it "worst storm of century." The National Guard patrolled the streets; all residents, with the exception of emergency personnel like doctors, were enjoined from driving. As it happens, of course, Villa-Komaroff's husband was a doctor. The couple also happened to live next door to the widow of James Curley, former mayor of Boston, whose posthumous perks seemed to include the swift arrival of snow-removal crews after a winter storm. And the work had to go on, so after being snowbound at home for two days, Villa-Komaroff got a lift to Beth-Israel Hospital with her husband, then took public transportation to MIT. The first entry in her lab notebook for February 9 acknowledges this epic journey: "Unsnowbound at last!" After two years of disappointment, and two feet of snow, finally, *finally*, it was on to the cloning.

The airport was still closed, and most streets in Boston would remain impassable for another four days, but on Thursday, Villa-Komaroff was in the P3 lab at MIT, slipping the insulin DNA into bacteria. She then spread the contents of the test tube on plates and incubated them; any bacteria which survived the procedure would proliferate and form pinhead-sized colonies, barely visible to the naked eye. They grew slowly, but two days later Villa-Komaroff returned to find some 270 colonies. She repeated the procedure a second time. On Valentine's Day, she hit the big payoff. That afternoon, after performing a big transformation three days earlier, she discovered hundreds of colonies. In all, she counted 2,355 of them, dotted on ninety-seven plates.

It was a promising start, but in no way conclusive. All Villa-Komaroff had determined up to this point was that about 2,300 hospitable microbes out of billions of potential recipients had taken plasmid pBR322 into their internal jelly. But did these bugs contain the insulin gene? Was the insulin insert aligned in such a way that it might even be read and expressed by the bacteria's transcriptional machinery?

To answer those questions, Villa-Komaroff and Efstratiadis would have to resort to more methodical biochemistry. They would cut the DNA with enzymes, run it on gels, analyze the patterns. For all these tests, the bacteria had to be killed and the DNA extracted in the P3 lab at MIT before Villa-Komaroff schlepped the material back up to the Bio Labs in her blue Saab. By now the streets were clear, and so too had the obstacles been removed from the cloning part of the project. Finally, fully a year after Axel Ullrich's work, the Harvard team had a promising colony of clones.

While Efstratiadis continued to run tests, Villa-Komaroff observed an odd pattern of growth in some of the plates back in the MIT lab. Rather than causing alarm, these aberrant colonies sparked great excitement. Some of the bacteria which had already been identified as having insulin inserts suddenly started growing on plates containing penicillin—something they ordinarily should not have been able to do. These bacteria were growing sluggishly, to be sure, but growing on penicillin nonetheless. As soon as she saw this, Villa-Komaroff rushed back to Harvard to report this startling development to Efstratiadis.

"I told him some of the guys with inserts were growing on penicillin," Villa-Komaroff recalls.

Realizing the significance, Efstratiadis's reaction was instantaneous. "Oh my God!" he cried. "Don't even tell the priest!" It was an old Greek expression: trust no one with your secret, tell not a soul.

This development suggested a new and, frankly, unexpected dimension to the experiment. If the bacteria were growing despite the presence of penicillin, then they had to be making penicillinase, the enzyme that neutralizes the antibiotic. And it was into the gene of that enzyme that the Harvard team had stuffed the insulin DNA. If those bacteria contained the insulin gene, they reasoned, insulin might even have been made and shipped out of the cell as cargo in the middle of the penicillinase protein. In other words, they might have jumped two steps ahead of themselves. Not only had they cloned the rat insulin DNA by getting it into bacterial cells, it appeared that the bacteria were reading the instructions and building the odd insulin-penicillinase hybrid protein ("expression") and shipping it out to the outer ramparts of the cell ("secretion"). It was a fantastic stroke of luck.

Efstratiadis insisted on secrecy; it was a hedge against premature scientific gossip. "We knew it was highly competitive," says Villa-Komaroff. "We knew people here at Harvard and MIT who knew people in California. Arg's feeling was 'Let's confirm it. It'll take a little time, but let's nail it down.' "

With Efstratiadis's injunction not even to tell the priest still fresh in her ears, Lydia Villa-Komaroff promptly rushed back to MIT and described the result to David Baltimore, her former graduate adviser, and Robert Weinberg, another faculty member and well-known cancer researcher. Villa-Komaroff burst into Weinberg's office with the words "Bob, I have the most wonderful news!"

Weinberg looked up from his desk and exclaimed, "You're pregnant!"

There was another bug that had entered the scene, however, and it effected the insulin work at Harvard in a significant way. It was the entrepreneurial bug, and after resisting it for a while, Wally Gilbert came down with a bad infection.

As early as 1977, Gilbert had received feelers from several private biotechnology companies. Molecular biology at Harvard was a "purist haven," he recalls, and the purists viewed industrial support with considerable suspicion. Only a few such companies even existed; Genentech and Cetus were probably the best-known examples, although a cottage industry had begun to grow up around such new-fangled lab supplies as restriction enzymes and linkers and probes. Gilbert viewed the division between academia and business as sharply as that between church and state, and he was constitutionally indisposed to blur the line. Still, since he was a world-class molecular biologist who would be the envy of any company, there surely would be more courtships.

Toward the end of 1977, the wooing resumed. Gilbert met informally with C. Kevin Landry, a managing partner of the Boston-based venture capital firm of T. A. Associates, and Daniel Adams, who represented another venture capital group. Landry recalls, "He clearly would not compromise his integrity in the academic community to join a commercial venture." That was not an unusual reaction. Unlike some other scientific disciplines in universities, such as chemistry or physics or electrical engineering, molecular biology had no history of collaboration or cooperation with industry; it was among the purest subcultures within the generally pristine biological sciences, where the principal attraction and primary motivation had always been to explain the intricate workings of natural phenomena purely for the sake of knowledge. To market a product, to make a profit, by contrast, seemed crass and intellectually suspect.

"The real skepticism," says Gilbert (who in speech often substitutes the past conditional for the simple past tense), "would have been, is there something useful for a company to do? Is there a useful way for the scientists to get involved in it? Does one *want* to get involved in it as a scientist, or is it going to be a distraction away from one's basic work?" The other main issue was independence: despite the reliance on federal agencies for funding, Gilbert treasured the fact that as a university professor he had "total independence." An industrial setting was likely to be quite different.

While Gilbert deliberated, not too seriously at that, the wheels were being set in motion for the formation of a biotechnology company with superstar aspirations. It would unite intellectual giants of the Harvard-MIT axis with the leading molecular biologists of Europe. The talent scouts for this enterprise were a pair of businessmen, the aforementioned Adams and Raymond Schaefer, who were associated with the venture capital arm of the Toronto-based International Nickel Company, Ltd. (INCO). INCO had been one of the earliest and most enthusiastic venture capital investors in biotechnology. When INCO learned that Kleiner & Perkins, the San Francisco venture capital firm, wanted to unload its share of the Cetus Corporation, a fledgling biotechnology company in Emeryville, California, INCO moved in. "We approached Tom Perkins," Schaefer recalls, "and told him we would be interested in taking over their investment in Cetus, in exchange for getting in on their next big project." That turned out to be Genentech. In due time, INCO invested $400,000 in Bob Swanson's company. But INCO wanted to get its own venture launched in what promised to be an exceedingly ripe area for industrial development. Cetus cofounder Ronald Cape, in fact, recalls providing Dan Adams with the names of some promising scientists in the field.

Schaefer and Adams, armed with the suggestions of Cape and others, spent a year and a half traveling through Europe bird-dogging scientific talent. Not unlike New York Yankee owner George Steinbrenner, the two business-men had a taste for the biggest names in the game, and pockets deep enough to pay them. After eighteen months or so, they formulated a "dream team" on paper—a preliminary list of all-star scientists they would like to recruit.

Just as in baseball, too, there were contentions that the venture capitalists stole Genentech's signs. In a dispute that was quickly and quietly buried and is rarely disinterred, Adams allegedly either obtained a copy of Bob Swanson's original business plan for Genentech or benefited from extensive business discussions with Swanson. Raymond Schaefer terms the allegation "absolute rot," and no litigation resulted from the charges, but according to Genentech spokesperson Deborah Bannister, the INCO venture capitalists "would not have invested in Genentech without looking at our business plan, and we were quite open about it. Then Dan Adams went on to start Biogen and was in fact the first president. Bob Swanson went to a [Biogen] board meeting in Geneva to talk to them about getting some payment from them, maybe a percentage of the company. . . . They were thinking about giving Genentech a percentage of the company, and then decided to make a payment." The payment, accord-ing to Genentech, was because "the business plan helped the guy from INCO get started." It is known that INCO paid Genentech an undisclosed amount of money relating to the charges. One source associated with INCO says, "It was an insignificant amount of money, corporately, and it shut them up"; Genentech's prospectus makes note of a $300,000 payment from INCO "in exchange for release of a claim by the Company to ownership participation, in exchange for services rendered, in a concern otherwise unaffiliated with the company."

The scientist Schaefer and Adams wanted most, Walter Gilbert, was the one playing hardest to get. In order to interest Gilbert, Ray Schaefer got in touch with Phil Sharp at MIT. Sharp, a young and widely respected researcher studying the molecular biology of adenoviruses, had worked once as a consul-tant to INCO. In fact, he'd been sent to California in the spring of 1977 to check out the credentials of the then-yearling Genentech, specifically to gauge the scientific feasibility of the somatostatin project. Largely on the basis of

Sharp's report, INCO made its sizable investment in Genentech stock. Obviously, there was a dismaying amount of incestuous financing going on in the early days of biotechnology; even while INCO owned part of one company, it was trying to set up a competitor.

Finally, over dinner in a Chinese restaurant in Boston with Gilbert and Sharp, Schaefer tried to interest the Harvard biologist in the idea. Schaefer's lone objective that evening was to lure Gilbert, without commitment, to an organizational meeting in Geneva. After considerable coaxing, Gilbert finally agreed. Following the Boston dinner, a second meeting was arranged in Cambridge, at which Adams and Schaefer unveiled their preliminary scientific list to Gilbert and Sharp. To anyone versed in molecular biology, it was an impressive roster. The group included, among others, Charles Weissmann of the University of Zurich, Kenneth Murray of the University of Edinburgh, and Peter-Hans Hofschneider of the Max Planck Institute for Biochemistry in Munich. Home-run hitters all. Gilbert and Sharp added a few suggestions of their own, and Gilbert agreed to chair the meeting. Once Gilbert was in the fold, Schaefer found it easy to lure the other scientists with a bit of name-dropping. "Wally Gilbert will be coming and chairing the scientific meetings . . ." he would say, and invitations were rapidly accepted.

Considering the fact that Genentech was launched over beers in a bar, the INCO initiative could not have been more formal or different. On March 1, 1978, wary groups of scientists and investors met for the first time at the Hotel Le Richemond in Geneva. The biologists, all high-minded academics and business naifs, suddenly found themselves discussed as commodities across the conference table. As Sharp recalls it, "It was sort of like undressing in front of your peers. You don't normally put a value on your activities in such a clear, upfront, and explicit fashion. You don't usually talk about money with other scientists." They quickly got the hang of it, though, and soon the biologists were enthusiastically banging their fists on the table, demanding 10 percent of the action. "The egos," Sharp was to recall later with a grin, "were *magnificently* large."

So, too, were the misgivings. Schaefer recalls the scientists worried that the businessmen would "pull something over on them," so they doggedly negotiated for a share of corporate control. They set up a corporate structure in which the power to approve product development lay with the scientific board. They put restrictions on the sale of stock so that none of the investors could bail out on short notice. They arranged for the company to fund ongoing research in their own university labs, so that they wouldn't have to leave their academic settings.

At Geneva, no one committed. Suspicions still ran high. But by showing up, the scientists had accepted the proposal as a real possibility. The motivation for becoming involved, Gilbert says, emerged from "a variety of reasons: wanting to do something socially useful, wanting to create an industrial structure, wanting to make something grow, wanting to make money. Although the desire to make money is not that high, generally, in scientists. Otherwise they wouldn't have been scientists in the first place."

Gilbert saw himself as a moderating force between the suspicions of the scientists and the interests of the business group. "In a sense, I became a spokesman for the scientists," Gilbert says. "I also had, in some ways, the greatest sympathy or affinity [for] the way in which the business side was structured." He expressed—"forcibly," to use Gilbert's own characterization—the scientific prerogatives, and at the same time represented the interests of the businessmen to the scientists. "Had I become a spokesman for the distrust," he says, "nothing would have ever happened."

With the INCO proposal to chew on, the scientists returned home from Geneva in early March. Another meeting was scheduled for Paris three weeks later. The scientists didn't know it, but at that meeting the businessmen wanted to get firm commitments on who was going to join. And what projects they were going to work on.

The advantages of a large lab with broad research tastes such as Gilbert's became apparent to Villa-Komaroff and Efstratiadis as they closed in on their insulin clone. Stephanie Broome, a young graduate student in Gilbert's lab, was independently making significant progress on her project, which dealt with the nominally unrelated topic of detecting proteins in small amounts. The results would intersect with the insulin work at precisely the right time.

Tall, blond, quiet, and bright, Broome had embarked on an independent project investigating the developmental growth of a slime mold called *Dictyostelium*. But in order to proceed, she had to overcome a considerable problem: how could you determine that a microbe was manufacturing a particular protein when everything was so small that even billions of bacteria themselves took up only slightly more space than a grain of salt? With Gilbert's encouragement, Broome devised a system for illuminating specific proteins in single-celled creatures, even if the proteins were present in minute amounts. Broome

195
. . . . . .

first tested the system out on that old bacterial standby beta-galactosidase. With serendipitous timing, however, Broome had worked out the kinks in the procedure just when Efstratiadis and Villa-Komaroff were ready to test for the presence of insulin.

By the time all the refinements were worked out, Broome's assay could detect the presence of as little as ten picograms of insulin—that is, ten trillionths of a gram—in a bacterial colony. That might be just a few molecules per cell. If among the clones made by Villa-Komaroff and Efstratiadis there were any colonies that were churning out insulin, the Broome-Gilbert assay, as it came to be called, would find them. And with the help of radioactive antibodies, they would bloom into light.

F o u r t e e n

''Interesting
Little Spots . . . ''

. . . .

He was, in Art Riggs's opinion, "the most productive young scientist I think
I'd seen." "Monomaniacal," said one colleague. The heart and soul of Genentech in the early days, said another. Competitive, clever, single-minded, mentally tough, physically inexhaustible—those were the qualities he showed as a
tennis player, mountain climber, and cloner. "He just has a greater capability
for hard work than most people and even most scientists have," Riggs would
say. ". . . some people have the knack of making the right decisions and the
work goes well not only because they work hard, but because they make the
right decisions. And he had the knack of making the right decisions." He was
a tall, restless splinter of a biologist. His name was David Goeddel. And it was
no accident that he came to be the key player in Genentech's race for human
insulin. Bob Swanson conceived of the early days of biotechnology as a series
of scientific races—for patents, for credibility, for attention. In order to run
races, you needed greyhounds. Goeddel was Swanson's greyhound.

In January 1978, Dave Goeddel was a twenty-six-year-old researcher at a small private laboratory in Menlo Park, California, called the Stanford Research Institute, where he worked with a molecular biologist named Dennis Kleid. Raised in a suburb of San Diego, Goeddel was tall and rangy, with a kind of teenage limberness about him; he had a high forehead, wavy brown hair that fell boyishly across his brow and covered his ears, large-lensed glasses that drooped down his nose. His was the standard wardrobe of molecular biologists—sport shirts, blue jeans, and running shoes—but somehow it seemed like slapping clothes over a coiled spring. There was a tiny scar on his chin which, although due to a childhood spill from a bicycle rather than any mountaineering mishap, left a telltale nick in this otherwise boyish appearance, just the hint of a hard edge.

Just as he came from the opposite end of the state from Goeddel, so did Dennis Kleid bring a diametrically different temperament to the working relationship. A native of Napa, north of San Francisco, Kleid effused a kind of loose bonhomie. Where Goeddel would speak in a deep monotone, Kleid's voice bobbed up and down with excitement and he built his sentences toward exclamation points; where Goeddel had that vigorous yet self-contained look, Kleid offered a kind of open volubility both friendly and challenging. With his wire-rimmed glasses, expressive, saucerlike eyes, bushy mustache, and wry expressions, he managed to project both excitability and serenity, like a character in a 1960s underground comic book. Perhaps his most endearing quality was a robust, humorous cynicism. "The minute a scientist thinks he's got everything figured out," he liked to say, "that's when he's in *real* trouble."

Kleid was the established scientist, Goeddel the apprentice. After doing his undergraduate work at the University of California–Berkeley and his graduate work at the University of Pittsburgh, Kleid went to Cambridge as a postdoc in the MIT lab of H. B. Khorana. He was one of the minions who chipped in part of the two hundred man-years necessary to create the first functional synthetic DNA gene, and he later collaborated with Tom Maniatis, of Mark Ptashne's lab at Harvard, on developing ways of sequencing DNA. (Those efforts were superseded when Allan Maxam and Wally Gilbert, one floor below, came up with the Maxam-Gilbert sequencing method.) After his stint in Boston, Kleid moved out to Menlo Park in 1975 to start up a molecular biology unit at Stanford Research Institute. DNA continued to be the topic of interest, but his particular focus was on learning how anticancer drugs might damage, or mutagenize, the DNA molecule.

Molecular biology's usual lines of communication brought Goeddel and

Kleid together: Goeddel had just received his Ph.D. in biochemistry at the University of Colorado at Boulder, where his graduate adviser had been Marvin Caruthers. Kleid and Caruthers had worked together in the Khorana lab. When Kleid needed an assistant, Caruthers sent along the curriculum vitae of his top student. Ergo Goeddel.

For his part, Goeddel wanted nothing to do with traditional paths toward scientific respectability, which obliged one to spend two or three years as a postdoc in as prestigious a lab as possible. "I didn't want to do an academic postdoc, mainly out of the feeling that I didn't need to do a postdoc," he says bluntly. "Here was a place you could go, have a job, and if you wrote your grants, you had a position. You got them funded. And I wanted to go to California. *Back* to California." Geography mattered to Goeddel, but in a somewhat eccentric way. He decided to go to the University of Colorado, he says, because it was close to the Rockies; at the time, he was "still more interested in being a climber than a scientist." Menlo Park was not that far from Yosemite; El Capitan and Half Dome, which he had climbed in the early 1970s, were reassuringly near. "I just like being near mountains," he would say.

When people asked Goeddel why he liked mountain climbing, he would reply that "in some ways it's like science—the chance to do certain things first. It's almost like being an explorer. And there're not too many exploring possibilities left outside of science and maybe climbing, being the first one to actually climb specific routes. It's very challenging, and close to being as tough mentally as science is, besides the physical toughness and difficulty of climbing."

Sometimes, with friends, the answer would be a little more explicit: "You get to a certain point," he would say, "and you're committed to finish. You can't just go home."

So there they were, living half a block apart in Menlo Park, setting up experiments at Stanford Research Institute; Kleid and Goeddel had themselves considered synthesizing genes, but hadn't drummed up any funding. SRI was a "soft money" institution: researchers like Kleid had to raise their own grant money to pay for research and salaries, and raising grant money was an odious full-time job in itself. "I was getting twenty thousand dollars," Kleid recalls, "and Goeddel was getting twelve thousand and I tried to get his salary up to fifteen thousand, and I was having an awful time." It was at precisely this moment that Kleid got a call from Herb Boyer and Bob Swanson. They had formed this new company, they said. Would Kleid be interested in joining?

At the beginning of 1978, Genentech was still just another paper tiger in the biotech business—its patent applications were on paper, its corporate identity was on paper, its main asset, the somatostatin experiment, just published in the December 9, 1977, issue of *Science,* was on paper. It had no products, no salesmen, no full-time scientists, not even its own lab. Bob Swanson maintained an office in the Wells Fargo building on Sansome Street in San Francisco's financial district, but it was difficult to convince anyone you were a leading-edge research company when you didn't even have a laboratory. "I had a desk, a telephone, and a rented secretary," Swanson recalls. The cloning continued to be done in Herb Boyer's lab, funded by Genentech's controversial grant to UCSF; the synthetic DNA work and protein chemistry was done at the City of Hope in southern California.

The work had reached a crucial juncture. The success of the somatostatin project had given the infant company a big boost. Much more was at stake with insulin, however. Both the Rutter-Goodman team at UCSF and Wally Gilbert's group at Harvard were thought to be closing in on the expression of rat insulin—that is, getting the bacteria to *make* the hormone. Having done that, human insulin would be almost a duplication of the rat work. True, the university labs were going about it in a different way scientifically by trying to clone an insulin cDNA while Genentech wanted to clone a synthetic gene; but commercial feelers were reaching the two groups, too, and the Harvard and UCSF teams in any event were in direct competition with the Genentech group for public attention, and that was crucial to any outfit with industrial aspirations. Public attention, particularly in the case of a young, unknown entity like Genentech, gave the company—and the technology it championed—a kind of credibility that money couldn't buy. Swanson was particularly attuned to this; financial analysts credit him with being extremely adept in those early days at orchestrating the media to bang the drum for biotechnology and, of course, for Genentech.

So the final push for insulin was on. Keiichi Itakura's team at City of Hope, led primarily by Roberto Crea, had been busy preparing fragments of synthetic DNA since the previous summer; Paco Bolivar, the former postdoc in Boyer's lab, was commuting from Mexico to Duarte to work with the City of Hope contingent, but he couldn't commute "often enough to really make the project go fast enough," Kleid says. Genentech, after going as far as it could with its itinerant researchers, had reached the point where it had to take itself seriously as an independent research lab. Now was the time to assemble the dramatis personae, not just for insulin but for all the important genes that would be chased next—growth hormone, the interferons and interleukins, tumor necrosis factor, animal vaccines, tissue plasminogen activator. Just as the feeling had been widespread in 1976 that only a few scientists were going to make the key discoveries with recombinant DNA, Genentech believed that only a few early-bird companies were going to clone the medically important genes and, more important, file the patents to protect those discoveries.

Kleid and Goeddel, however, were probably not Bob Swanson's first choices. Soon after the Rutter and Goodman labs had cloned rat insulin in 1977, Swanson mounted a talent raid on UCSF. He offered jobs to Axel Ullrich, Peter Seeburg, and John Shine; and the company also tried to get its hands on the rat insulin clones. There were high-level scientific meetings during which Genentech discussed the possibility of inducing mutations in the rat clones to convert the insulin gene to human. When Goodman learned about the discussions, according to Ullrich, he tried to get in on the deal; when Rutter learned that Goodman was negotiating with Genentech, he too wanted to join the discussions. Then the whole package collapsed. The reason, according to one source at Genentech, was that Goodman and Rutter wanted an equal say with Boyer in running the company (although Rutter denies this).

After this impasse, Herb Heyneker suggested bringing Dennis Kleid into the fold. Kleid had met Heyneker in Holland; the two scientists had become good friends when Heyneker moved to California to work in Boyer's lab at UCSF, and the two had even prepared a grant application together. Heyneker knew Kleid's work and had spoken enthusiastically about him to Boyer. So Boyer decided to give him a call. They agreed to meet at a French restaurant in San Francisco to discuss the matter.

Swanson and Boyer needed to lure Kleid to the insulin project. The insulin gene had already been partially cloned by Paco Bolivar, they told Kleid, but they wanted the project to move faster. "Now to me, the insulin project was, certainly that year, *the* project to work on if you were interested in this sort of thing, cutting and pasting DNA and trying to make things in bacteria," Kleid recalls. "Insulin was the project that was number one."

As the dinner went on and on (it would last four hours), however, Kleid remembers feeling somewhat dumbfounded as Boyer and Swanson described the grandiosity of their plans. "Swanson had analyzed that if we could make insulin," Kleid says, "we could then prove the credibility of the technology and get some royalties and get some contact with a big pharmaceutical company that knows what it's doing, and maybe that project could be the seed from which we could grow into a real organization. His plan was always to have a fully integrated company, all the way from science to process development to manufacturing to marketing—the whole gamut." When they asked Kleid to join Genentech to clone the synthetic insulin gene, he remained noncommittal. "To me," he recalls, "their idea of how it was going to evolve so quickly seemed *incredibly* naive . . . the idea of walking into some pharmaceutical house and saying 'I want a few million dollars to do these experiments' seemed to me to be really kind of naive."

Still, Kleid considered cloning insulin "the dream project of the day," and he recalls being particularly impressed by Boyer's argument that even in an industrial setting, Genentech's scientists could do basic research and publish their results. So Kleid went back and talked it over with Goeddel. Goeddel's response was immediate: "Well, if you can get me a job there, I'll go." Kleid had had something similar in mind. He invited Swanson down to Menlo Park for lunch several weeks later. Kleid said he wouldn't sign up unless he could bring along Goeddel. Like Boyer's decision to rely on synthetic DNA, Kleid's insistence on dragging Goeddel into the deal turned out to be a pivotal event in Genentech's early history—"one of the great gold strikes of all time," according to a scientist at a rival biotech company. In February, Kleid and Goeddel went to a scientific meeting in Colorado. When they returned to Menlo Park on February 24, job offers were in the mail for both of them. Swanson had bought the whole package, Goeddel and all. Now the ball was in their court. Should they make the move?

Biologists traditionally viewed the path from academia to industry as a one-way street, and a downhill one at that; they tended to look upon industry scientists as researchers whose scientific curiosity was conditioned by managers with one eye on the marketplace, whose freedom to publish was curtailed; and there even existed, just beneath the surface, a kind of snobbish machismo in basic research, an unspoken consensus that to tackle anything less than the stickiest, most confounding, most fundamental biological mysteries suggested a certain lack of intellectual ballsiness. In a field where intellectual rewards historically outstripped financial compensations, it meant giving up a lot.

Kleid had more at stake than Goeddel—he would be abandoning what every scientist works toward, his own lab. In exchange, he would earn the dubious honor of commuting to southern California frequently to do the work. For Kleid, the decision was "very chancy," and in the end, he cheerfully abdicated responsibility to Goeddel. "Basically I said to Dave, 'It's your decision. If you want to do it, let's do it. If you don't want to do it, we'll stay here,' " Kleid recalls. He knew that if he turned the project over to Goeddel, it would fly.

Indecisiveness is not one of Goeddel's weak points. He had no academic aspirations, so he didn't mind leaving that bridge aflame behind him. His salary at SRI was paltry compared to what Genentech was offering. Most of all, Goeddel liked the challenge. Here, he realized, was the opportunity to work on a hot project. Here was the opportunity to get in on the ground floor of a business that could have a huge future (although even Goeddel himself admits that personal goals exceeded any loyalty to the company when he decided to sign up). And, perhaps more important, here was the opportunity to go head to head with a scientist of enormous reputation, Wally Gilbert, on a highly visible project. "I was conscious, very conscious, of the competitive aspect," Goeddel says. "Actually, I found that very attractive. We knew we were in a race."

"Why don't we?" Goeddel finally urged Kleid. "You know, they have Itakura. They can make the DNA better than anybody. It's a good chance." Kleid agreed. Just like that, Genentech had its first full-time scientists.

Bits of synthetic DNA, ready to be assembled into a human insulin gene, were waiting for them when they began. Goeddel reported to work in mid-March; thus began a five-month period of work in which he claims to have taken only one day off—to attend the ten-year reunion of his high school class in Poway, California, outside San Diego. Kleid showed up, as he likes to point out, on April Fool's Day; that first day, according to his lab notebook, he signed out at four in the morning.

They worked against molecular biology's traditional deadline: yesterday.

Several miles to the north, the Goodman and Rutter labs at UC–San Francisco continued to work on expression of the rat insulin gene, and the

UCSF-Harvard competition continued to inspire much jesting in the *Midnight Hustler*. In one 1978 issue, two entire pages were devoted to "jokes"—the term is used somewhat conditionally—made at the expense of UCSF researchers. An example:

> What do you find written on the bottom of UCSF reagent bottles?
> "OPEN OTHER END"
> What do you find written on the top of UCSF reagent bottles?
> "SEE OTHER END FOR INSTRUCTIONS"

It was one of the more benign and less technical of many insults.

They may have been laughing at Harvard, but the UCSF work on rat insulin produced one of the classic papers in the early history of recombinant DNA work, and that was becoming apparent with each passing month. The tricks improvised by Axel Ullrich, John Shine, Peter Seeburg, and John Chirgwin were quickly incorporated into laboratory protocols throughout the world. The researchers were in demand to present seminars on the experiment and courted by biotechnology companies. Prizes were awarded; magazines and newspapers frequently mentioned the work.

Yet the Rutter-Goodman collaboration, by most accounts, traveled a fairly rocky road after the triumphant announcement of the cloning in May 1977. Like Efstratiadis and Villa-Komaroff at Harvard, Axel Ullrich was now attempting to demonstrate expression of rat insulin in bacteria. Using the very same plasmids that Genentech had employed for the somatostatin work, generously donated by the Boyer lab, Ullrich had patched in his rat insulin gene. Ullrich and Pictet thought they had expression, too, but they had less faith in their assay than the Harvard group had in Stephanie Broome's technique. They decided the data were not convincing, according to Pictet, and their only mention of expression appeared in an obscure Japanese publication after Ullrich alluded to the work at a meeting.

There were delays, too, ascribed to what Ullrich describes as the "political traumas and personal traumas" associated with the pBR322 episode. Many of the problems seemed to stem from competing claims of intellectual priority and scientific credit. These are always ticklish issues in a laboratory; the Goodman postdocs at least partially resolved matters among themselves.

"After some discordant times," Seeburg admits, "especially on that *Hae* III cutting business, we had an agreement—Axel and John Shine and me—that we would go separately for insulin expression and growth hormone expression, and whoever arrived first would include the other ones so that we could work in the same direction." Indeed, when Seeburg and Shine expressed rat growth hormone in bacteria and reported the result in *Nature* in 1978, Ullrich was one of the authors.

The schism between the postdocs and the lab directors was not so easily repaired. The high visibility of the work, the spotlight that patents threw on the origins of "discovery," and the random mosaic of ideas that contribute to breakthroughs all conspired to make UCSF once again an unhappy laboratory of science sociology. "It was only when [the techniques] began to be successfully applied that there seemed to be a couple of instances where the actual origination of the approach or the technique or the idea was viewed slightly differently between different people . . ." John Shine recalls, diplomatically tiptoeing around particulars. "In a rapidly developing field, it's an accumulation—someone begins to sound out a certain thought, someone builds on it, someone else adds to it, there's a lot of discussion about it, and a month or so later it's achieved. Then, of course, with three different people, they see it as 'Hey, I thought of that back then . . .' The truth is, they all did it in a sense, cooperatively, because you work intimately with each other on a day-to-day basis. But, you know, people always do see their input slightly differently than the other person sees it."

Even the most dispassionate observers describe this as a "very emotional" period at UCSF. The Goodman group in particular suffered from lingering ill-will. "There were a lot of emotions running around," recalls one worker in the lab, "because everything was so exciting. Basically, every week there was a new discovery. Sometimes when you work real hard, and close to other people, you get high feelings, and that's when greed and fame and ego, the grasping, enter in." Goodman's zeal to take credit for the work, in the view of numerous people in and outside his lab, created a bad impression. If recognition was the scientist's bread, then both Ullrich and Seeburg felt especially undernourished. John Shine, a model of Australian pragmatism, found himself consoling his two European colleagues.

"I can remember having many discussions with both Peter and Axel about this," Shine says. "It's always seemed to me that you can't rush the recognition you're going to get. If the work was worthwhile and you were very much responsible for it, then it comes to you in time." Shine told his German colleagues that there was no way they could "force that great acclaim from your peers," that the situation would change "as you establish yourself, show that it wasn't a fluke."

Goodman, for his part, no doubt felt he deserved some credit for creating a laboratory where the techniques were available to produce such noteworthy results in the first place. Shine believes Goodman's role may have been unfairly devalued. "I have to give Howard some more credit than I probably did back then," he says, "because now I have more appreciation [for] the other roles he had to play that were necessary in supporting, just financially, the grants for the laboratory." Nonetheless, the perception that the postdocs were not accorded sufficient credit was widespread and, in the opinion of one faculty member who asked not to be named, "legitimate." As a result, according to lab members, Seeburg and Ullrich became almost paranoid about discussing their progress with Goodman, a withholding of information that set the stage for even more traumatic consequences.

As the insulin expression work stagnated, Axel Ullrich began to feel totally isolated. Seeburg, at least, had a supportive and enthusiastic sponsor in John Baxter, a member of both the biochemistry department and the medical clinic at UCSF. Baxter promoted Seeburg's work at meetings and had been a constant source of encouragement during Seeburg's long and ultimately successful experiments on rat growth hormone, performed with Joe Martial. Ullrich got none of that. He traveled to exactly one scientific meeting during his period in the Goodman lab; not only did he have to pay his expenses out of his own pocket, but it was to the fateful meeting in Utah when the pBR322 crisis exploded. Ullrich says that his superiors treated him like a mere technician, "a pair of hands." Nor was he alone in this view; Raymond Pictet, a senior researcher in Rutter's lab, says, "Axel got cheated. I don't think Bill and Howard gave him the credit he did deserve for the way he worked on it. It must have been very hard for him."

Seeburg, who shared Ullrich's perception of the situation despite occasional tensions in their own competition, felt the problem had reached absurd proportions by the summer of 1978, when Howard Goodman was named recipient of the CIBA-GEIGY DREW Award in Biomedical Research for the rat insulin and growth hormone experiments, which carried with it a $2,000 prize. Ullrich provided champagne and cake—again, out of his own pocket.

"It was just ridiculous," says Seeburg. Ullrich doggedly continued to work with rat insulin—a little less naive, a little more protective of his work.

The Harvard researchers, meanwhile, were deep into a series of experiments trying to confirm that they had cloned rat insulin. They were engaged in molecular biology's version of finding a needle in the haystack: you fished out the one bacterium out of perhaps billions that contained the piece of DNA you wanted.

The key to this procedure, known as hybridization, was the creation of a kind of molecular magnet known as a probe. Probes were single-stranded pieces of DNA or RNA which had been chemically tagged with radioactive markers. Wherever these probes homed in on DNA with complementary sequences, their location would be revealed as black spots on film.

Efstratiadis borrowed from the techniques established by the UCSF work in the creation of his insulin probes, but when he searched for spots on the films, the screening technique seemed to work only too well. Villa-Komaroff examined some eight hundred colonies; more than a hundred lit up on film, suggesting that their probe had homed in on insulin sequences in the DNA of those bacteria. She identified the likeliest candidates by the intensity of the spots, grew up large amounts, and passed on the DNA to Efstratiadis for restriction mapping—the procedure of cutting DNA with a restriction enzyme and running it through a gel to get its "fingerprint."

And none of them was insulin.

Puzzled, they tried another tack. They selected nine of the positive colonies at random and gave them to Peter Lomedico. It was Lomedico who picked the first insulin needle out of the haystack. He screened each promising candidate using a technique known as hybrid arrest translations, or HARTs, which indicated which if any of the colonies contained insulin DNA. Lomedico performed HARTs, as it were, on all nine clones from Villa-Komaroff's Valentine's Day crop. As often happens, the most promising candidates flopped—no insulin there. It was a dark-horse candidate, clone number 19, which unexpectedly showed the most encouraging results. It had to be insulin. Peter Lomedico ran the tests on April 3; the next day, Lydia Villa-Komaroff was growing up a whole pot—three liters—of the clone, which was known as pI19.

Identifying a promising clone is like locating an object and centering it in the field of a microscope. It is then possible to increase the magnification and make the resolution even finer. Very quickly, the Harvard team took the necessary steps to sharpen their focus, but they did this by biochemical, not optical, means. Bacteria containing the rat insulin sequence were grown; Efstratiadis chopped up the DNA with restriction enzymes and ran the material down a gel. The "fingerprint" left by the migrating fragments matched Axel Ullrich's data. Gilbert was so eager to obtain the sequence of this clone that he gave the job to his technician, Richard Tizard, when Efstratiadis couldn't get to it immediately. Several days later, Tizard and Efstratiadis had the sequence of clone 19. More with relief than exhilaration, they saw that it was rat insulin. The long, frustrating two-year quest was over.

At this point, the Harvard team had essentially reproduced UCSF's result of a year earlier. The lone novelty was that they had used a different cloning site. They were poised, nonetheless, to make a great leap forward. Using the DNA from clone 19, they prepared an even more discerning probe; once again, the focus could be tightened even further. Within a week, Villa-Komaroff began to search through her clones all over again, through each and every one of the 2,355 colonies she'd created in February.

This highly discerning assay took place in Seal-a-Meal bags, the same ones used in nonscientific kitchens to keep food fresh. In fact, the whole procedure smacked of institutional cooking. You would dot out individual bacterial clones on filter paper, about thirty-five colonies per filter, allow them to grow into pustulelike colonies, and then rupture the cells and bake the DNA until it stuck to the filter paper. Into the Seal-a-Meal bags would go the filters, to marinate overnight in a fluid that contained the radioactive probes. Finally, the filters would be rinsed, covered with photographic film, and tossed into the freezer. Any spots on the film would correspond to numbered colonies on the filter paper where probes had congregated.

It took seventeen bags—four filters per bag, three dozen colonies per filter—to check through all the recombinant bugs created during those snow-bound days back in February. When all the films were developed, Villa-Komaroff identified forty-nine colonies that lit up. The single clone 19 probe had allowed them to find forty-nine in all that possessed the insulin gene.

From the moment Lydia Villa-Komaroff saw those peculiar bugs growing where they shouldn't have, she was convinced that the Harvard group had indeed achieved expression. She was right, as it turns out, but for the wrong reason. This supposition led the Harvard researchers down a long and winding road that grew into the New Jersey Turnpike of blind alleys. But Villa-Komaroff's initial conclusion was borne out, and Wally Gilbert headed off to Paris for another round of business meetings buoyed by the knowledge, based on false premises but heartening nonetheless, that his group had expressed and possibly even secreted insulin.

On March 25, the European scientists as well as the contingent from Boston, including Gilbert and MIT professor Phillip Sharp, convened in a hotel near the Paris airport to begin a decisive two days of discussion. Ray Schaefer was there, representing the INCO venture capital group, and so was Kevin Landry of T. A. Associates. Both businessmen had misgivings.

For the financial backers, the Paris meeting represented a go–no go decision point. Another meeting had been scheduled about six weeks later for Zurich, but the investment group made it clear that only scientists committed to the project would be invited. At the same time, they sensed continuing ambivalence, if not outright skepticism, among the ten or so scientists assembled. Schaefer believed Gilbert was "probably the most distrustful of all. He had a pretty low opinion, generally speaking, of how industrial companies were run." The scientists, meanwhile, worried that the businessmen would try to pull something over on them. "And that made things a bit difficult," Schaefer continues. "Everything had to be explained in minute detail. They wanted every line, every sentence, explained, because they thought there was something hidden in there."

Schaefer's view notwithstanding, Gilbert still viewed himself as the glue holding the scientists and the businessmen together; and if anything, his mind was pretty well made up by the time of the Paris meeting. He is not, by his own admission, "an extreme soul-searcher." He sees himself as "an activist. And so my reaction to getting involved in it at all was much more the sort of thing of 'Well, am I going to run this or not?' " No one, it appears, had any objection. The scientists were busy bashing out the articles of association for this new company and whittling down a list of potential products. But were they actually going to sign on?

To Sharp's mind, a crystallization of intent, if not an actual turning point, seems to have occurred during one of the recesses at the Paris meeting. "This particular afternoon," he recalls, "we broke around three or four o'clock and were taking a walk. Wally and I walked a little ahead of the others, and he started telling me about his experiments with insulin. They had gotten both expression and the secreted protein. I was excited when I heard about it and congratulated him on the achievement, and then I asked him if he was going to patent it, and we discussed the commercial possibilities. There was a rhythm, a nonspoken feeling, that people were going to try this. It was just a little piece of technology that had obvious commercial implications, part of a whole conceptual feeling. The feeling was that it was something to be done, and that it could produce something of interest."

As the Paris meeting drew to a close, Gilbert asked the business people to withdraw: the scientists wanted to hash things out among themselves. The businessmen, Schaefer and Landry, cooled their heels outside the meeting room. The scientists' caucus went on for an hour without any sign. Then for nearly two. "Looks like we're in trouble," Schaefer said to Landry. "Might as well pack our bags."

Finally, Gilbert emerged from the meeting room with, as Schaefer recalls, a "somber look" on his face. Schaefer immediately inferred the worst and asked, "Well, is it all over?" Gilbert nodded.

"How many are going to come with us?" Schaefer asked resignedly.

"All of us," said Gilbert.

"It was like the Pied Piper," Landry would say later. Everyone fell into line. Later on, the two venture capitalists, Ray Schaefer and Dan Adams, tried to think up a name. Schaefer wrote down a word on a scrap of paper and, passing it to Adams, said, "What do you think about this?"

Adams seemed to like it. The word was "Biogen."

It's not as though Wally Gilbert were not a frequent topic of conversational analysis in his lab group under normal circumstances, but this flirtation with a biotechnology company, when it became known, prompted much discussion. There were "some initial rumblings," in the words of one graduate student, but there had always been grumbling about his frequent travels and extended absences from the lab. People wondered if going into industry was a good idea in general; others wondered if it was a good idea for someone like Gilbert, whose scientific credentials were unassailable, but who didn't seem to exude the gregarious, hail-fellow-well-met, politic demeanor of a businessman.

The *Midnight Hustler,* in its inimitable fashion, weighed in with "Clonegate," an exposé by "Schwartzward and Broomestein"; photographs taken by a surveillance camera showed a hooded figure—"a paid agent of Biogen's dastardly rival, Genentech"—poring through Biogen files in Gilbert's office. The East Coast–West Coast rivalry was alive and well and residing now in the industrial sector.

But opinions were just opinions, and Gilbert, as usual, was interested in facts. The rat insulin work was beginning to yield some very delightful facts. By April 14, Lydia Villa-Komaroff had gone through all her Seal-a-Meal bags and identified her forty-nine promising candidates. Two weeks later, these clones would receive the Broome-Gilbert treatment.

The first insulin test of the Broome-Gilbert technique occurred on May Day, 1978. Villa-Komaroff prepared cell extracts from the forty-nine candidates in MIT's P3 laboratory. The same logistical obstacles pertained: since no living organisms could leave the P3 lab, of course, the bacteria had to be killed first and cracked open, each colony becoming a spot on a grid.

Broome had already obtained insulin antibodies from guinea pigs and penicillinase antibodies as well. The insulin antibodies formed the bottom layer of Stephanie Broome's analytical technique. Villa-Komaroff had taken these disks down to MIT and pushed them onto plates containing the forty-nine clones, so that each colony left a dot of its protein contents on the disk. Then she ferried these samples back up to Harvard. There, radioactive antibodies, insulin or penicillinase, were allowed to wash over the whole disk. A piece of unexposed film was slapped against the disk, and into the darkroom freezer it went for overnight exposure.

Early the following afternoon, May 2, Stephanie Broome came into the Bio Labs to develop the film. Normally Gilbert would hover around the darkroom, awaiting the results. "But that day he had a seminar over at the Science Center," Villa-Komaroff recalls. "And we pulled out this film and one of them was black, and in the right place." One of the clones, p47, showed a strong signal.

Expression.

Insulin.

The Harvard group had not simply inserted the insulin gene into *E. coli*. They had managed to switch the gene on. Out of the billions of bacteria that had unwittingly embarked on this man-manipulated odyssey, a handful of colonies contained the insulin gene; now, among those few finalists, they had found one colony, a single colony, that not only contained the insulin gene but actually appeared to be reading the instructions and making rat insulin. Bacteria making rat insulin! With the crucial addition of Broome's assay, the Harvard group had not only equaled the UCSF result of Ullrich a year earlier but had been able to leapfrog dramatically ahead. They were the first group to express a mammalian protein in bacteria with a cDNA "gene," and the second to express any mammalian protein at all, following Genentech and its synthetic DNA somatostatin.

Delirious with excitement, Villa-Komaroff raced down two flights of stairs and out past the rhinos, cut through the courtyard and across Divinity Avenue, followed the footpath behind the new biochemistry building, crossed Oxford Street, and dashed into the Science Center in search of Gilbert. She tracked him down in a stairway outside the seminar, just as he was heading for the men's room, and said, "You have to come and see this film." He knew exactly what she was talking about.

They both raced back to the Bio Labs. The scene there is indelibly etched in the memory of almost everyone who happened to be on the third floor at the time: Gilbert walking down the hallway, holding the photographic film up to the lights, and reacting with that most Gilbertian sign of approbation—the beaming grin—and typical understatement. "Oooh," he said, "interesting little spots!"

To Villa-Komaroff, the spots merely confirmed the suspicions she'd harbored since February, when she spotted colonies growing on penicillinase. "I knew then," she says. "I knew it for the wrong reason, but I knew it then." The May confirmation added to "a continuing euphoria." To Efstratiadis, the spots spelled relief. "I was happy," he says, "because we could have a paper that claimed something more than the simple cloning of insulin, which the people in California had already done."

The result could not have come at a more propitious time for Gilbert. Several days later, on May 6, he was in Zurich for another business meeting. This time, the scientists gathered to confirm the consensus that had been reached in Paris. Benefiting from sagacious counsel on the financial side, these biologists—naive in the ways of the business world only a couple months earlier—proved to have been quick studies. They headquartered their new company in the Dutch Antilles (to provide favorable tax treatment), and they set up the primary scientific facility in Geneva, Switzerland (to enjoy typically speedier European approval of new pharmaceuticals, a more relaxed legislative attitude toward gene-splicing, and the fact that in European labs, the scientist and not the university owns the research). The company later established a base of operation in Cambridge, Massachusetts, as well.

The businessmen, Schaefer and Adams, proudly revealed the name they had come up with for the new company. Had they run it by the scientists before formally selecting it, they would have learned that it wasn't exactly original. Biogen? Oh, that was the name of the company that used to make autoclaves and other scientific equipment.

Nonetheless, on May 6, 1978, Biogen was officially formed. Its scientific priorities had already been established at the meetings in Geneva and Paris. Alpha interferon and hepatitis B vaccine were the primary research targets, for they promised large and lucrative markets; but Gilbert returned to Boston with the knowledge that one of Biogen's three priority projects would be an attempt to duplicate the rat insulin expression. He would get the same people to do it and use the same experimental strategy as on the rat insulin experiment, which, in effect, had worked on the very first try.

This time, however, they would do human insulin. Heading into the summer of 1978, Gilbert was only slightly behind the two-year schedule he had invoked before the Cambridge City Council.

Fifteen

"It's Like They Were Whirlwinds"

· ·

If David Goeddel embodied the scientific soul of Genentech, his spirit became institutionalized in the motto on a T-shirt he later wore. "Clone or die," it said. That relentless, all-or-nothing determination was on abundant display from the very first day he showed up for work at the City of Hope. Art Riggs, a scientist of more contemplative demeanor, had planned to spend the morning discussing the work that would be done. Goeddel, however, is not what you would call a chatty scientist; he abruptly cut off the orientation session with what seemed to him to be a self-evident agenda. "Well, of course, we'll just clone these things," he said to Riggs. "Let's get going!"

By mid-March, when Goeddel arrived at the City of Hope in Duarte, the DNA synthesists had constructed most of the gene fragments for human insulin. Indeed, since the previous summer, when somatostatin was successfully cloned, Keiichi Itakura and his team had steadily created bits and pieces of the insulin gene. They had developed new tricks along the way, speeding up the synthesis procedure in general and improving purification. "We started immediately," recalls Roberto Crea, the organic chemist from Italy, who with Itakura took the lead on the insulin project. "We started working like crazy. And there was no other thing but making it, finishing it. We knew at that time that there was another group doing synthetic insulin. A group in Canada, Narang's group. But they were expecting to finish the project in several years."

Indeed, by the beginning of 1978, several other groups had inched into the insulin picture. Saran Narang, Itakura's former mentor in Canada, had teamed up with Cornell University biologist Ray Wu to attempt essentially the same approach as the Genentech team: synthesize the insulin gene from scratch, then express the hormone in bacteria. Donald Steiner's group at the University of Chicago, too, was known to be interested in cloning insulin by the cDNA method. Then, of course, there was the UCSF group headed by Rutter and Goodman and the Harvard group led by Gilbert. Eli Lilly had an in-house group working on the problem, and at the same time kept an eye on all other contenders. The field was beginning to get crowded.

The mandate to synthesize the insulin gene fell mainly to Crea, a short, stocky chemist with a broad, warm face who had grown up in the southern Italian city of Reggio Calabria. Son of a state railway worker, product of a southern Italian culture notorious for its poor education and limited opportunities, Crea began his studies of chemistry at the University of Messina in Sicily, moved on to do graduate work at the University of Pavia in northern Italy, and finally transferred to the University of Leiden's crack DNA chemistry group under Jacques van Boom in 1973. It was at Leiden that he crossed paths with Herb Heyneker; that connection ultimately paved the way for his migration to California, a place he considered "the Olympus of research." In addition to Crea and Itakura (and to maintain the international flavor of the enterprise), the group of DNA synthesizers included Tadaaki Hirose (Japanese), Adam Kraszewski (Polish), and Leonor Balce-Directo (Philippine-American). As if there were any further proof needed of the universal, Esperanto-like quality of the genetic code, here was a veritable foreign legion of scientists, all products of different mother tongues, all perfectly fluent in the language of DNA. Crea seemed a particularly apt choice for making a gene—his name, in Italian, means "he creates."

Each worked under a large ventilation hood, which sucked away most—but not all—of the noxious vapors created by the chemicals. Leonor Balce-Directo recalls how, each morning at eight-thirty when the five ventilation hoods were simultaneously switched on, "it was like jet engines." Inevitably, there were spills of pungent solvents such as pyridine, an essential ingredient to DNA synthesis but a substance to which the human nose is especially sensitive. Pyridine has a watery chartreuse color and smells like a combination of lime and ammonia; complaints from nearby patients—and finally the hospital administrator—began to rival the spills in frequency. The linoleum floor of the lab was pitted and eroded by chemicals. It was a smelly, hot, noisy, and generally unpleasant laboratory; one of Art Riggs's assistants refused to set foot in the place.

But DNA synthesis required an enormous amount of purification; and purification depended upon those unpleasant reagents. "Believe me, that's how we made holes in the floor," says Crea. "Because we were using cans, *gallons*, of solvents. We didn't even bother purifying the solvents. We were buying these five-gallon cans of solvents and plugged in the machine with big tubing and pumping solvents directly from the can. Imagine! It's a machine that can pump a half-liter of solvent per minute. So if you forget for a minute and you spill it, you have a half-liter of solvent on the floor. Now imagine that every purification step required four or five gallons of solvent. Imagine that laboratory when you did it four, five, six times in a row. . . ." In the rush to get the work done, there were occasional accidents or spills, and sometimes the entire wing of the hospital would smell of pyridine. Once, too, there was a small explosion; no one was hurt, but the glass left embedded in the ceiling reminded everyone that the material could be volatile. Both Crea and Riggs maintain that as unpleasant as the chemical fumes were, the solvents were for the most part non-flammable and posed no health threat either to the workers or to the hospital patients.

Itakura and Crea had streamlined the operation to the point where the construction of a gene, as long as it was reasonably small, became almost a mass-production proposition. During the previous autumn, they had painstakingly assembled a collection, or library, of trimers for the job. Trimers are subunits of DNA containing three letters (TGC, for example) which correspond to a particular amino acid; a string of amino acids, assembled in the order dictated by the bases encoded in the DNA, specifies a particular protein. What the City of Hope researchers did was to figure out in advance what three-letter "words," or codons, they were going to need to spell out human insulin. Once they decided they could handle the job with about thirty-two different codons (a "complete" library, with redundant chemical synonyms, would total sixty-four codons), they proceeded to mass-produce large amounts of these biochemical words.

It would be too technical to describe all the steps involved in producing one single trimer, but, as in the somatostation project, Crea and his team artfully added on phosphate groups, blocked reactions and protected chemical groups, and, after several weeks of arduous and demanding chemistry, managed to produce a few grams of material. After about three months of work, the refrigerator in the lab had a freezer full of glass jars containing faintly pastel powders—light yellows and pinks as well as whites. Each jar, as it were, contained one word, one unit, of the biochemical sentence that would ultimately spell out insulin. When that particular word was needed, the jar could simply be fetched from the fridge. It was not quite as easy as making freeze-dried coffee, for these powders had to be partially thawed and then dissolved in a reagent before use. But this stockpile of codon words afforded state-of-the-art convenience to the City of Hope synthesizers.

To speak of an "insulin gene" was not quite correct; Nature had never managed to produce what the City of Hope synthesizers had cooked up. In the human body, the gene for insulin produces a rather elaborate molecule, known as a precursor, which later matures into active insulin. This precursor, known as preproinsulin, has an initial segment (called a signal sequence) that serves as a mailing address within the pancreatic beta cell, allowing the molecule to thread its way to the proper destination for storage. When the signal sequence is snipped off, the molecule graduates from infancy to adolescence. This adolescent version, called proinsulin and discovered only in 1967 by Donald Steiner, contains a long segment between the A and B chains known as the C chain. The C chain manages to maneuver the other two chains into proper apposition, so that chemical bonds can form between them; it also holds the whole molecular structure in such a way that special joints are exposed, so that when insulin is needed by the body, enzymes come along to snip away the C chain. What remains is mature insulin—the A chain linked to the B chain, suddenly folded into the particular three-dimensional structure that gives it biological activity.

The Genentech team had no intention of imitating nature, and for very good practical reasons. Bacteria didn't know how to *process* insulin—that is, do all that aligning and snipping and girder-building to come up with an active molecule. "It wouldn't work directly," says Dennis Kleid, "which is what

217
. . . . .

everyone originally thought—let's just put the insulin gene in bacteria, and it'll make insulin. Bacteria *don't like* insulin." So in designing their gene, Itakura and Crea did the cutting ahead of time. They dispensed with the so-called "signal" sequence and the connecting C chain, and then constructed two smaller genes, for the A chain and the B chain. Each of these could be cloned separately. Like somatostatin, Genentech's insulin genes were designed with a link of methionine acting as a clasp between two different molecules (here the gods smiled rather benevolently upon the Genentech effort, because there were fifty-one amino acids in the two insulin chains, and not a single one was methionine). To complete this quiet *tour de force* in DNA synthesis, Itakura and Crea went on to incorporate a biochemical buckle in the middle of the long B chain to facilitate the cloning.

These innovative touches by the DNA synthesizers allowed Goeddel to clone the front half of the B chain in one batch of bacteria and the back half in a second, in case there were any mistakes in the DNA chemistry. During the fall, when this strategy was conceived, it seemed like a reasonable precaution; by the following spring, it turned out to have given the team an edge, saving perhaps several weeks of work. And at that point, even one week began to look crucial.

During the fall and winter of 1977–78, the City of Hope group created twenty-nine subunits of the insulin gene. The work was tedious and pressured, in smelly quarters. "It was kind of chaotic," recalls Leonor Balce-Directo, "and then it became more chaotic." When spirits sagged, Roberto Crea would sing old Neapolitan songs in Italian. As a device for venting frustrations, Crea brought in plastic packing material, the kind that comes in clear sheets with large air bubbles. Against the background noises of huge ventilation hoods sucking away the "pea green" fumes all day long, one could sometimes hear, from the laboratory at the end of the hallway, the sharp pop of ruptured bubbles.

Never for a moment were they unaware that speed was of the essence, that other groups were working with the same ultimate aim. Bob Swanson, Genentech's president and resident cheerleader, occasionally dropped in to rally the troops, but as optimistic as he characteristically tended to be, his mere presence was a reminder to everyone that much was at stake. The questions would tumble from his lips: "When? How? How soon?" Workers recall feeling "a sense of mission" and "almost a religious fervor."

"There *was* a certain urgency on his side to come up with some important results for the company," Crea says. "We knew that. But on the other hand, we were never denied technical support and financial support. We had to do it faster and better. So it was like, if we need it, we'll get it from Genentech. That was the word at City of Hope. There was no delay. There was no discussion about saving money here and there. It was just 'Do it. Do it fast. Do it before anybody else.' "

"Anybody else" was a euphemism for the Harvard team. As Crea recalls, "Definitely the name Wally Gilbert was in Swanson's mouth all the time. That we had to beat him. But I think Swanson used that as a management tool to keep pressure on Goeddel, knowing that Goeddel was so competitive."

When Dave Goeddel arrived in mid-March, he immediately began to assemble the gene fragments produced by Itakura's team. Dennis Kleid joined him several weeks later. Their arrival was viewed as nothing less than an invasion by some workers in Art Riggs's lab. Riggs tended to be meticulous, careful, and considerate; for example, after his technician, Louise Shively, became pregnant, he took over any tasks involving radioactive isotopes. The scientists from San Francisco were different. "It was unbelievable," Shively recalls. "It was clear from the first day that they were random and left a mess piled up behind them wherever they worked. It's like they were whirlwinds." Riggs would occasionally remind them to show some manners at the bench, but as Goeddel admits, "We were so concerned with the project that we weren't really listening too much." To everyone's relief, they did most of their work in a small lab across the hall.

Home for Goeddel was one of the trim, single-story, mission-style cottages in a small quadrangle known as the Rosenkranz complex, which had been originally built to accommodate the families of patients staying at the City of Hope hospital. Art Riggs's laboratory was right across "Auditorium Avenue," perhaps fifty feet away. But home was always a relative term for Goeddel: his real home, where he spent virtually all his time and felt most comfortable, was the lab. A work day for Goeddel ranged "anywhere from a twelve-hour minimum to, in some cases, twenty-four hours—working the whole day and into the next."

Work on the A chain went smoothly. Goeddel had that assembled and slipped into one of his cloning vectors by April 7. But something funny was going on with the B chain. As April passed into May, the problem proved to be more than a minor glitch. For some reason, the latter half of the B chain wasn't coming out right after cloning and sequencing. Goeddel repeated the sequencing procedure several times. Still, the sentence came out misspelled. "We said, 'Aw, half this B chain has a mistake,' " Goeddel recalls. " 'What's going on?' So we isolated another clone, did it again, went through it a few times, and then went back and checked the individual fragments, and one of the fragments had been made incorrectly."

A sentence in DNA suffers proportionally more damaged meaning from spelling errors than a sentence in English. The meaning, in biochemical terms, can change dramatically (a single substitution, T for A, changes one of the 574 amino acids in hemoglobin, and that one switch is sufficient to cause the disease sickle-cell anemia). After a bit of grumbling from the DNA synthesizers, they went back to the bench, isolated the mistake, and corrected it.

At just about that time, Herb "Bad News" Heyneker paid a visit to San Francisco. His return to Leiden the previous fall had been uneventful—too uneventful for his tastes, as it turned out. His last days in the United States in the fall of 1977 had been devoted to a triumphant seminar tour describing the somatostatin experiment to researchers in Washington, Cold Spring Harbor, Cambridge, Toronto, and elsewhere. Having tasted scientific life in the eye of the hurricane, in Herb Boyer's lab at UCSF, Heyneker found the Dutch academic setting a backwater by comparison. At the same time, recombinant DNA work in the Netherlands had come under vigorous attack, and the constraints were so entangling that it was difficult to get work done. Heyneker, for instance, wanted to clone and study the penicillin-resistant gene in a bacterium known as *Staphylococcus aureus*, the protagonist in what we call "staph" infections. "But at that time," he recalls, "I needed approval. First from the lab, then from the building. Then from the university, then from the county, and then from the country. You know, there were hundreds of layers of bureaucracy, because it was at the height of this debate on recombinant DNA and the dangers. I mean, I was creating monsters left and right." Heyneker exaggerates, but in an illuminating way, when he adds that he was considered "the most scary guy in the Netherlands."

So the Dutch scientist's ears were wide open and receptive to beguiling offers when, in February 1978, Herb Boyer and Bob Swanson visited him in the Netherlands. Their message was simple: come back.

Now it was Heyneker's turn to face a career dilemma. His family lived in the Netherlands. He had a lifetime appointment at the University of Leiden. His wife was at the time pregnant with their third child. But the lure of the work was too great to pass it up; to hear him describe it, it sounds almost like an addiction. "I felt that the progress we were making, and the excitement—you know, I was sort of *used* to it by that time," he says. "I mean, I couldn't live without it, basically. I saw there was too much going on, and I wanted to be part of that. Badly." His only condition was that any move would have to be delayed until after the baby was born and able to travel, probably by September.

In May, however, Heyneker found himself in the United States for a scientific meeting. Mixing domestic business with scientific pleasure, he headed out to San Francisco with the intention of looking for a house. He got as far as San Francisco, but that was when he bumped into Dave Goeddel and learned about the problems caused by the misspelling in the insulin B chain fragment. "People were up in arms and alarmed" about the mistake, Heyneker recalls. "It was sort of bad news."

Heyneker, of course, had cloned the somatostatin gene. He was intimately familiar with potential pitfalls in synthetic DNA, and it seemed as if he had known Goeddel for no more than a day or two before they were both on a plane headed for the Ontario, California, airport and nearby City of Hope. Swanson and Goeddel had persuaded him to go down. "We spent the Memorial Day weekend in Riggs's lab," Goeddel recalls. "We went down there, picked up the synthetic DNA fragments, and three days later came back to San Francisco with the clones already partially analyzed. He and I worked very well together. It was our first time working together. We were in this little room. Riggs only gave us one tiny bench. We both knew what the next step was, what the next enzyme would be, so when we were pipetting, one guy would be holding out a tube, the other person going in. It was about as efficient as it could be. When one of us got tired, the other one would take over. We'd go down and get a few hours of sleep and then come back up, so we worked around the clock the whole time and kept the experiments going." Heyneker considered it "the most fantastic collaboration . . . we were *extremely* efficient in slamming that gene together."

By the end of the Memorial Day weekend, the mistake had been corrected and the project was back on track. Goeddel popped the DNA in his suitcase and headed back north. The error, in Goeddel's opinion, cost the Genentech team "a month at the max."

Energized by this quick hit of science, Heyneker returned to Leiden. And did he find a house for his family? "No," he says, "that was not very successful, that part." After an embarrassed laugh, he adds, "Look, those were exciting times!"

The Dutch scientist merely reflected an occupational hazard of many biologists: they spent as much time married to their "systems" as to their spouses. Wally Gilbert was once asked how he described his work when he came home for dinner at night and his children asked, "Daddy, what do you do?" He paused for what seemed like an unusually long time, reflecting on the question, then finally said, "I think I probably didn't come home. I think I probably worked nights in the lab."

While Genentech was slamming its human insulin gene together, the Harvard group was rushing to prepare its rat insulin results for publication. They operated against a self-imposed deadline around the beginning of June, for that would still allow inclusion in the August number of the *Proceedings of the National Academy of Sciences*—the publication of choice when one had a hot result. Articles in *PNAS* could be published rather rapidly, with a turnaround time of as little as six to eight weeks. Wally Gilbert, elected to the academy in 1976, could at that time submit up to ten scientific articles a year that in effect would automatically be accepted for publication. The rat insulin paper would be one of them.

Yet no experiment is as neat and perspicacious as the paper describing it makes it sound. Harvard's rat insulin work was no exception. For one thing, all of the tests using Stephanie Broome's assay system had what was known as "high background"—it seemed as though there was enough diffuse radioactivity in the dish to produce a fuzzy, cloudy backdrop to the spots on film. "You could see spots where you shouldn't have," Lydia Villa-Komaroff recalls. "There was a haziness all over." They felt they needed a cleaner picture. The experiments were repeated.

A more pressing problem was that at the beginning of May, when the first "interesting little spots" had turned up, the Gilbert group had still established only that insulin had been expressed by the bacteria. They had not yet proved that the insulin was being secreted—packaged for export, as it were, by the bacteria and shipped outside the cell membrane. Aside from its immense biological interest, this process of secretion—if harnessed for industrial application—might prove to be an efficient way of collecting genetically engineered materials.

In order to prove secretion, they had to establish that insulin molecules had reached the periplasmic space, which is a kind of antechamber between the interior of the bacterium and the outer cell wall. Stephanie Broome provided the answer to this problem by adapting a technique first published in 1965. The idea was to shock it out of them. Not electrically, but with the equivalent of a cold shower—a procedure called osmotic shock.

Anyone with the slightest of anthropomorphic empathies will feel for the bacteria that underwent this treatment. The cells were first bathed in a high sucrose solution, which caused the cell membrane to contract away from the cell wall. "They shrink away from the walls of the periplasmic space," explains Villa-Komaroff. "The cells shrink. And then you hit them with cold water, and they go pfffft! They swell up and push against the wall, and anything that's in the space between the cell membrane and the cell wall gets shoved out into the water." If there were any insulin molecules in what came out in the wash, they too would leave spots on a piece of film. It turned out to be a difficult point to prove. When the Harvard group attempted its first osmotic shocks, around May 21, the debilitated 1776 strain, wimps to the bitter end, literally crumbled under the stress.

They actually began writing the paper while continuing to run osmotic shocks. Gilbert produced a first draft. The junior researchers promptly tore it apart and rebuilt it during a marathon, nonstop thirty-hour session in a first-floor conference room of the Bio Labs building. Villa-Komaroff, Efstratiadis, and Broome all took cracks at the cutting and pasting—of prose this time. When the last bottle of Perrier was finished, the last sentence reworked, there was a paper with eight authors. It began, like many Gilbert papers, with a question. "Can the structural information for the production of a higher cell protein be inserted into a plasmid in such a way as to be expressed in a transformed bacterium?" The answer, as to so many of Gilbert's experimental questions, was yes.

Next, they had to establish the order of authorship—an exercise that might be described as eight characters in search of a first author. The first author, by consensus, has made the greatest contribution to the project; it is also a convention of molecular biology that the last author listed is the senior scientist of the team, the group leader.

Usually, the retrospective view of each individual toward his or her contribution has a thematic richness and diversity worthy of *Rashomon*. It is useful to cite an account out of the nonscientific literature, no less an authority than the *Midnight Hustler,* to see how the Harvard scientists themselves made fun of what was clearly an important and often painful process. It was in the form of an "intercepted" letter from Jane Goodall to Louis Leakey making anthropological observations about "a tribe called Molecular Biologists." The letter observed:

> A central ritual common to this group is called the "Collaboration Dance." It may be performed by as few as two members or as many as 15 participants. When large groups perform the dance a well defined hierarchy exists, sometimes distinguishable from division of labor during the dance, but most often revealed at the close of the dance when a paper is written with the order of authors usually reflecting the hierarchy. Sometimes the order is alphabetical; however, experienced dance members often change their names so as to assure first position on such manuscripts. Occasionally this ordering process is equal in time to the dance itself. It is generally taboo to discuss the ordering process prior to the initiation of the dance. The dance unfolds with a preliminary gathering of participants. There is a ritualistic exchange of refreshments such as tea or coffee. Some groups share cookies or smoke together. During this early process participants pay homage to one another through verbal exchange of compliments regarding past achievements while carefully circling each other gracefully, occasionally making eye contact. The major part of the dance consists of each member performing a task in full view of all participants or privately. Another exchange, this time of small test tubes containing colorless liquids, occurs. To the keen observer the phenomenon we refer to as legitimacy and coercion is practiced with one member (the object of legitimacy) attempting to achieve compliance from another. One dance member attempts to make compliance with his/her wishes more attractive than the consequences of refusing to comply. The final resort I have noted among the Molecular Biologists is "Do what I want or I will kill you" followed by "O.K., kill me." I must close now, Louis.

Aware of these time-honored enmities, the Harvard group gathered at a long table to decide the issue. Gilbert sat at one end, Efstratiadis at the other, with Villa-Komaroff and Broome in the middle. "Let's talk about authorship," Gilbert said.

"Everyone was silent," Efstratiadis recalls. "Wally raised his hand toward me as if to say, 'Speak.' I said, 'I have invested more time than anybody else, but I didn't clone the gene. I think either Lydia or me should be first.'"

Villa-Komaroff, having contemplated beforehand what she would say, swallowed deep and said simply, "I need the paper more than you do." To her great relief, Efstratiadis did not object.

Indeed, after the long dry patch in her silkworm research, Villa-Komaroff needed a scientific byline. Stephanie Broome felt that her assay permitted the key point of the paper—expression—to be established; she, however, was relegated to third place on the paper, followed by Peter Lomedico, Richard Tizard, Stephen Naber, and William Chick. Gilbert's name, as group leader, came last, though its inclusion was not automatic simply because the work took place in his lab, as sometimes happened in other laboratories. Among his graduate students, Gilbert generally earned high marks for putting his name only on papers on which he had direct intellectual input.

While there were eight names on the *PNAS* insulin paper, there were only four names on another important paper submitted almost simultaneously by the Harvard group. It was their application for a U.S. patent on the cloning technique, which was submitted June 8, 1978; the four coinventors were Gilbert, Broome, Villa-Komaroff, and Efstratiadis. The younger Harvard researchers were distinctly blasé about this development, despite the fact that it was the first time any of them had applied for a patent. Efstratiadis found the idea "ridiculous." Gilbert, it seems, was keenest on the patent. "I think of all of us, Wally was the most convinced of the value of this as patented," says Villa-Komaroff. "He was thinking ahead, towards the companies and the impact that all of this was going to have on communication between scientists."

The patent was broad and ambitious: it did not cover simply the making of insulin, but the process by which *any* protein could be made and secreted by cells. All four "inventors" submitted to close questioning by patent attorneys in an effort to ensure that all the key people were listed on the patent, and that no one other than key contributors was included. It was a ritual, legalistic inquisition that many molecular biologists have subsequently endured.

Like many experiments, the expression of rat insulin did not come to a neat, unambiguous conclusion with the writing of the *PNAS* paper. In fact, the insulin team kept adding data up to the last minute. "Stephanie was

225
· · · · · ·

continuing to work on it all the way up to publication date," recalls one member of the Gilbert lab. "She did the osmotic shocks and developed the radioimmunoassay and got the data that said one hundred molecules per cell. All of that went in at the very last second it possibly could. . . ."

Timing was of the essence. The University of Chicago annually conferred honorary degrees on exemplary figures in the sciences; the institution took a particular pride in recognizing worthy recipients *before* the Nobel Prize committee got around to ratifying the same choice. It was Wally Gilbert's turn in 1978; the university planned to award him an honorary degree on June 9. That date may have served as more of a deadline than the *PNAS* date. It hardly seems a coincidence that Harvard's patent application was filed June 8 and the *PNAS* paper submitted June 9. The Chicago trip would provide a perfect opportunity to discuss the recent work in Gilbert's laboratory for the first time.

The speed with which the three teams raced toward solutions to what had seemed like intractable biological problems defied all but the most optimistic predictions. During congressional hearings in November 1977, Ronald Cape, chairman of the Cetus Corporation and one of biotechnology's more experienced and esteemed voices, predicted that insulin production by recombinant DNA technology "is at least ten years away. Commercial rewards are not around the corner." Scarcely half a year after Cape's testimony, all three groups were closing in on laboratory demonstrations of expression. What did the pharmaceutical companies make of this fast-paced academic progress?

Some observers may have been surprised, but not Irving Johnson. He had been telling colleagues at Eli Lilly & Co. that the work could be done in much fewer than ten years, perhaps as few as five. In fact, Lilly instantly appreciated the commercial significance of Genentech's somatostatin experiment. Somatostatin was only slightly smaller than the insulin A and B chains, so if Keiichi Itakura could build genes ever so slightly longer, insulin too would be within reach. Lilly had been in discussions with Genentech, even as the company was in the process of finalizing a research agreement with the Rutter and Goodman labs at UCSF.

Lilly had been synonymous with insulin for decades. Although it was not the kind of company to betray the slightest sign of outward concern, its most venerable product—and the marketing turf it occupied—faced unusual challenges. Lilly's market dominance in the United States was being challenged by a foreign manufacturer, its technical eminence was being challenged by both cloners and chemists, and it was being challenged by Federal Trade Commission investigators for possible antitrust violations in the collection of pancreases.

Within Lilly, there were doubts about human insulin. Could a cloned product be geared up to production levels? Could that process be done economically? There was some reason to doubt the economics of the situation, for supply-house insulin was one of the cheapest proteins available. There were, however, other factors at play besides economics.

There was, for example, the matter of prestige. As the recognized leader in U.S. production, Lilly could hardly be seen to be backward technologically, standing off to the side, while this new, state-of-the-art biochemical wizardry grabbed all the headlines and its aggressive young-turk practitioners invaded the company's most hallowed ground. "Also," adds a source familiar with Lilly's thinking at the time, "it's not that they don't make money on insulin. They do. You get the impression that they're giving it away, but they don't give it away. It still covers a lot of costs and it's not too far down the list in their big sellers." If it couldn't beat out the academic or entrepreneurial labs with its own in-house group, the company could at least be gilded by association with the winner of the race. Of course, no one knew who the winner would be. But companies don't amass net incomes of $277.5 million, as Lilly was on the way to doing in 1978, simply by being lucky or guessing right. Like any good company, Lilly covered its bets. Lilly talked to all three groups. In addition to Lilly's ongoing negotiations with Genentech and UCSF, Irving Johnson had held what he calls "corridor conversations" with Walter Gilbert about the Harvard approach. Their paths frequently crossed, in fact, because both spoke often to lay audiences about the benefits of recombinant DNA work.

The need for these multiple courtships was at least partially accelerated by foreign competition. The world insulin market up to 1977 in many ways retained the shape imposed upon it in 1922 by the University of Toronto when it apportioned rights to insulin production on the basis of the original Banting and Best patent. Lilly received the rights to market insulin in the United States, Central America, and South America. But the University of Toronto empowered the British Medical Research Council to oversee patent rights in Britain, the British Empire, and Europe, which effectively covered the rest of the world market. These rights did not grant long-term exclusivity, but they allowed certain companies to establish early and overwhelming market shares in various countries.

A particularly vigorous insulin industry grew up in Denmark, essentially because of one biologist's scientific curiosity. In the fall of 1922, August Krogh of the University of Copenhagen was in the midst of an American lecture tour, describing the work on capillaries which had won him the 1920 Nobel Prize in medicine/physiology. He kept hearing so much about the recent insulin breakthrough in Toronto that he decided to investigate firsthand. Krogh visited with both Fred Banting and J.J.R. Macleod, and received permission from the University of Toronto to make and sell insulin in Scandinavia. During the winter of 1922–23, Krogh established the nonprofit Nordisk Insulin Laboratory with an associate, H. C. Hagedorn. In the ensuing decades, the Nordisk researchers developed a number of significant refinements in insulin preparation, and these were picked up by the local insulin manufacturers. One of these, located right across the street from Nordisk, was Novo Industri.

Founded in 1925, Novo got its start with insulin, but its reputation later grew primarily on the production of industrial enzymes (the company's development of a brewing enzyme, amyloglucosidase, allowed the recent introduction of low-calorie "lite" beers). The company controlled the European insulin market by the 1970s (with the exception of Hoechst's dominance in Germany), and had also established a reputation for technical innovations. Novo had plans to implement several of its most promising recent insulin improvements in an attempt to break into the American market.

Novo's new products represented significant improvement over standard beef and pork insulin preparations. The company had perfected a method of purification that rendered pork insulin virtually nonallergenic, theoretically almost as good as human insulin. And to take matters one step further, Novo too was rushing to develop a human insulin. The company had originally considered the genetically engineered approach. In early 1978, in fact, company officials traveled to San Francisco and held discussions with Genentech. Unconvinced that the recombinant DNA approach to insulin would work, Novo instead opted for a very unusual chemical approach. Rather than bother with cloning and bacterial fermentation, Novo's biochemists devised a method to snip off the single amino acid in pig insulin that differs from human insulin, and to replace it with the human amino acid. It wasn't cheap, and it didn't address the problem of diminishing supplies of pig pancreases, but it proved an elegant chemical route to the human protein.

Novo's new insulins, human and purified pork, were formidable products. It is significant that most of the scientists working on insulin in California and Boston were not aware of these developments; Lilly was.

From Lilly's vantage point, the insulin landscape—once relatively serene and peaceful—began to show signs of ferment. Insulin was viewed with a certain proprietary affection and protectiveness back in Indianapolis. Recombinant DNA was the wave of the future, and that future included insulin. True, the recombinant DNA approach might never produce tankfuls of insulin and might not even be economically feasible. But could Lilly gamble on that failure this early in the game? There was no way to be sure. Novo's purified pork preparation would appeal to the insulin users who suffered allergic reactions. What was Lilly doing on that front? And since Lilly wanted to expand its insulin market into Europe, the company would similarly need a new, improved insulin to compete with Novo's strong product line.

So Lilly kept close tabs on all the research groups working on insulin. "We intended to keep all bases covered if at all possible," Irving Johnson recalls, "because we wanted to be there first."

Sixteen

"What We're All
Working For . . ."

By the summer of 1978, the question of how to clone human insulin took
second place to a more humdrum logistical question: *where* to do it.

Axel Ullrich had been collecting human pancreatic tumors—the equivalent of William Chick's rat insulinomas—in anticipation of going after the
human insulin gene at UCSF. Just as with rat insulin, one could make a reverse
copy of the messenger RNA (or a "gene") for human insulin and then clone
it in bacteria. But where? The NIH guidelines required that human genetic
material be handled in a P4 facility, the level at which germ-warfare research
was conducted. Practically speaking, there were none in the United States
available to university researchers.

William Rutter, orchestrating the UCSF effort, spent weeks trying to
persuade military authorities to permit use of a special World War II–era P4
lab at the Navy Bio Labs in Alameda, across the bay from San Francisco. After
considering the proposal, the Navy said no. Having been turned down in
Alameda, Rutter then looked overseas for a venue for his group's experiment.

Gilbert, too, was looking across the ocean. Inquiries had been made to gain access to the British military's top-secret research facility in Porton, England, for the Harvard group was preparing its final assault on the human gene. Between corporate interest in the work and personal investment of time and reputation, both research groups were willing to pay the price to find a place anywhere in the world where they could do the work—and, of course, do it first. What made it especially odd is that the major remaining obstacle to what both groups believed would be be a major breakthrough was not scientific, but simply finding the right laboratory.

While everyone else was scrambling all over the globe looking for that laboratory, Genentech's Dennis Kleid and Dave Goeddel puttered up Highway 101 each morning from Menlo Park in Goeddel's yellow Volkswagen Beetle, headed for the new Genentech labs—industry's first commuting cloners. Once again, the prescient decision to use synthetic DNA rather than human genetic material allowed the company to sidestep this most severe class of government regulation. "They didn't have to worry about the guidelines," Axel Ullrich would lament later. "That was the smart move." Genentech's synthetic gene was like a facsimile of a human gene; it looked and worked the same, but wasn't the same. Unlike a cDNA gene, a synthetic DNA gene was likelier to be pure, less likely to be contaminated by other bits of DNA. Furthermore, in a conscious attempt to ensure safety and minimize criticism, the insulin team essentially repeated the strategy used for somatostatin. They constructed a "buried," or fused, protein that would not be biologically active while still in the bacteria.

Finally, by the beginning of June, Genentech had a laboratory of its own. The previous January, Bob Swanson had leased the corner of a warehouse, 8,000 square feet in all, in an industrial park in South San Francisco. It was just north of the airport, a little south of Candlestick Park, on a lick of land called Point San Bruno, a public fishing spot poking out into San Francisco Bay.

Like the computer companies that started out in garages, Genentech conducted leading-edge biology in what was essentially an air freight depot. The architectural structure of the new offices seemed to reflect the fragility of the entire enterprise at that embryonic moment. Numerous guy wires, fastened to the high warehouse roof, held up the walls and ceiling, so many delicate strings forestalling imminent collapse. A sheet of polyethylene formed an impromptu roof over the room where two benches had been set up for the molecular biologists. At first there were no technicians, so Kleid and Goeddel cooked broth for their bacteria and washed their glassware as well as performed experiments; Kleid formed the local biosafety committee and drafted a secretary to fill out the group. Besides the main lab, there was a corner office for Swanson (which looked out across the street at a teddy-bear factory), two adjoining offices for the entire staff, a kitchen, several equipment rooms, and a small P3 facility. The rest of the 130,000 square-foot warehouse was given over to loading docks and freight storage.

Goeddel would swing by Kleid's house at about six in the morning; the insulin project leader would stumble out of his door shoeless, half-dressed, part of his wardrobe still a bundle in his arms, plop down in the front seat, and fall back to sleep. Thus would begin another fourteen-hour day of frontier science.

The problem with molecular biology, though, is that you can't really work fourteen hours straight. Growing bacterial cultures can take six to eight hours; enzymes may require two hours to do their job; incubations may last an hour or two. Biologists try to interweave these tasks, but inevitably they have time to kill as well, and it often gets killed in peculiar ways. This was apparent one day when a Genentech vice-president wandered into the cold room, which is basically a walk-in refrigerator where temperature-sensitive supplies are stored. There on the table was a huge striped bass. Dennis Kleid had caught the fish out back, during one of the endless little competitions with Goeddel that filled the dead time during experiments. Kleid meant to keep it as a refrigerated souvenir of his triumph.

Even when the scientists took a break from work, they never really took a break from their innate competitiveness. At least once a day, when the high tide rolled in and experiments were at an ebb, Kleid and Goeddel would sneak out to Point San Bruno and cast their fishing lines out into San Francisco Bay, vying to hook the biggest fish. The biggest, the fastest, the first—these were the parameters of everything they did, from cloning DNA to catching fish. When they took a walk out back, they would end up playing racquetball against the warehouse wall. When someone turned up with a Nerf basketball, an indoors version of Horse would be sandwiched into every spare moment; Goeddel was notorious for creating and perfecting exotic shots that bounced off the ceiling and caromed off walls before dropping through the net. "If there were five seconds of spare time, there'd be a game invented," Kleid says, "a constant series of competitions. But Dave always won. Dave Goeddel *always* won." The underlying hope at Genentech, of course, was that Goeddel's fierce desire to win would rub off on the real competition: the race to get human insulin first. The feeling, as Kleid recalls, was that "if we could pull this off, if we could make this insulin work and be the first ones to do it, then we would have ourselves a company."

The Genentech scientists viewed both Axel Ullrich at UCSF and Gilbert's group as estimable rivals; it was Gilbert's reputation, however, that seemed to billow and swell in their fearful imaginations like some white-coated Wizard of Oz as they toiled through the summer. Like many distant competitors, his rivalry assumed mythic proportions. He became a management tool. If there hadn't been a Wally Gilbert, in fact, Bob Swanson might have had to invent him for motivational purposes.

No one at Genentech knew for sure where Gilbert's group stood, but even their ignorance did not preclude typical scientific paranoia: they were *sure* he was closing in on the problem. Goeddel describes the anxious uncertainty this way: "It was like each day it was almost a relief when you didn't hear word that someone else made an announcement that they'd finished the work." "Someone else" was really a euphemism for the Harvard group; Roberto Crea recalls that during the summer Goeddel could be heard to say, "We have to beat Gilbert. . . ."

Competition, as it often does, forced the Genentech scientists to modify their plan of attack. They needed results as soon as possible, of course, but they could ill afford any time-consuming mistakes. Be aggressive; play it cautiously. Cut corners; plan back ups. Yin; yang. Goeddel; Kleid. These philosophical oscillations between risk and prudence perfectly suited the temperaments of the two scientists. Goeddel, the straight-ahead player, tried to think up every quick-and-dirty shortcut possible to speed up the work. He would hurriedly yank films out of the darkroom prematurely, attempting to perceive exposure patterns from mere smudges of development. He would cut incubations short by ten minutes; if the reaction still worked, he'd try to trim off more time. Instead of doing overnight preparations, he and Kleid worked through dinner and stayed late. They stole an hour or two here, ten minutes there, and didn't, as Goeddel puts it, "let the time of day interfere with the work."

Kleid, as counterpoint, played the intellectual cynic, the worrier; at each step of the way, he had designed backup experiments and fallback positions, a safety net in case Goeddel's angle of attack turned out to be too steep and the project stalled. "Dave always worked on, more or less, the thing that had to be done *that day,"* Kleid recalls, "and I worked on the backup experiment in case something went wrong." A number of times, they had to go to the backup.

The two, however, basically fended for themselves. Herb Boyer was available for consultations, and sometimes Kleid and Goeddel would go up to UCSF to hash out a problem, but Swanson, the nominal scientific director of the company, really couldn't provide any scientific guidance. That Genentech was not topheavy with big scientific names in the early days, in fact, seems to have played a determining role in its identity. It was, if anything, a company of postdocs.

"It gave us a feeling that if we accomplished this, we would get recognition ourselves," Kleid says. "And every postdoc's dream is to work in a lab filled with postdocs and *no boss!* So you had a critical mass of people to work together, but then nobody took the credit from you for what you did. The postdoc system is really *terrible* that way. They pay you very low and the lab guy gets all the credit."

Kleid freely admits he has problems with authority, and that became almost an institutional attitude during the early days at Genentech. To borrow on the Gilbert group parlance, they didn't want any Big Guy Scientists on the premises. "Because," Kleid explains, "you won't be as creative, you won't look to yourself to solve really tough problems. When the going gets tough, you'll run to somebody and he'll give you a fifteen-minute discourse on it. You ask any scientist for advice and he'll give it to you, whether he has any good advice or not. What you need to do is to depend on yourself to figure out the answers, and if it's a really tough thing, it should be *your* ass on the line to solve it."

For one brief moment, during the second week of June, it seemed as if all that brave scientific self-reliance might have been for naught. There was a bolt of news which sent a quick, devastating chill through the researchers. Someone heard it on the radio, possibly Dan Yansura (a new recruit who had worked alongside Goeddel at the University of Colorado). It was a fragmentary news report out of Chicago. A group of Harvard scientists, the story went, had produced genetically engineered insulin.

Hearts sank in South San Francisco. Gilbert! they thought instantly. It must have been Gilbert and the Harvard group. Had they been beaten out?

Goeddel and Kleid immediately called down to Keiichi Itakura at City of Hope. Itakura, the foreigner, somehow had better contacts through the grapevine. Maybe he knew what the news report was about. As usual, he did. The Harvard group had cloned and expressed insulin, all right.

"But they didn't advertise the fact," Kleid points out now, much more relaxed than on that day in 1978, "that this was *rat.*"

On June 8, 1978, a patent application entitled simply "Protein Synthesis" was filed with the U.S. patent office, with Walter Gilbert, Stephanie A. Broome, Lydia J. Villa-Komaroff, and Argiris A. Efstratiadis listed as coinventors and the assignee specified as the "President and Fellows of Harvard College." One day later, on June 9, Wally Gilbert was in Chicago to accept his honorary degree from the University of Chicago, and during the course of his stay he gave a seminar in which he revealed that the Harvard group had made rat insulin in bacteria. It was the first public disclosure of the work, and a science reporter for the *Chicago Tribune* happened to be there. The group back at Harvard, still working to improve the results on the experiment, knew something was up when Gilbert phoned that afternoon. "Wally called me from Chicago," Efstratiadis remembers, "saying he had presented the work and therefore it was in the public domain, and furthermore there was a journalist in the audience." Gilbert instructed Efstratiadis to call *Boston Globe* science writer Robert Cooke and give him the story. It was on page one of the *Globe* the following day.

Once again, the word "insulin" worked its magic. Press reaction was swift and overwhelming. The Harvard experiment received widespread coverage in newspapers, on the radio, and in the scientific press. The *Wall Street Journal* acknowledged that it was "a major step toward producing human insulin for diabetics by bacteria"; *Chemical and Engineering News,* an authoritative trade magazine, reported the result as if handicapping a race. "Recombinant DNA research advocates have been promising insulin production as one of the practical achievements that the new field might bring," began the account. "And several research groups in the U.S. and Canada are speeding toward that end of making insulin in the test tube with a little help from some microbes. A team of scientists led by Dr. Walter Gilbert of Harvard University has just edged into the lead." The Harvard researchers, however, learned the same hard lessons as the UCSF postdocs had a year earlier. "Everyone was very disappointed by the *Globe* article," Efstratiadis says, "because although I had organized the research, Wally's name was all over the place and our names were at the bottom of an inside page."

All the attention was slow to subside, and the Harvard researchers beat a retreat to the hills of New England, where each summer scientists from a variety of disciplines congregate for a series of informal colloquies known as the Gordon Conferences. There was the sense that momentum was building, that the race for human insulin was heading into the home stretch.

There was also a sense that perhaps all the scientists were taking themselves a bit too seriously. So one night that summer, a group of Bio Labs graduate students and postdocs gathered in the third-floor Tea Room to regress in style.

Electrical power supply units, the ones used to drive DNA through gels, were arranged like amplifiers along one wall; heat lamps became microphones, sonicators became audio equipment. Greg Sutcliffe, in sleeveless T-shirt, wielded a guitar; Karen Talmadge, all punked up, had a cigarette dangling from her lips and paper clips in her ear; Allan Maxam sucked on a bottle of Scotch, and a postdoc named Stephen Stahl scowled fiercely with greasy, slicked-back hair; Debra Peattie was dressed like a counterculture debutante, the bearded Peter Lomedico appeared as a sheik, and Argiris Efstratiadis stood off in one corner with a flat, wide-brimmed hat that made him look a bit like a Spanish dancer. On the blackboard, behind the group, someone had written "Pee 4." And in the middle background, both totemic and incidental, rose a six-foot-high atomic model of a DNA molecule that twisted up from the floor toward the ceiling.

Numerous photographs were taken and developed and compared; two were blown up into posters. One represented the Introns, the other the Exons. The posters were smuggled over the border into New Hampshire, and were exposed one morning in July.

As it turned out, Wally Gilbert was chairman of a session of the Gordon Conference that summer. The topic was DNA sequence organization in eukaryotes. As the scientists filed into the first session on the morning of July 17, however, they were greeted with a poster showing two hoodlumish rock bands and the legend "The two greatest rock bands in molecular biology . . ." and "For the most MIND-BLOWING meeting EVER, bring THE INTRONS and THE EXONS to your next Gordon Conference! Contact our agent at the Harvard Biological Laboratories." This was terribly juvenile and inappropriate at a gathering of serious scientists, of course, but the students were sending a message to their leader. Gilbert, who enjoyed a good prank (this, after all, was the man who'd worn a purple lab coat for years), suffered the humor in it. It was an inside joke, and it pivoted on perhaps the most interesting biological revelation of the period, and one in which Walter Gilbert played a leading theoretical role.

Biologists had for a long time been puzzled by an inexplicable mystery in the packaging of DNA in the chromosomes of humans as well as other organisms. There were something like fifty thousand to one hundred thousand genes, most of them about 1,000 base-pairs in length and some considerably longer, in the total human complement of DNA (a hereditary total known as the genome). Yet the total amount of DNA in each cell nucleus was much larger; it amounted to perhaps 3.5 billion base-pairs. Genetically speaking, it didn't add up. As much as 80 percent of the DNA seemed to be genetic padding. Why all the extra DNA?

A totally unexpected answer began to take shape in the MIT laboratory of Phil Sharp, one of the founding scientists of Biogen. Back in the spring of 1977, a postdoctoral researcher named Susan Berget had been studying the way genetic information is transcribed—that is, the step where DNA is copied into messenger RNA. The biological window which provided Berget with a view of this process was a much-studied virus known as adenovirus, which causes certain respiratory illnesses. Biologists had long accepted as dogma the fact that for each base along the DNA helix a complementary base of messenger RNA would be made, so that one could imagine the molecules as two threads of equal length lying side by side.

Electron microscope views of the two molecules, however, revealed an absolutely astonishing anomaly. One did not see two threads of equal length lying side by side, as expected, but rather one straight threadlike molecule and one longer molecule scrunched up so that nucleic pleats stuck out like loops. Those loops turned out to be the padded parts of the genetic manuscript— segments of the DNA molecule that somehow dropped out and disappeared from the messenger RNA. At this most fundamental point of information transfer, in other words, some significant genetic editing was going on.

The same phenomenon was beginning to turn up in other laboratories. Allan Maxam had seen evidence of it in experiments he was doing that spring, researchers in the NIH laboratory of Philip Leder were busy investigating this curiosity in mouse globin genes, and a group at Cold Spring Harbor was also hot on the trail.

The mechanism described by Sharp and Berget affected almost every gene in higher organisms, including the insulin gene. They called it "RNA splicing," and it seemed to work like this: RNA polymerase II, the enzyme that plops down on DNA and transcribes the genetic instructions for a protein into messenger RNA, was actually a discriminating reader of biochemical texts. It faithfully copied every base of DNA into its complementary RNA base. But before this long strand of messenger RNA left the cell nucleus, a dramatic bit of genetic editing took place. Somehow, the "padded" segments of the gene got spliced out, bringing the functional segments of the gene together. It was this shortened, edited piece of RNA that carried the message out to the cell, and this explained the mysterious pleats when the RNA and DNA strands intertwined, forming those telltale loops that showed up so nicely in electron micrographs. Obeying signals in the DNA sequence, it seemed, some agent in the nucleus actually performed this cutting and pasting operation on the RNA. Hence there were portions of the DNA that were expressed into protein (that is, surviving as messenger RNA) and sections that remained unexpressed (spliced out of the text). Some enzymatic editor in the nucleus had the capacity to glue all the expressed texts together into a coherent RNA message. In other words, Sharp and Berget had discovered that the genes of higher organisms, unlike bacteria, were strung together as a series of interrupted segments, and these segments were ultimately stitched together into RNA.

The next obvious question was why. Gilbert's lab had not done any of the experiments involved with RNA splicing, but it is typical of Gilbert's aggressive scientific demeanor that he leaped forthwith into the theoretical fray. He sent to the British journal *Nature* a speculative letter which attempted to answer the "why" of RNA splicing. The one-page contribution—a famous and much-cited communication titled "Why Genes in Pieces?"— appeared in the February 9, 1978, issue of *Nature*. It immediately created a tremendous stir for reasons ranging from evolutionary theory to scientific etiquette.

Gilbert theorized that this mechanism was nothing less than a powerful and radical way in which evolutionary change could occur. New genetic hands—including potentially advantageous ones to the organism—could be reshuffled when the RNA spliced new segments together. A single mutation at the splice point, Gilbert proposed, could mean that entire blocks of DNA— not just words, but entire paragraphs and chapters of heredity—could at a single stroke be brought into play; the padded portions of the genetic manuscript were "both frozen remnants of history and . . . the sites of future evolution." He argued that this could be a swift and expeditious way of introducing variety into species; natural selection could then pluck out the most successful of these serendipitously dealt genetic hands.

In perhaps the most scientifically innocuous but controversial aspect of his commentary in *Nature,* Gilbert went so far as to propose nomenclature for RNA splicing. The sections of DNA that were transcribed into RNA he called "exons," because they were expressed; the sections of DNA that were spliced out were called "introns," because they were the intragenic regions. In his zeal to participate in the discussion, however, Gilbert appears not to have observed all the professional courtesies; when Phil Sharp was asked some years later if he had advance warning from Gilbert that the terminology for his MIT lab's discovery was about to be established by somebody else, a look of chagrin came across his face. "No," he said. "He saw the opportunity to do it, and I didn't. That's how you have to do it."

To put it mildly, there was a tremendous amount of resistance to the terms "exon" and "intron." During that first summer of 1978 in particular, almost endless discussion and argument swirled around the appeal of Gilbert's terminology. It was this boiling controversy which inspired the elaborate prank by the Gilbert group. But beyond the controversy lay biological questions of great import, and in insulin, Gilbert and his colleagues had a molecule which could provide answers to the puzzle posed by introns and exons. In retrospect, one can even view this as an early but significant juncture for Gilbert the entrepreneur. Would he concentrate on the expression of human insulin with the kind of focus such commercially minded aspirations seemed to require, or would his intellectual fascination with introns and exons prove so great that he would divide his attention and pursue basic research as well? It didn't seem like an either-or question to the biologists at that moment; they immediately began attacks on both fronts.

The Harvard group made the rounds of several different Gordon Conference sessions that summer to describe the findings of their insulin work, but the intron-exon controversy seems nearly as memorable to many as the scientific breakthroughs. At one session, Argiris Efstratiadis acted as democratic impresario, demanding a vote on the issue: did the scientists prefer the Gilbertian "intron" or the Lederish "intervening sequence"? (Efstratiadis jokingly withheld "exon" from a vote on the grounds that it was "capitalistic propaganda.") As an amused Gilbert himself looked on, "intervening sequences" won out over "introns." But for the rest of the meeting, Lydia Villa-Komaroff remembers, "everybody used 'introns' because it was easier."

Even though the Harvard researchers got away from Boston, they couldn't entirely escape aspects of the insulin race. In delivering a talk on the rat insulin work, Efstratiadis found himself following Keiichi Itakura, Genentech's gene synthesizer, to the podium. Here were exemplars of the two main pathways to cloning, the synthetic DNA wizard and the master of the cDNA approach; Efstratiadis described his preferred method as "the old way" and wondered "if reverse transcriptase is as competent as Itakura." He went on to point out that Harvard's method of making rat insulin had so small a yield that "if you make a fermenter run, with gallons and gallons of *E. coli,* you will produce only enough insulin for one dose to cure a *very* small diabetic albino rat." Referring to the biologists who were to follow him to the podium, and scheduled to speak about increasing just such expression yields, Efstratiadis concluded his talk with a theatrical prediction. "So if the next speakers do not solve the problem of overexpression," he said, "Genentech will go brooooookkke."

The rest of the summer passed quickly for Gilbert's insulin group. Arg Efstratiadis had been hired as an assistant professor at Harvard Medical School, and he was in the process of establishing a laboratory there. Lydia Villa-Komaroff had been offered an assistant professorship at the University of Massachusetts Medical Center in Worcester, about thirty miles west of Boston, and was similarly involved in the process of setting up shop there. Gilbert, however, wanted to keep the rat insulin team intact for the run at the human gene. There was no pretense about commercialization: Biogen planned to underwrite the entire venture. He asked Villa-Komaroff, Efstratiadis, and Stephanie Broome if they would participate in the experiment. They all accepted (Villa-Komaroff had to request a leave-of-absence after one month on her new job).

But there was still the problem of where to do the work. Gilbert had entered serious negotiations with administrators at the British army's Microbiological Research Establishment in Porton Down, England, for access to the facility's P4 containment facilities. The request came at a particularly delicate time, for the fate of the institution was being weighed by Britain's Medical Research Council (similar in scope to the NIH). Porton was nominally engaged in some biological warfare research, and there was concern about press and public reaction. The director of the Porton lab, Dr. Robert Harris, however, expressed enthusiasm about the insulin project, and "to give the Gilbert group an air of respectability," according to one Porton official, the insulin cloning project was billed as a collaborative venture between Harris, Gilbert, and Kenneth Murray of the University of Edinburgh, a highly respected molecular biologist who also happened to be on Biogen's scientific board. Gilbert's application went all the way up to the Ministry of Defence. The ministry, after subjecting all the scientists to "negative vetting" security checks, finally issued its approval.

As in the United States, the hypothetical scenario of an insulin-producing microbe loosed in the gut had gained some currency, so there was concern as well that the insulin work might antagonize the public because of safety reasons. Harris approached an outside consultant to assess the risk of the experiment. Finally, the proposal was steered through the Genetic Manipulation Advisory Group, which was England's national equivalent of a local biosafety committee. British officials undertook an immense amount of logistical work on behalf of the Gilbert group, an indication of how much pull Gilbert had in the international biology community.

So Gilbert had scored a coup. While other American researchers were scrambling around to find a P4 facility, the Harvard group had slipped into Porton. They were given exactly four weeks, beginning in September, to complete the experiment.

The month of August was thus filled with last-minute preparations. Anyone wandering into the Tea Room was treated to the unusual spectacle of a portable cloning, hybridization, and radioimmunoassay lab taking shape in three Woolworth's trunks on the floor. Lydia Villa-Komaroff was in charge of organizing the evacuation to England. She made checklists of essential and even marginally useful ingredients for the experiments. Every ingredient in every recipe was prepared and packed, along with the cookbooks. All manner of emergency was foreseen: into the trunks went velvet, India ink, Saran Wrap, and toothpicks; sterile needles, Ziploc bags, X-ray films, and squirt bottles; a yellow magic marker and, yes, a purple magic marker. They took gel plates, they took test tubes, they took petri dishes, they took a battery of enzymes; they packed bacteria, and they packed DNA. They even packed some Cuban cigars, a gift sneaked in by Allan Maxam, which were to be presented to Gilbert upon completion of the successful experiment.

Nothing was left to chance, not even English water. "I had everybody make up buffers that they felt would be critical because we didn't want to take any chances," Villa-Komaroff says. "Different water can often kill you in the enzymatic reactions. We made up everything, from salt solutions on up."

In a repetition of the rat procedure, William Chick had obtained some human insulinomas, and Efstratiadis used these to prepare the cDNA for human insulin in mid-August. Villa-Komaroff had the material ready for cloning by August 22.

There was a *Mission Impossible* air to the whole thing, which is how the *Midnight Hustler* ultimately described it, reporting on the "unassuming, mild-mannered biochemist" who slipped into a phone booth and turned on a tape recorder.

"Hello, Wally," went the story. "[GC] is a seemingly innocuous graduate student in your laboratory. In reality, he is a member of a world-wide conspiracy, bankrolled by a large, oil-producing country, to create international chaos through the insidious induction of the disease syndrome known as diabetes. Your mission, if you should accept it, will be to go to England with your Impossible Mission Team, to work in a top-secret military laboratory and to clone the gene for human insulin. If you fail, the Secretary will deny all knowledge of your actions. This tape will self-destruct in ten seconds. Good luck, Wally."

It was a different television show that preoccupied Genentech's biologists as they moved toward cloning their human insulin. Working alone in the lab late on weekend evenings, the wind whistling over nearby Mt. San Bruno and rattling the freight doors of the vast abandoned warehouse, Dennis Kleid tuned in *Saturday Night Live* to keep him company while cutting and pasting DNA. "If you're here around midnight on a Saturday night, and you're the *only* one here, and you hear these doors rattling all the time, it's *scary*," says Kleid, eyes widening to register the fear. "So I brought my television in. . . . At least I could hear somebody's voice." Night sounds, not recombinant DNA, struck fear into the heart of this biologist.

July turned out to be a pivotal month for Genentech. Kleid and Goeddel kept the experiments going twenty-four hours a day on the two insulin chains, but the work was never fast enough to relieve Bob Swanson's anxieties. Dave Goeddel had already spliced the A chain gene into a plasmid, a customized version of good old pBR322, and successfully cloned it. The longer B chain was proving to be trickier.

This was the gene in which Itakura and Crea had designed a kind of biochemical buckle right in the middle of the B chain gene. By such engineering, they could in effect unbuckle the gene in the middle, creating a front half (confusedly designated "BH") and a back half (designated "BB"). Each could be cloned and sequenced separately, and BH and BB could be joined together later on to form an intact, complete B chain gene.

While Goeddel worked on the B chain, Kleid nurtured his garden of backup experiments. Among these was something called the mini-C chain. This synthetic gene was designed to mimic the B-C-A chain structure of real proinsulin—the B chain DNA continued to read right on into the C chain DNA. Due to variations in genetic syntax, the back half of this B chain—the BC—was slightly different at the end. It did not contain the bacterial signal sequence in BB that said "full stop," but rather a few biochemical letters that said "Continue reading on into the C chain." It was only a difference of a couple of nucleotides, but if the two B chain fragments ever got mixed up, the confusion would be enormous. That, predictably, is exactly what happened.

The mixup may have arisen even as the original mistake in the B chain was being reworked over the Memorial Day weekend. "We ended up cloning both of those down there, bringing the clones back up to San Francisco, sequencing them, and showing everything was fine," Kleid says. "And then we made some big plasmid preps of BB and BC to be used in the subsequent step the next month. I made those plasmid preps, and the possibility is that we got one of them mislabeled or something like that, because we couldn't tell the difference between BB and BC until you put the whole thing together." The difference, unfortunately, was between a grammatically correct sentence (in the biochemical sense) and the genetic equivalent of a run-on sentence.

By mid-July, the entire B chain had been assembled. Paco Bolivar, the biologist who had worked in Herb Boyer's lab, came up from Mexico for a week to help out on the project. Both Kleid and Goeddel were "sick of sequencing," so they drafted Bolivar to sequence the freshly cloned B chain to confirm that it was right. Meanwhile, Bob Swanson was informed that the second of the two insulin chains was in the bag. Swanson, who enjoyed the reputation of being especially encouraging and supportive of his scientific staff, felt this benchmark worthy of celebration. He also wanted his two scientists to meet Thomas Perkins, the venture capitalist who was Genentech's key early backer, so a celebratory dinner was scheduled for the following Saturday evening, July 24.

Kleid believed that the sequencing would be "just a confirmation." Bolivar spent the entire week sequencing the B chain, but everyone expected the results to be pro forma, to set the table for the Saturday celebration. They planned to check the film at the end of the week.

"And that morning," Kleid says, "that Saturday morning, we went into the lab to check out Paco's sequence and do the next step, whatever it was. And we read the gel and the goddam thing was wrong."

It wasn't immediately clear what had happened. In the initial shock, panicked thoughts raced through Kleid's mind. They might get fired. They couldn't tell Swanson. Certainly not that evening. "But the thing I remember saying is that I thought it was a really stupid mistake," he says. "Anybody who knew anything about this field would say, 'You mean you don't even keep track of your DNA pieces? Here are these synthetic pieces and you can't tell them apart. You should have labels on these little tubes, you know.'" He knew. That's what made it so embarrassing.

Although the source of the problem was uncertain, it *was* clear that the evening's celebration, at the boss's invitation, would have a premature and somewhat forced gaiety to it. Mum was the word when Kleid and Goeddel and their wives showed up that evening. Not even Swanson knew that the unexpected error had occurred and that his two molecular biologists were planning to return to the lab after dinner. "All four of us kept quiet and had a great time," Kleid recalls. "Tried to, anyway."

Before going out to dinner, Thomas Perkins escorted his visitors through his house and grounds. It was a striking home, located in the wealthy Marin County suburb of Belvedere, and had been designed by Julia Morgan, the same architect who had built the Hearst Castle at San Simeon in southern California. There were tapestries on the wall and a little bar with a view of the Golden Gate Bridge. There was a scale model of Perkins's yacht—"and this model," Kleid said later, "cost almost as much as the yacht." There was a swimming pool out back and a garden tumbling down the hillside and a greenhouse where Perkins was installing a hot tub. Perkins regaled the two young couples with stories about the companies he owned, about how he had recently attempted to trade in his Ferrari and instead ended up buying another one. "And you know, here we were, living in *apartments*," laughs Kleid.

It was a comment made by Bob Swanson that evening, however, that still stands out in the memory of both biologists. Six months earlier, they had been struggling to raise grant money at a small research institute in Menlo Park. Now, standing outside Perkins's home, they listened as Swanson motioned toward the house and said, "This is what we're all working for."

Kleid and Goeddel returned to the lab that very night. They worked all day Sunday and all day Monday to fix the fouled-up B chain. By Tuesday, the problem had been licked. The engineered bugs were sent down to Art Riggs for the final steps.

How did Eli Lilly view all these developments? Cagily. It is unclear how much the various scientists knew about Lilly's thinking at the time, but there were plenty of reasons for them to be outwardly confused by the signs Lilly gave. Lilly continued to negotiate with Genentech, as it had throughout the summer, but no deal had been signed. At the same time, Lilly was negotiating with UCSF to support research by the Rutter and Goodman groups. Swanson, meanwhile, renewed his efforts to raid the UCSF biochemistry department by again offering jobs to Peter Seeburg, Axel Ullrich, and John Shine. And no one in California knew what Lilly thought about the Harvard initiative.

To all outward appearances, however, Lilly seemed to be backing the UCSF researchers. They needed to do the experiment, as did Gilbert's group, in a P4 facility, and the Navy had refused. Where to go? Eli Lilly came to the rescue.

If the Rutter-Goodman group couldn't do the human insulin cloning experiment in the United States, then Lilly would foot the bill to do the work overseas. If the group couldn't get access to the right kind of laboratory, then Lilly would build them one. Europe was perfect. "We knew that the French regulations were a little more relaxed," Ullrich recalls. "You needed P3 to do it. So Eli Lilly had a subsidiary, a small company close to Strasbourg. It was basically a plant that was making gelatin capules to package drugs, and they built a P3 lab within three weeks. Within *three weeks* they built a P3 lab." In fact, Lilly had briefly considered building a P4 lab back in the United States until it became clear that the expense would be exorbitant.

Lilly, however, actually leaned more toward Genentech. According to Irving Johnson, the synthetic DNA capability of Itakura's lab gave Genentech a unique advantage. The two-chain approach, though a little unwieldy, was the likeliest to be done first, and Lilly wanted very much to be first. At the same time, the investment in UCSF was not misplaced, because Lilly's long-range plan was to use the kind of cDNA cloning of proinsulin and growth hormone being developed by the Rutter and Goodman labs. The secretion system developed by the Harvard group left Lilly unimpressed. "Wally, as I recall, was using an expression system which hooked the gene for whatever desired product you wanted to the gene for penicillinase," says Johnson. "And we just didn't think that would be a very powerful promoter. If I remember rightly, he was suggesting that this would allow secretion to occur as well. And no one, not even today, has a very good secretion system in *E. coli*. So we felt it was probably flawed, from our own knowledge of microbiology."

Genentech's best business move at this point was purely scientific, and Swanson knew it. "I've always been a believer that you should show people it's doable, and if it's valuable to them, the deals are easy." By the beginning of August, the molecular biologists at Genentech had managed to hitch both the A chain and B chain genes to the end of the beta-galactosidase gene and to induce bacteria to make the corresponding fusion proteins. They sent the clones down to the City of Hope and waited a few days. Nothing happened. They sent Dennis Kleid down for a few days. Still, the project wasn't moving along fast enough. People at Genentech still remember the afternoon at the beginning of August when Thomas Perkins roared up to the door in his red Ferrari. "He talked to Bob for a long time," Goeddel recalls, "and right after he left, Bob came in and in effect said to me, 'You're going down to the City of Hope. Don't come back until it's done.' " Goeddel was on a plane the next day, on his way to Duarte. With an open return.

As usual, Goeddel stayed in one of the Rosenkranz cottages, and, as usual, he didn't do much sleeping there. At the City of Hope, they harvested the insulin chains from their clones and snipped the chains away from the molecular dross with cyanogen bromide, just as had been done with somatostatin. So far, everything was proceeding according to the same plan that had worked for somatostatin. In the final step, however, the insulin experiment branched off from the somatostatin protocol. This was the step most laden with unpredictability—the step about which Art Riggs experienced some late-blooming reservations.

The final procedure—conceptually obvious, technically difficult—was to bring the A chain together with the B chain to make a hale and hearty insulin molecule. This process was known as reconstitution, and Nature's way of doing it was extremely elegant. The C chain, which came between the A and B chains in the precursor molecule proinsulin, held the two chains in proper molecular position, so that atomic girders could form to give structural integrity to the combined A and B chains. Once it had performed this function, the C chain was lopped off by enzymes and proinsulin became insulin. This was the mature, three-dimensional molecule that circulated in the bloodstream and controlled carbohydrate metabolism.

Nature knew where to put those girders, but there was nothing natural or routine about tossing the two chains together in a test tube and retrieving insulin out of the soup. Several different configurations of bonding could occur between the A and B chains; only one of them was correct. And Riggs began to hear skeptical grumbling from protein chemists, who believed that the reported yields of such reconstitutions had been exaggerated on the high side. "Over a period of a year or so, we did talk to a lot of people, and we began to worry about it," Riggs says.

The recipe Riggs had chosen for the experiment, more than a year earlier, dated back to the 1960s and had been worked out by a scientist named Panayotis Katsoyannis. The Katsoyannis strategy called for an excess of A chain to drive the reaction, and predicted a 50 to 80 percent success rate in rejoining. "I was naive enough not to have much doubt," says Riggs. "The others in the field had tried to repeat it and had limited success. They didn't get the efficiencies that he had obtained and claimed."

Riggs didn't know that others had tried the technique and doubted its efficiency. No one was willing to guarantee the yields claimed by Katsoyannis. Riggs himself had thought that if there was any potential stumbling block in the project, it would be in the assay, in not being able to prove the presence of insulin in a test of cellular extracts. "And in this business, you know, if you run into any problems, sometimes it can take years to work out," he says.

For the first week or so, Goeddel and Riggs and Robert Crea fiddled with test runs of the joining procedure. Riggs and his technician Louise Shively, meanwhile, prepared the radioactive assays which would allow the researchers to prove that insulin was being made. Crea, having synthesized the DNA, now turned to purifying the cloned insulin chains produced by the bacteria. Goeddel did the cloning and expression, harvested the A and B chains, toyed with the reconstitution reactions.

Purity stood as the final obstacle. To test the joining procedure, Goeddel and Riggs would take human A chain made by their bacteria, for example, and join it to highly pure B chain from bovine insulin (this insulin could be purchased from chemical supply houses). The reactions worked, but they didn't work particularly well. The insulin chains made in *E. coli* weren't totally pure, and minor impurities could disrupt, in a major way, the ability of the two chains to form bridges.

Goeddel and Crea wrestled with this problem in the dog days of August. They would stay up all night, running A chains and B chains through a purifying device known as an HPLC (for high-performance liquid chromatography). A typical HPLC run would take half an hour and produce tiny amounts of insulin chain, between one and five millionths of a gram; the procedure had to be repeated again and again. The odds against success were directly proportional to purity. The dirtier the chains, the lower the yield.

By the fourth week in August, the City of Hope team had worked up a quantity of reasonably pure A chain and partially pure B chain, both of which had been made in *E. coli*. For two weeks, Goeddel had been doing dry runs with store-bought insulins. They didn't always work. Crea and he had to go back and repeat purification steps and try again.

No one needed to remind them of the pressure they were under, although it is unlikely that they realized just how sticky things were getting. Lilly had just concluded a research agreement with UCSF for insulin, and Axel Ullrich was rumored to be heading to Strasbourg, if he wasn't there already, to attempt to clone a human insulin cDNA. There were rumors about Gilbert's group going to Europe as well. Swanson was still trying to arrange a deal with Eli Lilly; Lilly still wouldn't commit itself. "My feeling then," says Goeddel, "was if we didn't do insulin first, we might not have raised any more funding and that might have been the end of this industry for a couple more years, until someone else came along and got something going. I feel that really was the case. A lot of people didn't believe this stuff was going to work."

Late in the evening of August 23, Goeddel figured he had enough pure material to shoot for a real molecule or two of human insulin. The time for dry runs was over. "I had a feeling," Goeddel recalls, "it might work."

Seventeen

"Three Acrobats
and a Magician"

On the evening of Friday, September 1, 1978, at seven forty-five, Walter Gilbert's team of molecular biologists from Harvard boarded a plane at Logan Airport in Boston to fly across the ocean to clone the gene for human insulin. It was an experiment which four months later, due to relaxed governmental regulations and the availability, finally, of the controversial P3 lab at Harvard, could have been performed in Cambridge with a minimum of expense and, to put it mildly, a good deal more convenience. It was the nature of the project, however, that a four-month wait would have seemed an unconscionable delay.

Why had these reputable scientists been obliged to squeeze a state-of-the-art molecular biology laboratory into several Woolworth's trunks and fly to England to do an experiment? One reason had to do with the NIH guidelines governing recombinant DNA research that had been in effect since June 1976. "The attitude that was embodied in the guidelines at that moment," Walter

Gilbert recalls, "was that human material might be *uniquely* dangerous in some unknown, unforeseen fashion. So any work with human material was done in P4 laboratories, and P4 laboratories are designed to contain the most virulent viruses. All the techniques we used [were the ones you'd use] if you *knew* the thing you were working on could be totally lethal."

The other main reason had to do with competition. Human insulin was the ultimate prize in the long scientific journey that began with rat pancreases; and more was at stake than its considerable scientific value and potential medical applications. The dawn of biotechnology brought with it the realization that traditional scientific restraint was at odds with the public relations value of important discoveries made by young and unknown companies like Biogen and Genentech. Biogen, after all, was paying for the trip.

There was another issue associated with the trip, although it remained almost superstitiously unexplicit. It became accidentally clear to Lydia Villa-Komaroff on the flight over. A prodigious reader, she could be counted on to travel with a small cache of science fiction. Gilbert, who hadn't brought along a book, asked if she had anything to read. Villa-Komaroff reached into her bag and produced a volume called *Nobel Lectures in Molecular Biology,* and Gilbert practically recoiled from the offering. "He said something like 'It wouldn't do for me to be seen reading *that,*'" she remembers.

The fact of the matter was that Gilbert had been considered a likely Nobel Prize candidate for years, certainly for the work on the repressor, more recently for the rapid sequencing technique. He had collected many of the signal honors leading up to it: election to the National Academy of Sciences in 1976, the Freedman Award of the New York Academy of Sciences in 1977, the honorary degree from the University of Chicago in the summer of 1978, with more soon to follow. The *Midnight Hustler* made frequent teasing allusions to Stockholm, and one year the lab group presented Gilbert with "The Nobel Prize Game," complete with board and instructions. In the view of some (but not all) of the students in the lab, the successful cloning of human insulin could be the kind of project that pushed Gilbert over the top. "Whether he thought this would do it for him or not," says one former graduate student, "at least it was one more notch in the gun. And something to keep him in the public eye."

At first, the fact that Gilbert's group had gained access to the top-secret British facility at Porton appeared to confer a spectacular advantage. As a biological warfare research facility, the Microbiological Research Establish-ment (or MRE) clearly was not the type of place where foreigners, to say nothing of nonmilitary British scientists, cavorted freely. Yet the Biogen-sponsored group appears to have slipped in just at a time when the British scientific community was still puzzling its way through the public furor over recombinant DNA work. The MRE "was a typically stuffy and conservative British research institute," according to Peter J. Greenaway, current head of the molecular genetics lab of the now-public laboratory at Porton and at the time specially appointed liaison with the American group. "It was steeped in tradition and had not yet woken up to the realization that a revolution was happening in the biological sciences. It also had a very close association with the army, so life was very formal. It was against this backdrop that four obviously very with-it Americans arrived to take the place by storm."

It is unlikely that the Harvard group had pondered all the ramifications of the Porton arrangements—the military milieu, the security clearances, the unusually strict procedures demanded in a P4 ("Category IV," in British parlance) facility—as they winged their way over the Atlantic that Friday evening. The mood of the group was relaxed and expectant. After obtaining expression of rat insulin in the spring, as Lydia Villa-Komaroff recalls, the team felt it "had a real shot at getting the human sequence then, getting it first."

Wally Gilbert was expansive and optimistic; he seemed to enjoy the sheer swashbuckling adventure of the trip. His young graduate student Stephanie Broome, certainly the shyest of the four, found the prospect fun, but admits that "perhaps I didn't feel I had as much at stake in the whole thing as those guys." Efstratiadis was excited by the prospect of getting his hands on the human gene for scientific reasons. Villa-Komaroff organized, intervened, mediated, entertained; she liked to sing, and often invented the lyrics.

A rental car awaited the group at Heathrow Airport in London. After loading it up with their suitcases and paraphernalia, and after cleaning up the mess left by a shattered bottle of sodium chloride solution, they drove some eighty miles from the airport to Salisbury Plain, where the Microbiological Research Establishment in Porton receded fairly anonymously into a land-scape more noted for the ancient circle of Stonehenge and the towering cathedral at Salisbury. In the footsteps of druids and men of the cloth, at a discreet historical distance behind superstition and religion, came these new high priests of science: the cloners.

"I remember being enormously impressed by their arrival . . ." Peter Greenaway would remark later. "They jumped out of the car, introduced themselves, and then began pulling out what appeared to be endless numbers of suitcases and trunks, all of which contained reagents necessary to do the work that they had planned." The Harvard researchers remember being impressed, too, but for entirely different reasons.

On their first day at Porton, the laboratory was closed (work was not conducted on weekends), so they could do little more than unpack their scientific gear and set it up in the appropriate spot. But all four were issued little pink identity cards, to be carried at all times, which—in impeccably cordial, restrained, and unalarmist tones—warned physicians to treat the bearer with caution and care: "In cases of P.U.O. [fever of undetermined origin], especially when associated with respiratory or CNS [central nervous system] symptomatology, the possibility of an accidental laboratory infection cannot be excluded and you are asked to contact as soon as possible . . ." As if that did not give sufficient pause, each of the scientists was asked by Greenaway to sign a "code of practice," which outlined work procedures in the containment facility; one of the conditions imposed by British authorities, Greenaway points out, was that "any breach of MRE's safety code for the facility should immediately lead to their becoming *persona non grata.*"

Because sensitive military work was going on at the laboratory, Greenaway was required to accompany the Harvard researchers at all times on the premises, and although this role seems to have begun as a cordial chaperoning, it did not take long before it warmed into friendly comradery. He was a tall, thin, personable fellow who immediately won flattering reviews from his American guests for everything from courtesy and scientific knowledge to skills of diplomacy. "He didn't watch the clock and he didn't wear a tie," Villa-Komaroff recalls approvingly, as if those were the most telltale traits of a good biologist. It was not merely that he "baby-sat" the Americans, as Efstratiadis put it; he also proved to be a helpful and unflappable ambassador who moved easily between the two very different scientific cultures. Just how different was revealed on a daily, if not hourly, basis. "Work for normal people at Porton started at eight-forty and finished at five P.M.," Greenaway dryly observes.

"The fact that the American group wanted to work after hours was considered really strange."

Time was tight, however; and the Gilbert group had a human insulin gene they wanted to plug into *E. coli*. The members of "Operation Lollipop"—as Villa-Komaroff unofficially christened the enterprise—knew exactly what they wanted to do and immediately launched into the work. On the Sunday after their arrival, according to an informal ledger kept by Villa-Komaroff, the group prepared an overnight culture of HB101—the Herb Boyer strain of *E. coli*. The following morning, on September 4, they made their first attempt at transformation. Villa-Komaroff used one-tenth of her cloning material—pBR322 plasmid into which Efstratiadis's insulin gene had been spliced—and smeared the transformed bacteria out on plates. On Tuesday, she performed a second massive transformation of *E. coli* using another tenth of the DNA. Then they waited. The nice thing about bacteria was that you didn't have to wait long.

Unfortunately, bureaucratic and logistical interruptions constantly intruded. For example, on their first full day at the lab, they were obliged to get fitted for gas masks, which in Villa-Komaroff's opinion may have been the most harrowing aspect of the work—certainly more frightening than exposure to any recombinant organisms.

"This little man gave us a gray raincoat, sort of gave us a look, and then found a mask that he *thought* would fit," she recalls. "Then he waved an ammonium salt around, asked if we smelled anything. Everybody said no. So then he said, 'Well, shake your head and nod up and down.' We did that. And then he put us all in a little room, along with a guy from Porton who was not sure his mask was intact and wanted to check it. So there we were, standing there with these gas masks and these long gray British raincoats, in this glass room, and the man drops these two pellets into this little cup and this *gas* rises!

"He's having us nod our heads up and down and side to side, and we're standing there taking deep, huffing breaths. The guy who was checking his mask—it turned out his was *indeed* not intact. He started gasping and choking and scratching at the door, and they let him out. When we got out ten minutes later, the guy was still sitting there, tears streaming from his eyes." The test substance, of course, had been tear gas. The scientists did not have to wear the gas masks continuously, but they did have to put them on during such routine bench operations as running the centrifuges or transferring material from test tube to test tube with the use of pipettes. The fear was that either operation could create "aerosols," invisible droplets of material that could be accidentally inhaled or ingested by a researcher. The precaution seemed excessive for work involving insulin and *E. coli*, but such standard P4 protocols were mandatory for experiments involving other research pathogens at Porton, which at the time included the deadly hemorrhagic fever viruses (Lassa fever, for one) and the botulism bacterium, *Clostridium botulinum*.

Then there was the lab. Considering the general casualness of today's molecular biology laboratory, and given the relative freedom with which Genentech's tennis-shoed, T-shirted researchers bounced through their insulin work in 1978, the extraordinarily restrictive conditions at Porton made the Harvard team a potato-sack entry in the insulin race. Indeed, the futility of the entire regulatory picture could be seen in pathetic perspective from Porton's P4 lab—while Genentech's scientists worked with their synthetic human material under P2 conditions and Axel Ullrich was cloning the human gene in a P3 lab in France, the Harvard researchers were wearing gas masks in England. In his inimitably straightforward fashion, Argiris Efstratiadis says, "The working conditions were the pits, okay?"

Merely *entering* the P4 lab was an ordeal. After removing all clothing, each researcher donned government-issue white boxer shorts, black rubber boots, blue pajamalike garments, a tan hospital-style gown open in the back, two pairs of gloves, and a blue plastic hat resembling a shower cap. Everything then passed through a quick formaldehyde wash. Everything. All the gear, all the bottles, all the glassware, all the equipment. All the scientific recipes, written down on paper, had to pass through the wash; so the researchers slipped the instructions, one sheet at a time, inside plastic Ziploc bags, hoping that formaldehyde would not leak in and turn the paper into a brown, crinkly, parchmentlike mess. Any document exposed to lab air would ultimately have to be destroyed, so the Harvard group could not even bring in their lab notebooks to make entries. After stepping through a basin of formaldehyde, the workers descended a short flight of steps into the P4 lab itself. The same hygienic rigmarole, including a shower, had to be repeated whenever anyone left the lab.

At the bottom of the steps, through a door opening to the right, was the warm room, which maintained a steady 37°C (98.6°F) temperature. Here bacteria replicated and the biologists dozed. To the left was the main labora-tory area. Under a bank of windows stood three glove boxes—sealed boxes into which an experimenter could stick his or her hands to handle materials, pour reagents together, pipette material, and perform other operations deemed potentially dangerous in the open air. After each day's work, the glove boxes too were flushed out with formaldehyde.

At the far end of the room stood a metal table and metal countertops (the easier to wipe up spills), three huge autoclaves, and a door that led to the kitchen, where media were prepared and glassware cleaned. The auto-claves, where everything from utensils to growth media is "cooked" under pressure to ensure sterilization, swallowed all manner of jetsam in the Porton lab: not just the usual lab refuse, but also waste water from the showers that each researcher had to take before leaving the lab and even paperback books, from science fiction to *All Things Bright and Beautiful,* that the Harvard workers read in the laboratory to kill time between experiments. Into the autoclave they would go, and then, sterile and sopped and ruined, into the trash.

As if the physical containment procedures were not sufficiently trouble-some, the highly routinized way of doing things at Porton came as a rude shock to researchers accustomed to the twenty-four-hour-a-day, open-access, come-as-you-are atmosphere back at Harvard. "The people were pretty friendly, to the degree they could be friendly," Efstratiadis says. His reaction was partially conditioned by the epic journey he took one day to procure a rotor, which looks somewhat like an imploded bowling ball but is in fact the circular object with holes that holds test tubes in a centrifuge.

"They were so funny, those British people," Efstratiadis recalls. "I mean, they said person X had a rotor for the centrifuge, so I walked a *mile*— it's an *enormous* building, okay?—to find that, and I said, 'Can I take your rotor?'

"The guy said, 'What time do you want it?' I said, 'At five o'clock in the afternoon.' And he said, 'Fine. Sign for the rotor.'

"And so I did, and I started going, and he said, 'Hey! Where are you going? You didn't sign for the key!' I said, '*What* key?' He said, 'The key that opens the locker where the rotor is. But then you have to sign for *another* key that opens the *door* to go to where the locker is.' They had three keys or something, [and] you had to sign for each and every one of them." In the midst of this military milieu, it is difficult to imagine four less militarily inclined individuals than the Harvard researchers.

Even Wally Gilbert, he of Olympian scientific repute, found himself relegated to the role of bottle washer and microbial saucier. Porton provided no kitchen staff to prepare culture plates; in fact, the Harvard group did not even have access to kitchen-scale autoclaves for such routine operations as preparing growth media. As Villa-Komaroff recalls, "Wally ended up making the plates. We couldn't use the autoclave because it was very hierarchical. Only autoclave people could use the autoclaves, and they turned them off at five o'clock. So he would make media in hundred-milliliter bottles in a pressure cooker. Usually you make media by the liter, three liters at a whack. So he'd make one hundred milliliters [a bit more than three fluid ounces] at a time. One hundred milliliters makes three plates. We were going through *lots* of plates. He was," Villa-Komaroff adds, "very good about that."

The work began well. They looked at the plates from Monday's cloning run and discovered a number of colonies growing up through the antibiotics in the media. A good sign: the insulin-bearing plasmid had made its way into some of the bugs. A sense of excitement began to build. Within a day or two, the positive clones could be grown into colonies and tested by hybridizing probes designed to zero in on insulin. That would nail down one step: cloning of the human gene. Positive colonies from that procedure would then be turned over to Efstratiadis for analysis and to Broome for a check to see if any of the bugs were reading that gene and making human insulin. The analysis would take a few days, but on that Tuesday, September 5, Lydia Villa-Komaroff made a one-word entry in her ledger: "Whoopee." Things were going exactly according to plan.

Despite, it should be added, continuing logistical problems. On the first full day, the Americans made a disastrous excursion to the officers' mess hall, which was open only for an hour during midday—not the kind of flexible schedule preferred by scientists working with bugs on a microbiological, not military, schedule. Male patrons were obliged to wear ties. Efstratiadis claims to have owned only one tie at the time, black, to wear to funerals (which, he says, "I didn't care to bring with me"). But Gilbert, with characteristic color panache, had prepared for such a contingency by packing two ties, one green and the other bright orange. Both Gilbert and Efstratiadis shared a certain disdain for formality in convention and dress. (One Harvard graduate student tells an amusing story about the time Gilbert went to an exclusive, highly formal restaurant in Boston with a purple tie tucked in his pocket; sure enough, the restaurant pointed out that ties were required, and so Gilbert took great pleasure putting on this loud purple cravat over his customary orange turtleneck. "And really," he is reported to have said, "it only made *them* look silly that I had to wear this tie.")

The Harvard group did not realize, however, that only recently had women been allowed access to the officers' mess, and only if properly attired in dress or skirt. When Villa-Komaroff and Broome strode in, wearing the standard-issue slacks and shirts of biology's front-line troops, and the unshaven Efstratiadis bounced in on sneakers, it was the Boston Tea Party all over again. The place experienced the kind of silent uproar only the British can manage. "The presence of distinguished American guests in flagrant violation of mess rules caused obvious embarrassment to the gentlemen scientists reclining in their comfy armchairs for a quick nap after dinner," observes Peter Greenaway. "I believe I am right in saying that this was the only time the American group went to the officers' mess."

Next, they tried picnics. When the morning work seemed fairly well along and a natural break could be anticipated, one of the scientists would shower out of the lab and drive back to the Pheasant Inn in Salisbury, where they all stayed, to pick up a prepared picnic lunch. But this seemingly innocuous plan, too, ran afoul of the authorities. "The picnics were initially taken on the grass directly in front of the building," Greenaway recalls. "However, complaints were made to the director concerning the bad image that this presented, particularly when we were drinking beer directly from the bottles." They were tactfully asked to remove themselves from the front of the main building and to picnic behind the cycle sheds.

Food presented a different kind of problem in the evening. The Harvard group characteristically worked well beyond typical closing hours at Porton, often staying until ten o'clock in the evening. By that hour, however, most of the eating establishments in provincial Salisbury were dark and shuttered. After one of the first late nights in the lab, they drove all through the city and finally happened upon an open restaurant called La Gondola. It was run by Italians, and it maintained Italian, as opposed to British, hours. The scientists gratefully invaded what was to become their regular culinary hangout in England.

The arrival of four odd-looking foreigners at so late an hour, of course, provoked curiosity on the part of the proprietors that first night. One of them asked the group if they worked with the circus currently performing in Salisbury.

"He's a magician," confirmed Efstratiadis, pointing to Gilbert, speaking in his deep resonant voice, "and we're all acrobats."

This brief exchange was the inspiration for a song composed by Villa-Komaroff that became an unintentionally ironic anthem for the entire trip. It started out like this:

> *Three acrobats and a magician*
> *Went off on a dangerous mission*
> *To clone and express the insulin gene*
> *And thereby to thwart Genentech's schemes . . .*

Gilbert, forever apostrophized by colleagues and rivals alike, was the magician; Broome, Efstratiadis, and Villa-Komaroff were the acrobats. The rest of the words, naturally, would depend on how well they pulled off this scientific high-wire act in England.

The magic held for a few more days. On Wednesday, September 6, the group returned to the laboratory to find that the second batch of transformed bugs had somehow been contaminated. But the first set of clones looked good. Now they were ready for further analysis using a method with the intimidating name "Grunstein-Hogness hybridization." It would show which, if any, of the clones contained insulin DNA.

At this point, the Harvard researchers knew only that their pBR322 had made its way into some cells. To find out if any of the bugs contained human insulin, they had brought along some radioactive rat insulin DNA, from the famous pI19 plasmid, as a probe. This hot rat DNA was mixed in with the DNA of Villa-Komaroff's clones. Just as in the rat insulin experiment the previous spring, filters dotted with DNA from each of the Porton clones were bathed with the radioactive probe for a time, then removed, placed against X-ray film, and left overnight in a freezer for exposure.

At Harvard, this would mean a few short steps into the third-floor darkroom. At Porton, it meant that Efstratiadis turned these filters over to Greenaway, who climbed into special protective garments—a "space suit," the

Harvard people called it—and took the filters, sealed in plastic bags, out of the P4 lab. The films had to be exposed at $-80°C$, and the only freezers at Porton that got that cold were in the animal wing. That wing was off-limits to all visitors, and also to any employee who did not have the requisite immunizations, from bubonic plague to Venezuelan equine encephalitis. If any of the colonies lit up, the group would have a very strong signal that they had cloned human insulin. The film lay dormant overnight. While the biologists slept, electrons emitted by the decaying radioactive probe, clinging like glue to any similar cloned DNA, fired out from the filter paper, striking and toppling silver halide crystals in the film. If enough electrons hit the same area of the film, they would again produce interesting little spots. By Thursday morning, they would have an answer.

The next morning, the researchers arose as usual around seven o'clock in the morning; Lydia Villa-Komaroff, the only one of the four who had remembered to bring an alarm clock, would knock on the doors to wake up the other three. They straggled downstairs and enjoyed a typically immense British breakfast of eggs and bacon at the Pheasant Inn, then were off on the short drive to Porton. By eight o'clock they were in the lab; a short time later, they were poring over the films Peter Greenaway had retrieved from the darkroom.

Even the most cursory glance was exhilarating: there were dark, telltale spots on the films. Thirty-two colonies in all lit up during the assay. It had to be insulin. *Human* insulin.

Villa-Komaroff immediately set out to grow overnight cultures of the chosen thirty-two. More tests would need to be run, more characterization undertaken, more verification. But, as Greenaway later observed, "the isolation of clones that hybridized with the rat insulin clone was certainly greeted with excitement and celebration." Lydia Villa-Komaroff employed a red pencil, in fact, to highlight the event in her ledger. "Red Letter Day," she wrote. "Good spots."

Plans were made for an official celebration that evening at dinner. Even as preparations were underway to celebrate, Arg Efstratiadis anxiously awaited a sufficiently ample amount of DNA to analyze more closely. Two obvious steps remained. Once Villa-Komaroff's bacterial colonies grew sufficiently large, Efstratiadis could crack open the cells, excise the insulin insert, and run some preliminary "fingerprinting" tests on the DNA. Meanwhile, Stephanie

Broome could scan those same colonies, checking to see if any of the microbes were making insulin.

Euphoria undeniably outdistanced caution that Thursday. Allan Maxam, back in the Harvard Bio Labs, received a call from the Gilbert party reporting the news; like a lot of scientific gossip, before long it had made the rounds of the Bio Labs in Cambridge. Back in England, the group brimmed with optimism. Broome had yet to run one of her assays, but everyone was hopeful. It had worked the first time with rat insulin. Why not with human? Peter Greenaway was dragged along to the celebratory dinner; and at the end of the meal, Arg Efstratiadis produced one of the Cuban cigars which Allan Maxam had given him and presented it to Wally. With Churchillian gusto, Wally Gilbert lit up a cigar.

"But then," as Efstratiadis puts it, "there was no cigar."

Even as the Harvard scientists savored their apparent success over dinner in England that Thursday evening, the scientific community in the United States was digesting the first news reports out of Duarte, California, where an important press conference had been held the day before, on September 6, 1978. Scientists at the City of Hope and at a small, little-known company called Genentech had successfully created human insulin with the help of bacteria. It made the network news shows Wednesday night, and by Thursday morning it was front-page news in national newspapers.

Word of Genentech's triumph, telephoned to the Harvard group by Tony Komaroff, was bad enough; the late-breaking news coming out of the P4 in Porton was suddenly even more distressing. First, on Friday, Villa-Komaroff discovered she had inadvertently mixed too much tetracycline into the growth media while preparing individual cultures of the thirty-two promising clones. All the bugs had been wiped out by the antibiotic; a whole new set of cultures had to be grown up overnight. Broome, meanwhile, ran her first expression assay; nothing turned up. The Friday results were negative, but not ominous.

On Saturday and Sunday, they began to look ominous. Efstratiadis chose to take a close look at twelve of the promising colonies. He had extracted the DNA from these clones, each of which had given a "good spot," and then used a restriction enzyme to snip out the inserted DNA. He then prepared to run a gel with each clone's DNA insert to make sure they were roughly the same size.

Everyone expected the insert to be the human insulin gene, and if it was like any other experiment of its kind at the time, the little DNA fragments Efstratiadis cut out would not be absolutely uniform, but rather would show some disparity in size. Cloning produced this kind of effect: some of the inserts, for a variety of reasons, came out a little longer, others a little shorter. When they were run side by side down a gel, they didn't all stop at the same spot, so they would leave a pattern like this:

When Efstratiadis ran this DNA on a gel that Sunday, the group had its first whiff of fiasco, because the pattern looked like this:

"The inserts of those positive clones started coming out exactly the same size, which was fishy," Efstratiadis says. Indeed, all the different cloned inserts left the same sharp, clean, well-defined band in each lane of Efstratiadis's gels. Not at all what one would expect. It became apparent that, as Greenaway puts it, "there was something very badly wrong with the clones isolated."

The euphoric mood dissolved. Efstratiadis worked feverishly to resolve the mystery. A molecular detective sorting through nearly invisible clues, he ultimately solved a mystery that spelled out the group's own experimental demise. After staying late one night, Efstratiadis reached an inescapable and devastating conclusion. Somehow, some way, the plasmid the Harvard group had plugged into their bacteria had been contaminated. Worst of all, the contaminating agent appeared to be the *rat* insulin gene. In effect, the group had spent tens of thousands of dollars and flown thousands of miles across the ocean, doffed their clothes and donned their gas masks, to repeat by a terrible accident the same experiment they had done by design in Massachusetts earlier that year.

"The contamination," Greenaway recalls, "was discovered within days of the news that Genentech had already cloned the human insulin gene. Wally was most philosophical about it, claiming that first was not always best."

How the contamination occurred was "the subject of much debate," to cite Peter Greenaway's understated characterization, in the ensuing discussions. In all likelihood, an infinitesimally small scrap of rat insulin DNA may have clung to a plate of glass used during gel electrophoresis. DNA sticks as stubbornly to glass as egg yolk to dirty dishes. An improperly cleaned gel plate can invisibly harbor such contaminants, and that's all it takes to foul up weeks and sometimes months of work.

Such insidious contaminants have since become recognized as gremlins that can inhabit any laboratory. Nadia Rosenthal, one of Efstratiadis's graduate students at Harvard Medical School, later coined a word to explain the phenomenon. Playing off the term "transposition," which refers to a jumpy patch of DNA (or "transposon") that hops from place to place within a chromosome, Rosenthal invented the term "transtubation." It denoted an element (or "transtubon") that hops from one test tube to another. It is not a naturally occurring phenomenon, of course, but rather gallows-humor jargon suggesting that anything less than the most rigorous of lab technique can result in experimental failure.

Almost every molecular biologist is familiar with the effects of "transtubation" these days. In the fall of 1978, however, in Porton, England, when cloning was still an experimental art and the experiment at hand was so highly competitive and pressured, it was a novel, unwelcome, and shattering development. "We were all very depressed," says Efstratiadis. "And at a certain point, actually, we started becoming rather hostile to each other."

The main issue, of course, was the contamination—or, in Villa-Komaroff's words, "Who did it? When did it get in there?" Neither Broome nor Gilbert was culpable. It was something that turned up in the cloning, which pointed to either Villa-Komaroff or Efstratiadis. The two scientists, not unexpectedly, inquired as to each other's possible responsibility for the mishap.

Villa-Komaroff does not remember an argument, but she does recall thinking, "It must have been during the cloning, the synthesis of the double-stranded DNA, because *you never put your pipettes away, you're always using more than one of them.'* Arg probably thought I put it in when it got tailed, because my terminal transferase was contaminated. . . . At the time, he thought it was a horrendous mistake and he certainly didn't want to have been responsible for it," she continues. "He wasn't that sloppy. And the only other person would have been me. And I was probably pissed at him for being so obstreperous, because I *knew* how sloppy he *could* be, as well as how sloppy *I* could be. But I don't remember the specifics of an argument." Villa-Komaroff and Efstratiadis had had their share of arguments over the years, and this history of intimate disagreement helped them in this moment of great despair. "We *knew*, at least the two of us, when to yell and when not to yell, and when we were pushing too far."

Gilbert—the fire-breathing taskmaster, according to legend, the take-no-prisoners biologist, to whom in matters scientific you were either in first place or no place—the supposedly ferocious Wally Gilbert played peacemaker. While his reputation was a bit on the line, not to mention the embryonic aspirations of Biogen, he kept his temper, and those of his colleagues, on an even keel. "Wally was the only one of the four of us who never lost it there," says Villa-Komaroff. "At different times, Stephanie and Arg and I just lost it. Got either very angry or very tired or both, and just had a fit. And Wally was the only one who never did. . . . He was the mediator of that trip. And one reason we could all get along was that when things got too tense, he would mediate." As hostilities festered, Efstratiadis recalls, "We sat down one day like adults and said, 'This is nonsense. I mean, it doesn't matter whose fault it is—if it's anybody's fault. This is life, and we have to bite the bullet and face it.' And everybody relaxed, I think."

Relaxed emotionally, perhaps, but not professionally. In the short time remaining, the four researchers dove back into the wreck and tried desperately to salvage the experiment anyway. Time and Category IV conditions were the ultimate enemies.

They had to start from scratch, which meant they had to prepare plasmids and do other steps that had previously been performed in the relatively unrestrained environment of Cambridge. At Porton, the most trivial task—a trip to the centrifuge, the preparation of media—became an exceedingly complex operation, full of unexpected pitfalls. To prepare plasmids, for example, one normally placed the DNA in a heavy solution known as a gradient, such as cesium chloride, and ran it through a centrifuge; the plasmid DNA would migrate to a specific band in the gradient and could be collected. Routine at Harvard, this simple procedure became disastrously complicated in England.

"At Porton the ultracentrifuges are down the hall and outside the P4," says Villa-Komaroff. "So we had to put stuff in the gradient, wrap the rotor in plastic, and dip it through the formaldehyde. Somebody had to go get undressed and dressed and showered and all of that. Then you'd go to the ultracentrifuge, which was surrounded by a sealed container, and you would pass the rotor in through another formaldehyde bath and out onto a table and unwrap it and put it in the centrifuge. . . . Which would have been okay," she continues, "except that you had to repeat that coming back, and we just couldn't hold it steady enough." Because of the jostling, in other words, the DNA band sloshed around and was no longer concentrated in one spot. "We never got good yields. We finally started making big preps on the gels, which we could do in the P4. And with the hybridizations—we were screening with the rat clone. And we just never got anything convincing, and the month ran out."

On September 28, Gilbert, Broome, and Villa-Komaroff returned to Boston (Efstratiadis had left a day earlier). For all their effort, they headed homeward with a tin can sealed on the premises of Porton containing the contaminated DNA. In a "To Whom It May Concern" note accompanying the tins and intended for customs officials, Peter Greenaway wrote, "These samples are for research purposes only and have no commercial value." For Biogen in particular, that sentence was depressingly unequivocal.

The exhausted researchers returned to Boston. It was around the time of year when Nobel Prizes were announced, and there was a feature in the latest issue of the *Midnight Hustler* that took a subtle poke at Cambridge scientific egos. On page six of that issue, there was a graph entitled "Bottles of Dom Perignon (1964) Purchased from Cave Atlantique (per month)." Cave Atlantique is a Cambridge liquor store, and there was a huge spike for the month of October. At the bottom of the page was the addendum: "Next issue: Graph B. Bottles of Dom Perignon Returned Unopened to Cave Atlantique (per month)."

The Harvard team returned to discover one final stunning reversal. The entire library of insulin clones, the 2,300 colonies created during the February blizzard, had inadvertently been left in the MIT incubator when the group left for England. An urgent message to remove the clones either got lost or unheeded, for when the dejected Harvard researchers returned, all that was left of the library was "these thin dry flakes of dead cells."

Lydia Villa-Komaroff's informal ledger of scientific activity came to an abrupt end after her September 10 entry, when the contamination by the rat insulin gene showed up in the gels. The famous song, too, remained unfinished.

> *Three acrobats and a magician*
> *Went off on a dangerous mission*
> *To clone and express the insulin gene*
> *and thereby to thwart Genentech's schemes.*
>
> *To do that they flew off to Porton,*
> *Met Peter, who proved quite important.*
> *They made lots of clones the very first day* . . .

"And that's all I can remember," she says now. "Only had a few more lines to go, but I never really could bring myself to put the rat insulin stuff in."

The trip, in Walter Gilbert's own crisp two-word summation, was a "total disaster."

Eighteen

"Jumping Around a Little Bit, Even"

· ·

September 6, 1978, was a warm, humid, smoggy day in Duarte, California, with the temperature climbing to the upper 70s by midday. It was much hotter inside a conference room at the City of Hope, where, under a bank of television lights and surrounded by a large group of inquisitive journalists, scientists from City of Hope and Genentech were in the process of announcing to the world that human insulin had been made for the first time by recombinant DNA techniques. Art Riggs, Keiichi Itakura, Roberto Crea, and diabetes expert Rachmiel Levine of the City of Hope as well as Dave Goeddel, Daniel Yansura, and Bob Swanson of Genentech sat behind the dais. Dennis Kleid stood at the microphone, making a few comments about the molecular biology involved and the safety of the technique.

"Okay, and now about the ethical considerations . . ." said Kleid. There was an audible groan down at the end of the table, where Bob Swanson awaited Kleid's comments with pained anticipation. "Swanson *was* groaning,

266
· · · · · ·

yeah," confirms Goeddel. " 'Get him down!' " Thus were the details of Genentech's remarkable achievement first revealed to the public.

The September 6 press conference was the last time Genentech held such an affair, although not because of anything Dennis Kleid said. Announcing scientific news in public had become more controversial than the news itself, and one heard "Gene Cloning by Press Conference" used as an increasingly pejorative term. Researchers, particularly those affiliated with new biotechnology companies, were accused of making end runs around the normal self-regulating apparatus of communication within the scientific community. Slick, manipulative, hyped-up, unscientific—many scientists viewed the press conference as a way of going over the heads of experts in the scientific community, as it were, to make a grandstand play in public. Journalists were unlikely to ask the same tough kind of scientific questions that scientists themselves would pose. Careers were advanced, companies trumpeted, products hyped. It was tub-thumping science.

And yet "science by press conference" appeared much more controlled and manipulative in theory than it ever was in reality. It almost always exacerbated personal tensions among team members over such issues as credit. Nor were private companies always the instigators; aggressive educational institutions, eager to publicize work and attract greater funding, pushed equally hard for visibility. Genentech's press conference, as it unfolded that Wednesday, was not exactly a model of the genre, but it did serve an important role that no refereed journal article could. It got Genentech's name out in the public, and hitched it up to a much more recognizable name: Eli Lilly & Co. That was one small step in Bob Swanson's business plan, one giant step for the world of biotechnology.

As early as January 1978, when Boyer and Swanson recruited their insulin team, Dennis Kleid had come away from their first meeting with the distinct impression that a deal with Eli Lilly had already been nailed down. There was an object lesson in that, too. "It was *not* the case," Kleid says. "It was that they were *going* to approach Lilly. You know, whenever you're talking to business guys, there's a gap between what is going to happen and what has happened, and it's a very gray area."

While Kleid and Goeddel worked on the project, only slowly realizing that a deal had not been worked out, Swanson and his associates danced across

267
. . . . . .

that gray area with Lilly in search of a deal. Early on, around the beginning of 1978, Swanson held discussions with Lilly's archrival, Novo Industri. The Danish manufacturer never expressed great interest, according to Swanson, because "Novo didn't believe it could be done. They didn't believe the technology would work, basically." Lilly's interest remained steady but non-committal.

Details of an agreement with Lilly were roughed out over the summer. It had been Swanson's plan from the very beginning to hitch Genentech's wagon to a powerful, profitable, and well-regarded pharmaceutical company. Genentech's scientists knew nothing about processing, fermentation, scale-up, testing, shepherding a drug through clinical trials, winning FDA approval, marketing, and sales. Swanson wanted to do his apprenticeship with an established company. When it came to insulin, there were only two U.S. companies as major players in the market, Lilly and Squibb, with the lion's share going to Lilly. Genentech wanted to side with the lions.

There were several hitches. First, Swanson was determined to participate in all the planning and scale-up activities; he regarded any potential deal with Lilly, associates say, as a kind of business seminar, and he wanted to get the most out of it. "He didn't have any idea what it took to develop these products," says a source familiar with the negotiations, "but he wanted to be part of that. He wanted to use these agreements to leverage Genentech up on the learning curve. . . . At the very least he wanted codevelopment, so that as Lilly learned how to make these products, we'd learn, too. And Lilly was adamant that they weren't going to do that." Second, Lilly clearly occupied the high ground in the negotiations, because even if Genentech had figured out a way to make insulin by gene-splicing, they didn't know how to scale up production, assure quality control, pass the regulatory hurdles, and market it—and they really didn't have anywhere else nearly as good to take their clones. Third, Lilly had no intention of signing any agreement at all until Genentech had clearly and unequivocally demonstrated that it *could* make insulin. And so the shrewdest business move Genentech could make at that point was strictly scientific: make insulin.

Two big questions loomed over the final stages of the insulin work. Could the A and B chains be put back together to make a molecule of functioning insulin? And would the yield of that operation be sufficiently high?

The Genentech and City of Hope scientists worked sleeplessly toward answers during the last weeks of August. Dave Goeddel had spent three weeks

268
. . . . . .

down at City of Hope making the final preparations of A chain and B chain. For all his immense confidence and scientific self-assurance, Goeddel nagged himself with obvious questions. "Are we going to make any material?" he would ask himself. "Are we going to make enough? Is it going to be stable? Is the protein going to be degraded in *E. coli?* When you finally purify it out of this form, are you going to be able to reconstitute it into active insulin?" Doubts about the Katsoyannis procedure had freshly arisen, and the embarrassing, anticlimactic absence of somatostatin in the June 1977 experiment surely weighed on their minds as well. Swanson held Lilly at bay with promises of imminent success; meanwhile, Goeddel called up to San Francisco almost every day with updates on what was happening.

The initial reports, in fact, were not terribly encouraging, although neither were they grounds for serious alarm. Goeddel had tried several times to join the A and B chains, but the insulin chains produced by the bacteria were not sufficiently purified, and that seemed to impair proper joining. "You could get the partially pure B to work with the pure A that was derived from native insulin, and the reverse," says Goeddel. "But then when you had to make them both from *E. coli,* we had to go through some additional purification steps." At this point, Roberto Crea played a major role in purifying the proteins. It was a slow, laborious process, but Crea and Goeddel worked around the clock until they felt the two chains were sufficiently pure to allow reconstitution to occur. The magic hour fell in the middle of the night, sometime between August 23 and 24.

"The final experiment was to put the *E. coli* A chain and the *E. coli* B chain together," Goeddel says. "I remember staying up all night doing all the experiments, the appropriate controls, and the next morning putting them all in the scintillation counter to look at the radioimmune activity.

"All the vials were in the scintillation counter, and Art arrived at work that morning with almost perfect timing. We went over what I'd done the night before, and then we went into the instrument room and watched the counts right as they came off the scintillation counter." If their jerry-built insulin molecules, forced together chain by chain overnight, had folded up into the proper three-dimensional shape, radioactive antibodies would glom onto the molecules in some of the vials; those antibodies would fire off radioactive scintilla, an invisible fireworks show, that would register on the machine. "And this was an antibody that didn't react to A chain or B chain alone," Goeddel says. "You had to have insulin correctly folded."

First the control vials were processed, to make sure the assay worked. Then the experimental samples went through. Fourteen months earlier, they had all gathered around a similar machine in a similar ritual and watched as the somatostatin experiment blew up in their faces. This time, however, the numbers looked good right off the bat. There was real insulin, *human* insulin, in their vials. "As soon as the counts came out, we knew we had it," Goeddel recalls. "I think we were actually jumping around a little bit, even." Adds Riggs: "He was probably happy just to get some sleep."

The celebration, at least initially, was low-key. Riggs and Goeddel shook hands; Crea and Itakura were informed and came over to celebrate. Goeddel immediately called Swanson and Kleid to give them the news. "Knowing the personalities of both of them," Goeddel would observe later, "my guess is they must have been even more excited than I was at the time." Back in Itakura's lab, Goeddel loosened up enough to exchange exuberant high-fives with Roberto Crea, and they celebrated boisterously enough to earn a reprimand from a hospital guard.

As a rule, scientific celebrations tend to be short-lived. But for everyone involved, August 24 was a red-letter date. Kleid and Goeddel, having quit more established jobs for what could have turned out to be fly-by-night biology, had pulled off the project. Swanson had drawn the ace he needed in his poker game with Eli Lilly. Riggs, Itakura, and Crea, the City of Hope's contribution to Genentech's critical mass, saw their faith in synthetic DNA chemistry vindicated once again—perhaps not as dramatically as with somatostatin, which was the seminal achievement in Riggs's mind, but perhaps in a way that could more easily be appreciated by the public.

So they celebrated briefly. Then they got down to the real work.

There was a world of difference between the laboratory creation of human insulin and commercial production. The total amount of insulin resting in Goeddel's test tubes was later estimated to be about twenty nanograms—twenty *billionths* of a gram. Could those scant few molecules actually serve as the bedrock upon which to build an industrial empire?

The view from Indianapolis tended to be skeptical. After all, a Lilly official, J. Paul Burnett, was at that very moment sitting in an office outside a laboratory in Strasbourg, France, patiently waiting as Axel Ullrich, working inside the lab, attempted to clone human insulin by way of a cDNA gene. Lilly had actually invested time, money, and energy into the Ullrich initiative; the company had tossed not a farthing in the direction of Genentech, according to Swanson. Taking nothing for granted, Lilly may have tried to hold out until the last moment, hopeful that Ullrich could pluck the human insulin gene out of his clones.

But when Goeddel called Swanson with news of success, and heard Swanson's emphatic "Okay!" over the line, the young scientist knew the result would give Genentech a boost in its negotiations. "I think," he would recall later, "he could use that to say, 'Okay, now we can have a news conference.' And then he could pressure Lilly by saying, 'We're going to have a news conference. Do you want to be part of it or not?' "

It has been suggested that Genentech scheduled a press conference within two weeks as a ploy to back Lilly into an agreement; either the pharmaceutical giant could participate in the press conference, and share in the achievement, or appear to have been technically outmaneuvered by a much smaller company. According to Irving Johnson, however, Lilly signed its agreement with Genentech on August 25, 1978—one day after Goeddel completed the confirming experiment, two weeks prior to the press conference. "From the date of the contract with Genentech, it seems unlikely that we were pushed to the wall in September," says Johnson. "We don't get backed into very many agreements, and I think that would probably be a specious statement." Indeed, Genentech undoubtedly needed Lilly's several million dollars at that point more than Lilly needed Genentech's twenty nanograms.

There is no question, however, that Swanson wanted to hold a press conference as soon as possible to establish Genentech's priority. Indeed, if any pressure was being applied, it was not on Lilly, but on Genentech's own scientists to agree to the news conference; there were, as Goeddel politely puts it, "slight discussions" about that.

Art Riggs, as he had during the somatostatin project, argued strongly that a scientific paper should be written and accepted for publication before there was any public discussion of the results. He opposed the idea of an instant press conference, but Swanson, he recalls, was terrified of a similar announcement by competitors. As a compromise solution, Riggs arranged for Keiichi Itakura to give a seminar at UCLA on the insulin work. "I got argued into that," Riggs concedes. "I was arguing that we had to wait. Swanson and others said we couldn't wait, that the press conference had to be yesterday." They also had to have a paper ready within two weeks. Goeddel, a maven for pressure and deadlines, said, "I think we can do it this fast."

To the races once more. The last week of August saw a furious burst of literary activity in Duarte and South San Francisco; two papers would emerge from the work, one on gene synthesis with Roberto Crea as first author and the other on cloning and expression with Goeddel topping the team. Goeddel lingered for a few more days at the City of Hope, repeating the experiments and verifying the results, then returned to the Bay Area. Riggs worked on part of the paper down south, Goeddel started on his portion of it back at Genentech, and then on the Saturday of that Labor Day weekend, most of the authors—an army that included Goeddel, Kleid, Yansura, Crea, Hirose, Kraszewski, Itakura, and Riggs (Bolivar and Heyneker, also authors, were out of the country)—gathered to work on the draft. Riggs and Goeddel then spent the next two days polishing up the paper. By Monday, September 4, the paper was finished. And a good thing, too. Swanson had scheduled the press conference for September 6 at City of Hope.

Most scientific journals, of course, do not accept manuscripts on two days' notice, so it was going to be difficult to make the claim that the papers had already been accepted for publication. The insulin authors did the next best thing, however. They planned to submit both papers to *Proceedings of the National Academy*, because *PNAS* papers were automatically accepted when submitted by an academy member. A scientific colleague of Art Riggs's at City of Hope, Ernest Beutler, sent the cloning paper in, although almost a month passed before the work was officially received at the journal: "Expression in *Escherichia coli* of Chemically Synthesized Genes for Human Insulin" reached the National Academy on October 3. That meant the paper had not technically been accepted at the time of the press conference. Rather, as Goeddel puts it, "it was submitted to the person that was going to submit [it] to *PNAS*."

There was no backing down on the press conference. By setting a date and sending out announcements, Swanson in a sense had issued polite ultimatums both to his scientists, who were under the gun to produce a paper in less than two weeks, and to Lilly personnel. The Lilly contract was already signed, but the company's presence at the affair would lend inestimable credibility to Genentech's achievements.

There was another, less obvious reason for the sense of urgency that Swanson felt. Word that Wally Gilbert had taken a group of scientists to Europe had ripened, though somewhat sourly, on the molecular biology grapevine; rumors reached the West Coast that Gilbert planned to make a major announcement at a scientific meeting in Munich on September 7. Keiichi Itakura cited this, to the trade journal *Medical World News*, as a partial reason for the rush to go public about the insulin work on September 5 at his UCLA seminar. Once presented in such a forum, the scientists could claim that their peers had had an opportunity to critique the work before the press learned of it the following day.

The insulin press conference, in box-office parlance, was a monster. The TV networks showed up, along with newspapers, radio, and trade press representatives. Rachmiel Levine, the City of Hope's deputy director of research and a well-known insulin researcher, introduced the successful cloning team. When all was ready, Levine—a short man with a round, bobbing head and wry, avuncular manner—began to describe the historic experiment. It is perhaps typical of the generation gap between the cloners and the non-cloners that neither Kleid nor Goeddel recognized the name of one of the leading diabetes researchers of his day.

"We'd never seen him before," Dennis Kleid recalls. "And all of a sudden he is standing up there, talking about insulin and how this is very valuable and all that. Then somebody asked him, 'Well, how did you do it?' And he started answering *those* questions: 'Well, we synthesized this gene, and we did this . . .' Then they asked him *more* detailed questions about it. He was getting pretty deep into the science of this whole thing, and Goeddel started elbowing me there and said, 'Hey, *we* should be answering these questions. What's *he* doing up there?' "

There was, in fact, a fair amount of elbowing going on behind the table. As usual, the issue was credit; the reaction was almost reflexive. Goeddel remembers Swanson saying, "Why don't you say something?" Goeddel, the same fellow who thought nothing of scrambling up the sheer face of El Capitan, dangling thousands of feet above the valley floor, found the prospect of addressing the assembled press "pretty scary." He leaned over and nudged Kleid, urging him to say something. When someone asked a question that seemed too technical even for Levine, he hesitated for a moment. Kleid rose to his feet and said, "Well, I'll answer that question." And he continued to answer a few more.

"Then they started asking me more detailed things. The first one was about the gene itself, and I answered something about [that] it was the synthetic DNA. And then somebody else asked, 'Well, what about . . .' Something about the ethical implications of this. 'What about the ethics of doing this? The safety and the ethics?' " Kleid replied that the NIH guidelines on recombinant DNA research had been followed; he did not know that, as he puts it, "what you're supposed to do is ignore the questions that are on an ethereal level. You know, the ethical considerations. I said, 'Okay, and now about the ethical considerations . . .' And I started hearing this groan down there by Swanson, who was now going crazy. He was praying that this guy Levine was standing up there instead of me." As Goeddel would note later, "Bob was always so nervous about scientists saying anything about the morality of the work. It was a very new industry, and he was very conscious of image. He was probably worried that the scientists would go off and say some wild kind of thing."

Kleid fielded one more question. Someone asked about the potential danger of escaped bacteria making insulin in the gut of humans. This was one of the historic concerns of critics, and Kleid attempted to explain Genentech's use of fusion proteins—the concept that each insulin chain was fused to a larger molecule to form an inactive protein. The subtlety, he felt, was lost on the audience, and he almost created the impression that no insulin had been made at all. "Well, we really didn't make insulin . . ." he said at one point, startling the reporters. "What we did was make these fusion proteins in two different organisms. If we say we made insulin, that's a lie. We didn't make insulin. What we made was these fusion proteins."

"And 'lie' was kind of a poor choice of words," Kleid says now, "but I was very nervous."

At that point, Levine formally introduced everyone and Bob Swanson rose to say a few words—and, perhaps, engage in his first assay into damage control. He made a brief statement, and then somebody asked a question. "He was *just* about to answer the question," Kleid says, "and all the lights went out. The place was getting very hot. This was the typical summer afternoon in Duarte. It was boiling in the sun. And it was hot in there. So the minute the lights went out, in about sixty seconds they were convinced that they weren't going to come on right away, so everybody picked up their stuff and streamed out. That was it." End of press conference.

There was still time for one more mishap. A television news crew wanted to film the historic vial of insulin, so Kleid and Goeddel headed over to Art Riggs's lab, where Kleid was stage-directed to walk to the laboratory refrigerator and open it up. As the cameras rolled, he walked as directed to the refrigerator and pulled the door open. About a hundred small tubes flew out of the refrigerator and went skittering all over the floor. End of photo opportunity. Grateful that it was over, the scientists adjourned to a local restaurant for lunch before returning to San Francisco.

For all the groans and spills and ill-chosen words, it did not take long for the success of the news conference to make its mark. In the time it took for Kleid, Goeddel, and Yansura to eat their lunch, go to the airport, and board a northbound plane, the presses of the *San Francisco Examiner* had begun cranking out the news, so that when the three scientists walked through the corridors of San Francisco International Airport upon their return, an early edition of the afternoon paper was already on the stands, a huge banner headline across the front page proclaiming: "New Insulin for Diabetics." The kicker, typically provincial, added: "Bay Area labs lead the way." Bob Swanson eventually had paperweights made out of a replica of that front page.

The success was repeated in newspapers across the country the next morning: there were front-page stories in the *Los Angeles Times, Washington Post,* and *San Francisco Chronicle,* and quite likely there would have been one in the *New York Times* as well if there hadn't been a newspaper strike in New York. The story received prominent coverage in the *Wall Street Journal,* the *Boston Globe, Time, Newsweek,* trade journals, science magazines—it was a clean sweep.

Almost as important, as far as the mores of science are concerned, it set a standard to which biotechnology companies aspired for years to come. Even as "science by press conference" began to be excoriated by scientists, it moved toward the top of the business syllabus for start-up companies. The reasons were eminently clear. Following Genentech's somatostatin press conference in December 1977, the company easily raised $1 million to do the insulin project. Following the insulin press conference, the company ultimately raised $10 million from the Ohio-based Lubrizol Corp. to pursue such projects as interferon. And when it went public in 1980, Genentech was, if not a household name, certainly well known to investors large and small.

Hardly anyone in the press, however, bothered to ask a rather fundamental question. Did Genentech's insulin actually work? Was it biologically active, and did it perform up to the metabolic standards expected of such a crucial hormone in a living organism? Did it reduce high blood sugar in test animals such as rabbits? Among all the reporters who described the work, only a handful of publications—among them *Nature, Medical World News,* and the *Los Angeles Times*—raised this important point.

And the answer was: they didn't know. They hadn't tested the substance in animals because Kleid and Goeddel's bugs simply were not churning out enough insulin to conduct a meaningful test of biological activity at that point. Kleid maintains it was "a silly criticism." "That biological assay was kind of hokey anyway," he says, "because sometimes the rabbits would respond without insulin. It was a statistical kind of thing." Riggs, too, insists that the chemical evidence was convincing. Eli Lilly apparently was convinced, because Dennis Kleid recalls handing the bugs over to a company representative at the press conference.

Lilly's involvement provided a strong pharmaceutical industry endorsement of the recombinant insulin work. In announcing the agreement at the press conference, Swanson noted that Genentech would help in continuing research and development on insulin, leading up to actual industrial scale-up. The agreement, never officially disclosed, reportedly involved several million dollars and a licensing deal whereby Genentech's royalties—amounting to less than 10 percent, according to a company source—would be shared with the City of Hope and UCSF. What Swanson neglected to say, and what no one—not even Genentech's scientists—knew at that point, was that Lilly had negotiated an extremely tough deal. With no commitment to produce human insulin commercially, Lilly in effect was still saying to Genentech, "Show us." Show us that the yields from the bacteria could be increased. Show us that quantities of *pure* A and B chain could be produced by this method. And show us that it could be done within a certain time period. Otherwise, it was just another nice laboratory demonstration.

It was only after the deal was signed and sealed, Kleid recalls, that the scientists found out that Lilly had stipulated a series of deadlines, or benchmarks, for this work. To a pessimist, they looked for all the world like escape clauses out of the deal if the technology was not sufficiently sophisticated to go beyond mere demonstration. When Kleid and Goeddel learned about the conditions of the deal, they immediately concluded that it was impossible—especially the benchmark on yields farther down the road. Herb Boyer's attitude prevailed during these moments of doubt. "By then," he told the troops, "we'll figure out how to do it."

This marked a rather dramatic fork in the road for molecular biologists, a point where industrial work veered off from academic pursuits. If science indeed revolves around the choice of problems, then product-minded industrial biologists had a much narrower range of questions to ask. Genentech's molecular biologists were among the first to learn this lesson. In academia, a single smashing success at the postdoctoral level could land one a good job, a good laboratory, and a good system to explore throughout one's career; in industry, the smashing result in a model system merely initiated a kind of decathlon requiring improvements, amplifications, modifications, and adaptations. Managers demanded, deadlines loomed. The sole aim was to produce a commercially viable product, and pure intellectual curiosity could be a liability.

Kleid says, "There was never any satisfaction making the goals, as far as I could see. The goal would instantaneously self-destruct and there'd be a new one to take its place." Indeed, judging by some comments Bob Swanson made to a group of financial analysts in 1979, he had pretty high expectations for the living things employed by Genentech. "We should all have people working for us who work like microorganisms," he said, tongue in cheek. "They divide every 20 to 30 minutes and work 24 hours a day—just for room and board. What's more, they make the ultimate sacrifice—they die making the product for you." And Kleid was complaining about getting home late.

Axel Ullrich was in France when Paul Burnett, the Lilly research official, took him out one night to buy him a drink. The gesture wasn't coincidental, in Ullrich's mind, because Burnett used the occasion to break the news about Genentech: the company, with Lilly's knowledge and encouragement, had successfully cloned human insulin. The achievement surprised Ullrich in one sense, since he was among many who thought the synthetic DNA approach would prove difficult. In another sense, it was about par for the course for 1978, which had started out as a year of bewilderment and frustration, and was ending up on that note, too.

Ullrich's continuing attempts to express rat insulin never went beyond ambiguous results. Next, Ullrich attempted to isolate the chromosomal rat insulin gene, an experiment that was permissible under certain conditions and would be highly revealing. A "chromosomal gene," unlike a cDNA copy of a gene, corresponds to the exact genetic information as it is laid out in the chromosome, introns and all. The cDNA copy, by contrast, corresponds only to the information in messenger RNA, which is in effect an edited, shortened transcript of the chromosomal gene (after the introns have been spliced out). Ullrich's experiment, unfortunately, suffered much the same mishap as the Gilbert group's England experiment had: contamination of the material with plasmid containing the rat insulin gene. "Transtubation," as many pioneer cloners learned the hard way, was a distressingly common occurrence.

While Ullrich's attempts to express rat insulin foundered, enthusiasm and excitement in the Goodman lab shifted toward the growth hormone work. Peter Seeburg, Ullrich's German compatriate, had discovered that the growth hormone gene, though nearly four times as large as insulin, was somewhat easier to manipulate in bacteria; it did not require the same elaborate fusion proteins engineered by the City of Hope and Harvard groups on insulin. You could, in effect, hook the gene up to a promoter without having to trick the bacteria with bits of bacterial genes. Seeburg's growth hormone supplanted insulin as the hot project in the lab. At least that was the way Ullrich saw it.

"We were sort of competing," he says. "He had growth hormone cloned, and growth hormone, as we know now, works for everything in expression systems. So Peter started working on that, and it looked better than the insulin. . . . We had problems proving that we had really expressed insulin." Finally, in the summer of 1978, Seeburg achieved the successful expression of rat growth hormone genes in bacteria. It was a monumental achievement that once again focused worldwide scientific attention on the Goodman lab. Ullrich was still determined to get the insulin work untracked again. "So okay," he told himself, "Let's get human insulin. . . ."

The lab group, already sharply polarized, headed toward a fateful rupture. Goodman, according to some lab members, felt his two German postdocs were trying to exploit the insulin and growth hormone work to their own, and not the university's, advantage. Seeburg and Ullrich, lab members suggest, felt wary of what they perceived as Goodman's efforts to take credit for their work. "Both Axel and Peter were very possessive about their projects," recalls one Goodman postdoc, who spoke on condition of anonymity. "There was a strong effort to exclude Howard from whatever was being worked on," says this postdoc, specifically because of the issue of credit. This smoldering conflict, exacerbated by what the postdoc called "a fairly big personality clash," totally ignited over the issue of patent credit. During the summer of 1978, Eli Lilly was putting the final touches on its funding arrangement with both Howard Goodman's and William Rutter's labs. Lilly wanted to support general research not only on insulin, but on growth hormone. On August 11, 1978, the university filed a patent application on a process for making proteins in bacteria; the three inventors listed were Goodman, Rutter, and John Baxter. On August 17, 1978, Eli Lilly signed its basic research agreement with the

University of California–San Francisco. Roger Ditzel, the university's patent officer, says the timing of the two events was "purely coincidental." Ullrich was unaware of the new patent filing.

Soon after the Lilly agreement, with the company's P3 facility outside Strasbourg completed, Ullrich traveled to France and finally got started on the human insulin gene. This time, the work proceeded smoothly. With Lilly's Paul Burnett overseeing the project, Ullrich successfully isolated and cloned the human proinsulin gene. Meanwhile, disturbing news drifted overseas from San Francisco.

First, Burnett told Ullrich about Genentech's feat. This was not nearly as distressing to Ullrich as it might seem, since the Rutter-Goodman team's interests were in isolating the human gene—glimpsing the array in its entirety, with all the exons, introns, and control regions. Genentech's synthetic DNA approach couldn't reveal those biological nuances. More troubling, Ullrich learned from Seeburg about the new patent filing by UCSF. The maneuver was complicated—and, in the view of the postdocs who felt they contributed to the work, controversial.

When the rat insulin and rat growth hormone clones had originally been obtained in 1977, the University of California–San Francisco had applied for several patents on the techniques. On June 9, 1977, for example, UCSF had applied for a patent on sequence purification, based on the original rat insulin cloning, in the names of Rutter, Pictet, Chirgwin, Goodman, Ullrich, and Shine. A little more than a year later, a week before signing the Lilly agreement, the university filed its application for a much broader patent on the expression of proteins in bacteria; it was entitled "Synthesis of a Eucaryotic Protein by a Microorganism," and this time the list of inventors included only Rutter, Goodman, and John Baxter, on the advice of the university's patent attorney. None of the postdocs was included. In a department already riven by rumors and rivalries, stories began to circulate about million-dollar royalties, of patents being typed up in the dead of the night. Rutter insists that "it was *never* the intent" of the lab chiefs to exclude the postdocs from royalties, but in fact the postdocs didn't learn of the revised patent until after it was filed, at which point they protested bitterly.

"I heard about that while I was in France and got really angry," says Ullrich. "Peter was angry, too. Peter said he had been approached by Swanson again, and Swanson offered him a job and reactivated the offer to me, to all three of us—John Shine, Seeburg, and me. So I accepted. I said, 'Okay, this is it. I don't want to have anything to do with these people [at UCSF] anymore.'" Later that September, at the annual gathering of the UCSF biochemistry department at Asilomar, Ullrich informed Howard Goodman that he would leave UCSF at the end of the year and move on to Genentech. Ullrich admits offering to bring his insulin clones—with Goodman's prior knowledge and consent, he says—to Genentech, and a source at Cetus, who asked not to be named, says Ullrich and Seeburg both shopped their clones there as well.

The new patent application was the final straw for Seeburg as well. The German postdoc felt he had pushed the growth hormone project along in the face of active discouragement, then found himself deprived of patent recognition. "What shattered me," he says, "was when we told Howard that we had expression of growth hormone. The next thing we knew, they had called in patent lawyers from the university and patented the [expression] between Goodman, Rutter, and Baxter, putting themselves on as inventors, and they hadn't even known about it [until] we had told them. We had all the data." The university's position was that the three faculty members had organized and directed the research; the postdocs had merely done what they were told.

"The issue is whether they made a patentable contribution to the invention as claimed," says Ditzel, who joined the UCSF Patent office after this application was filed, but spoke in general about the university's patent position. "I do know that very often people think their names should be on a patent because their names are on a paper. The ground rules on a patent are *very* different from names on a paper. We always try to make sure we get the right names as inventors." The issue became sufficiently sticky that later on there was at least one meeting in which the postdocs were asked if they wanted to be listed as inventors; they said no, as Seeburg recalls, but signed an agreement by which they would share in any monies that derived from the patent.

Once again, the dispute revived the issue of credit—this time shifting to the legal sphere—between the lab chiefs and the postdocs. "Sure, Howard was very much opposed to starting the work," says John D. Baxter, one of the three faculty members named as an inventor on the patent application. "The work got started anyway. Howard didn't initially plan the experiments. So you could argue, 'Okay, Howard doesn't deserve credit.' But that's really unfair to Howard, because Howard set up what was probably at that time one of best few labs in the world to do this technology. Howard did that, nobody else—

Seeburg, Axel, Baxter, nobody else did that. . . . You know, if Seeburg and Axel think they're so great, why did they come to Howard's lab to do it? Why didn't they do it back in Germany or somewhere else? They came there because of that intellectual environment." Discouraged by this turn of events, Seeburg decided to switch from an academic lab to an industrial one. And, without Goodman's prior consent, to take his clones with him.

There was one final confrontation, however, and it seemed to symbolize all the discord and jealousies that had welled up in the superb biochemistry department at UCSF. It took place in November in Goodman's lab and has been known among insiders ever since as the "freezer incident." It stemmed in part from the fact that despite announcing his imminent departure for Genentech, Peter Seeburg had begun to work on the expression of the human growth hormone gene at UCSF. (The account of this episode comes from Peter Seeburg and was independently confirmed by two other members of the Goodman laboratory; Howard Goodman declined to be interviewed.)

As he freely admits, Seeburg was taking aliquots—portions of the total amount—of his human growth hormone clones to Genentech. Professors and postdocs normally reach agreement ahead of time on the transfer of materials, but that apparently was not done in this case. Against a backdrop of strained personal relations in the lab, industrial considerations may have exacerbated the situation. Since Eli Lilly had by this time begun supporting Goodman's work on growth hormone, both Lilly and a potential competitor, Genentech, could conceivably find themselves working with, and competing on, the very same clones.

"I had gotten," says Seeburg, "a clone, [had] cloned cDNA from a pituitary tumor for human growth hormone, and I was just sequencing that. Then Howard said I wasn't allowed to sequence on it anymore. I was standing there. I showed him the film and I was standing there, and he said, 'You're not allowed to sequence anymore.' And I said, 'Well, Howard, I'll close my eyes, and I won't look at the film anymore.' "

Disgusted, Goodman stalked out of the lab, yelling as he went, "Get out of the lab! I don't want to see you from tomorrow onwards!"

That same night, Seeburg returned to the laboratory. John Fiddes, a postdoc from Scotland, was in the lab, and he told Seeburg, "Peter, something terrible happened."

"What?" Seeburg asked.

Seeburg had been in charge of the restriction enzymes for the lab, and they were kept in his freezer. But when Fiddes had gone to the freezer to get a restriction enzyme for an experiment, he opened the door and discovered to his surprise that it was empty. All the restriction enzymes had been removed. All of Seeburg's test tubes from his growth hormone experiments were gone as well (they were for the most part unlabeled, their contents known to Seeburg only by their location in his test-tube rack).

Fiddes said, "Do you know what happened?" Seeburg did not. He immediately called up Goodman.

"He had taken all my material," Seeburg recalls, "and locked it into his freezer. And then he came in, because Fiddes did need that enzyme." Goodman arrived sometime between nine and ten-thirty; Seeburg had been waiting for him. Goodman unlocked his freezer and opened the door. Nothing had prepared Seeburg for the sight. "There's all my stuff. You have all these experiments in racks and little tubes and labeled and stuff, and instead of taking these racks, they had just been thrown into his freezer. They were all lying upside down, tubes all over the place. All the experiments which I'd been working on, which he got so much credit for and everything—I was totally shattered. I just sat there. I couldn't believe my eyes . . . I just sat there, stunned."

To tamper with another scientist's work—for whatever reason, with whatever motive—is considered an unusually grave transgression. As Goodman's own technician, Fran DeNoto, would say later, "You should *never* do that. A person's experiments are a person's personal property. Scientifically, it's like committing a mortal sin." Not everyone would agree that a scientist's experiments are personal property, especially the universities that pay for the labs and equipment, but rarely does it reach such extreme confrontations.

Here, unfortunately, was the ultimate rupture in that archetypal conflict between lab chief and postdoc. In the best of circumstances, it was always a difficult relationship. In 1978, at UCSF, after three years of groundbreaking research, the relationship utterly disintegrated under the novel pressures of patent recognition, commercial interests, priority, personality, and the vagaries of scientific temperament. It was not the kind of thing that universities seek to publicize, but in an odd way perhaps it should have, for it might have served as an object lesson to all molecular biologists that in the passage from the old biology to the new, from basic science to applied, in abandoning Medawar's "dire equation" where uselessness equals good, the loss of innocence could come at a very high price indeed, and sociological thresholds every bit as invisible as the biological ones could be crossed, never to be regained again.

The biochemistry department at UCSF remains superb to this day, but the freezer incident brought down the curtain on a successful and memorable period of work. As one disgusted member of the Goodman lab later concluded, "That was the end of an era, yes."

Nineteen

"The Bacteria
Really Went Bonkers"

· ·

If the teenager is father to the man, then Herb Boyer and Wally Gilbert (and, to an extent, the companies they founded) came to fulfill prophecies one could read in their personae as high school seniors. Boyer, the football and basketball player, junior and senior class president, was thus saluted in his yearbook: "Likes almost everything. . . . Dislikes hardly anything. . . . Ambition to become a successful businessman. . . ." Gilbert, in filling out his application to the Westinghouse Science Talent Search, listed as "hobbies" the following: photography, astronomy, heliography, bacteriology, photo-micrography, minerology, and chemistry. It is hardly surprising that one matured into a scientist with down-to-earth entrepreneurial talents, and the other into a biophysicist with unusually broad and insatiable scientific curiosities. These attitudes—one sociable and practical, the other powered more by abiding intellectual fascinations—spilled over into the companies each man formed in his own image.

As Genentech and Biogen revved their engines for the initial races in biotechnology, an unexpected calm settled over the political landscape. Opposition to recombinant DNA work had lost much of its momentum. It was with the creaking sound of Pandora's box being opened in the background that Ruth Hubbard, in March 1977, at the National Academy forum in Washington, predicted that gene-splicing research would never be contained or controlled if allowed to proceed without stiffer regulations. Eighteen months later, this prediction stood confirmed, thanks to aggressive lobbying efforts by scientists and to highly visible achievements like the somatostatin and insulin projects. It should be noted, too, that none of the hypothetical dangers materialized, and studies indicated that crippled versions of *E. coli* were incapable of colonizing the human gut.

For the second straight year, Congress failed to pass legislation; the NIH's guidelines had been revised, meanwhile, with more relaxed rules going into effect January 2, 1979. Environmental groups such as Friends of the Earth and the Natural Resources Defense Council mounted spirited opposition to recombinant DNA work, but suffered severe credibility problems when several scientists trusted by the lay public—including René Dubos of Rockefeller University, Paul Ehrlich of Stanford, and Lewis Thomas of Sloan-Kettering Cancer Center—publicly distanced themselves from the groups' perception of hazard in the work. Indeed, the climate had changed so significantly from the heady *vox populi* atmosphere of the Cambridge City Council meetings in 1976 that Maxine F. Singer, in a *Science* editorial three years later, essentially asserted that "elitist criteria" should prevail in allowing the work to go forward.

The state of the opposition, both its predictive insights and flagging momentums, was summarized in an appearance by MIT scientist Jonathan King at a public debate in Great Britain on genetic manipulation in October 1978. King claimed that risk-assessment research in the United States, aimed at establishing the relative safety or danger of genetically engineered organisms, had been blocked; in King's estimation, "it's been blocked because some risk is going to emerge." The MIT biologist speculated that recombinant organisms might become increasingly pathogenic or interfere with a host organism's immune response, and insulin continued to be invoked as a potential hazard. "I don't know what would happen if a newborn infant picked up an *E. coli* meningitis infection and that bacterium was [spewing] out human insulin; but it seems to me there's every reason to be concerned . . ." King told his British audience. "And any endocrinologist not in the pay of the company producing it would have to give pause to the uncontained DNA."

King, however, sarcastically predicted a rosy economic future for insulin. "A fortune is going to be made from the cloning of insulin in bacteria," he claimed. "Four million doses are sold three times a week in the United States. They are not going to sell that insulin cheap; they are going to sell it expensive, because it's human insulin. They are going to *produce* it cheap."

Even as King spoke, Genentech's molecular biologists were shamelessly working toward precisely those goals. Genentech's success in putting together the A and B chains of insulin was widely perceived as a landmark in the history of recombinant DNA—but it threatened to be a short-lived one. Genentech's scientists, still only months out of their academic settings, faced the entirely novel experience of taking a laboratory result, their mere smidgen of feasibility, and trying to prove you could make large fermenters full of the stuff, reliably and profitably. This was the real work of biotechnology, as opposed to molecular biology, and it took place only after the TV crews packed up their lights and the reporters closed their notebooks.

By the end of that first week in September 1978, after Genentech's whirlwind press conference and swirl of attention, Dave Goeddel disappeared into Rocky Mountain National Park in Colorado, climbing in El Dorado Canyon. When Goeddel took off on these mountain-climbing jaunts, the people at Genentech held their collective breath. "We were all scared to death that he might hurt himself," recalls one early employee. "And if anything had happened to him, we would have been up a creek." When he returned, Goeddel concentrated on Genentech's next races: cloning the genes for human growth hormone and interferon. One of Biogen's big guns, Charles Weissmann, simultaneously went after interferon, so the insulin rivalry resumed, this time over a new biological substance. It fell to Dennis Kleid, meanwhile, to take Genentech's test-tube demonstration of insulin and turn it into a convincingly viable product.

Lilly had "tremendous doubts"—in the words of one manager—that the technology would prove commercially feasible. Under terms of the agreement, Genentech had agreed to provide Lilly with a minimum quantity of purified A chain, purified B chain, and reconstituted insulin by fixed dates over a period of several years, as well as improving the yield of these proteins from bacteria. The first of these benchmarks came due during the following summer. No one on Genentech's scientific staff was particularly sanguine about meeting the deadlines.

Nonetheless, Robert Swanson and Herb Boyer had anticipated this next phase of development by expanding their scientific staff. Roberto Crea, who had taken the lead in synthesizing the DNA fragments for insulin at the City of Hope, was hired to head up the company's synthetic DNA operation in September 1978. Herb Heyneker rejoined the company later that same month. Two young scientists from the East Coast, Michael Ross and Ronald Wetzel, were hired that fall, and Peter Seeburg and Axel Ullrich were due to join by the beginning of 1979.

Wetzel, a slender, dark-haired chemist with glasses and a scholarly air, had taken a Ph.D. in organic chemistry at UC-Berkeley, had studied enzymology during a postdoc at the Max Planck Institute for Experimental Medicine in Germany, and had begun a three-year postdoc in the Yale University laboratory of Dieter Söll in 1975. Mike Ross came to Genentech by way of Harvard, where he had worked in Mark Ptashne's lab. Wetzel was brought in to help Roberto Crea synthesize nucleic acids, and Ross was a protein chemist who learned, upon his arrival in the fall, that Eli Lilly expected to see a seemingly impossible forty milligrams of A chain and twenty milligrams of B chain by the following June. It wasn't just a scientific goal; as Brian Sheehan, a former Genentech vice-president, put it, "as you rung the bell, you got cash."

This time the race was not against other competitors, but for long-term credibility, and Genentech staggered out of the starting blocks. Crea returned to his native Italy for the Christmas holidays, had his visa declared invalid on the very day of his return, and spent three months untangling himself from the bureaucratic mess. Ron Wetzel reported to work on November 1, helped Crea set up the lab, and then sat around with little to do while his boss and Genentech officials negotiated with authorities in Italy and the United States. But Mike Ross would stop by now and then to discuss progress on the first insulin benchmark, and it wasn't long before the two divvied up the labor: Wetzel concentrated on purifying the A chain and Ross concentrated on the B chain.

The problems they encountered illustrate precisely the kind of obstacles that make biotechnology, despite its presumed high-tech glamour, a biological crapshoot. Molecules are finicky, for all sorts of reasons. The ability to make an important molecule, like insulin, in minuscule test-tube amounts is no assurance that the molecule will cooperate when more ambitious levels of production are attempted. Each chemical step of manufacture and purification, for example, invariably produces some undesired side effect; molecules stick together, get gummed up, "behave poorly," as the chemists put it. As they attempted to meet Lilly's deadline, Wetzel and Ross struggled to impose order on their unruly molecules.

Wetzel quickly discovered that the amount of A chain produced by Genentech's bacteria was well below expectations. Part of the reason, it turned out, was the sheer size of the hybrid protein they had devised. By tacking the gene for human insulin onto the very end of the long bacterial beta-galactosidase gene, Genentech had produced a white-elephant protein so large that the molecule literally seemed to fall off the ribosome (the cellular site where proteins are assembled) before the tail of insulin A or B chain could be finished. Mike Ross's problems were even trickier, because of the greater length and biochemical "stickiness" of the B chain. All the while, Swanson's imprecations rang in their ears, pushing them to hit the benchmarks.

And so it was, on the morning of June 25, 1979, only a short time before their plane was to depart for Indianapolis to deliver this first installment, Ron Wetzel was in his laboratory at Genentech, packing a frozen solution of A chain in dry ice for the journey to Indiana. The samples safely stashed in a little box, they headed off for the airport and hand-delivered the consignment to Lilly officials in Indianapolis. For the first furlong, at least, Genentech had made good on its word.

Still, day-and-night work over a period of six months had produced sixty milligrams of insulin chains—by weight, less than one-fourth of a single aspirin tablet. That trifling amount was unlikely to dazzle officials at a company where freezers were stacked high with beef and pork pancreases. The reaction at Lilly, Wetzel recalls, was guarded and skeptical. Lilly biologists were working with the same bugs, and Irving Johnson, who headed the Lilly scientific task force, says, "The levels of expression were almost nil in the organism that we received from Genentech. We clearly had to improve that."

Could the yields be increased? Could the bacteria be induced to churn out enough of the stuff to justify the project economically? Despite the glib predictions of Jonathan King, there was considerable doubt about this. Back in South San Francisco, Dennis Kleid and Herb Heyneker and the rest of the scientists put their heads together and tried to figure out a way to tickle the bacteria's machinery into making more insulin. It was quality problem-solving, but of a more practical, less basic sort than in academic labs.

Given the hoopla surrounding Genentech's cloning of human insulin, it came almost as an afterthought, an anticlimactic footnote, when the Rutter and Goodman groups—still collaborating, but with undiminished rivalry—sifted through the 3.5 billion base-pairs of human DNA and, by the end of 1979, found the human insulin gene. Not an abbreviated version, like the one created by Genentech's synthetic chemists. The real human gene.

Anthropomorphic biases aside, continuing work on the rat insulin gene actually cracked the genetic structure of the hormone. Once again, the UCSF and Harvard groups found themselves dueling on the same project. The California group isolated one of the two chromosomal rat genes, but it was Peter Lomedico of the Gilbert lab who, searching through eight hundred thousand clones, isolated both rat insulin (actually preproinsulin) genes. And in comparing the two sequences, Lomedico discovered a dramatic difference. Both genes possessed an intron, a chunk of "padded" DNA, in the region flanking the informational portion of the gene, but the rat insulin II gene revealed a large and surprising intron smack in the middle of the C chain, which posed a question of immense biological importance. "An evolutionary test," the Harvard paper concluded, "is clear: the comparison of different preproinsulin genes will show whether new introns arise or old ones are lost."

Suddenly the Harvard group's insulin investigations, reflecting Wally Gilbert's broad interests, shot out in several directions. Efstratiadis set off on an evolutionary detective hunt. After collecting and cloning insulin genes from a variety of species—rat, human, chicken, guinea pig—he closely examined the A-T-C-G sequence of each gene. By comparing the order of these nucleotides, he could pinpoint the exact location, letter by letter, where mutations had occurred over tens of millions of years. These changes, in a sense, were the footprints left by evolution. Tracing these changes nucleotide by nucleotide, Efstratiadis tried to impose mathematical order on the patterns he was seeing. Almost inevitably, his epiphany came at an odd moment. "We were struggling with this, to understand the phenomenon, for a very long time," he recalls. "One night I was at home watching Johnny Carson, and during a commercial I was playing with the figures and I found it! I found that you could plot the numbers and surprisingly enough—actually, *very* surprisingly enough—get a straight line. And this is now millions of years of evolution, okay?"

Once convinced of the data, Gilbert retreated to a room of his home one day, Efstratiadis recalls, and in half an hour recorded his impressions on a dictaphone, which were published virtually intact as the "Discussion" portion of a 1980 paper in *Cell.*

Karen Talmadge, a graduate student in the Gilbert group, took off in another direction. She used the insulin gene as a biological keyhole to peek at the mysteriously precise process of secretion, the way in which cells ship proteins to the membrane. In an elegant series of experiments, she tinkered with the signal sequence, the address on the envelope that tells bacteria where to deliver a protein within the cell, and performed an artful series of molecular grafting operations. By splicing DNA together, she mixed and matched signal sequences from bacterial and mammalian proteins, and then determined whether these hybrid addresses permitted the protein to be delivered outside the cell. The surprising outcome of these experiments demonstrated that cellular zip codes have remained essentially unchanged for 3.5 billion years, since the first life forms appeared on the planet. If you patched the signal sequence from a rat onto a gene and cloned it into bacteria, the microbes knew not only where to send it, but how to clip off the signal once it got there. Cells, Talmadge concluded, recognize a "very general (and very ancient) aspect of structure."

When the human gene was finally identified, across the continent in San Francisco, it provided an excellent up-to-date understanding of our ideas about genes. The work was led by Graeme Bell and Raymond Pictet in Rutter's lab, Barbara Cordell in Goodman's. These were the heirs apparent who stepped in after the diaspora that saw Axel Ullrich and Peter Seeburg go to Genentech, and John Shine move back to Australia. Ironically, Genentech's bugs were producing human insulin for nearly a year before scientists isolated the real insulin gene from the human chromosome.

What exactly did a gene look like?

To judge from the scientific account of the work that appeared in the British journal *Nature,* the lay reader might conclude that the insulin gene was a monotonous string of bases—As and Gs and Ts and Cs—that stretched on for about eighteen hundred letters. Only a few spelled out the amino acids for insulin's precursor molecule, preproinsulin.

It is probably more useful, however, to view the gene the way an enzyme might, which is to say topographically, almost as does a pilot approaching an airport. The sequence of letters, in a way not totally understood even today, seems to create recognizable landmarks and bumps. About 250 bases before the start of the insulin gene's informational, or coding, portion, the enzyme spots a sequence that reads TATAAAG. To us, it is nonsyllabic nonsense; to a polymerase enzyme, it is a kind of landmark, a guiding beacon, which aligns the enzyme so that it will land with great precision on the gene itself. Shortly after this aligning beacon, there is another important landmark, known as the CAP site; this marks the point where the enzyme, having touched down, actually begins copying the DNA's information into messenger RNA.

The UCSF researchers discovered two large introns in the human gene, one just prior to the B chain and the other in the middle of the C chain. Like boulders on a roadway, these chunks of extraneous DNA broke up the contiguous flow of useful, encoded information; yet the polymerizing enzyme, picking up certain encoded signals, ensured that these bumpy patches would be edited out, literally picking up the thread of the insulin story on the other side of the interruption. Finally, at the end of this long genetic runway, the UCSF researchers identified the point where the enzyme finished up its copying work.

The insulin sequence reflects our deeper understanding of the gene. At the level of molecules, where physics and biology converge, information in a sense is topology. Just as an airplane cannot plop down or take off vertically from a runway, an enzyme reading a gene does not interact merely with a discrete patch of DNA pavement. The gene must also permit some maneuvering room, some biochemical airspace before and after the actual coding sequence, in order to allow the transcribing enzymes to move into position. Hence, biologists have increasingly pored over these approach areas, known as flanking regions, to see if they contain any hitherto undetected signals or landmarks for the enzymes. And a whole new order of control regions, known as enhancers, is now being investigated. Each gene, in short, is a small stretch of genetic landscape; the sequence of bases gives it particular landmarks, and these landmarks function as signals. Recombinant DNA has permitted us to see—to *begin* to see—how ingenious Nature is in giving DNA such harmonious and coordinated expression.

The UCSF authors concluded their paper with a discussion of diabetes. It was one of the few times in the entire course of the insulin work that molecular biologists spoke at any length about possible origins of the disease, not just expression products. Given the fact that both Type I and Type II diabetes can be genetic disorders, it followed that an understanding of the disease—and thus hopes for a possible cure—would be greatly abetted by the information contained in the gene. One of the ironies of all the mad dashes is that once the main genes were cloned, the cloners went on to other problems and longtime insulin researchers, so fearful in 1976 that their territory had been trespassed, could study the biology of the gene in relative peace.

It was during the spring of 1980 that Dennis Kleid searched around in vain at Genentech for a fairly standard piece of lab equipment. He had been working on improving the expression yields of the insulin clones, and had a hunch he wanted to check out. By that time, some of the most advanced biology in the world was being undertaken in the Genentech labs. Still, Kleid had to go out and buy this piece of equipment himself. It was called a microscope.

The Genentech scientists had been quick studies when it came to syllo-gisms of the marketplace. If the bugs didn't produce a lot of insulin, Genentech would fail to produce an economically viable process; and if the process wasn't economic, Genentech's work would simply remain an interesting laboratory demonstration of principle. Kleid and his colleagues worked furiously to boost yields. The effort concentrated on reducing the size of the huge beta-galactosidase molecule, more than a thousand amino acids in length, with the insulin chains trailing behind.

"Lilly was working on the beta-gal thing," Kleid recalls, "and trying to see how they could purify those peptides and link them together, and it was working okay. But the economics were lousy. It was maybe a hundred times more costly than regular insulin. So we had this economic problem. I remember our deal with Lilly said that they would have to make a certain amount of insulin per liter in order to meet the next benchmark. We had to figure out a way to make it economically."

It was Giuseppe Miozzari who provided the crucial clue. Yet another international member of Genentech's all-star cast, this Swiss biologist had studied in the Stanford University laboratory of Charles Yanofsky and had been recruited by Genentech in 1978 to join the scientific staff. At Stanford, Miozzari had done experiments with a suite of five bacterial genes—the so-called trp operon. These five genes made three enzymes which, one after another, synthesized the amino acid tryptophan. One of the enzymes, *trp* E, was much smaller than beta-galactosidase, so a cell's biochemical economy would not be nearly so monopolized by its production as by the huge beta-gal molecule. And because of some peculiarities controlling the tryptophan gene, its on-off switch could be artificially manipulated. These appealing features led Kleid and Heyneker to engineer a new insulin system. Instead of attaching the insulin chains to the tail end of beta-gal's thousand or so amino acids, they hooked up the A and B chains to the tail of *trp* E's 190 amino acids. It performed spectacularly.

"This *trp* system," Kleid recalls, "was driven by a promoter we could turn on and off with the density of the bacteria. You grow the bacteria up and then, when they got to a high density, the bacteria would run out of the amino acid tryptophan, and the bacteria would say, 'Let's make tryptophan!' And they would turn on, making tryptophan in our plasmid. The bacteria really went bonkers. The bacteria would just fill up with this protein." Every molecule of the tryptophan enzyme, of course, tugged along a chain of insulin in its wake, and the yields soared. "The amount of fusion protein, instead of being five or ten percent, was twenty or thirty percent," Kleid says. "It was really dramatic. So we introduced the chains into that system, got it cleared through the RAC committee, and sent those bugs to Lilly *just* in time to meet the yield requirement."

Kleid was so convinced the bacteria were bloated with insulin that he thought it might even physically distort the microbes. Everyone in the lab told him to forget it. "We didn't even have a microscope in the lab," he remembers, "so I went out and bought one, and sure enough, you could see protein precipitated in the bodies." Indeed, there is an allusion to the process contained in a brief report in the February 5, 1982, issue of *Science*, along with some dramatic photographs. The bacteria, distended with huge globules of protein, look like sausages primed to burst, almost disfigured by the pharmaceutical stuffing inside them. That stuffing, Dennis Kleid is pleased to relate, is human insulin. Even now, Kleid cannot resist the engineered beauty of those obese cells. "Look at this," he will say, grabbing a copy of the journal off a shelf. "This cell is just solid protein. Now let me tell you, Wally Gilbert never came close to making anything like that."

Lilly learned of the *trp* system in early 1980, when Kleid delivered the customized bugs to Indianapolis. The rest of the scientific community did not learn of the details until a patent application was filed, because neither Genentech nor Lilly ever published it. That was another side of industrial science. "We didn't tell anybody about this for years," Kleid says.

Meanwhile, Axel Ullrich, now working at Genentech, had not been idle. By the end of 1979, he had successfully cloned human preproinsulin in *E. coli*. Ullrich's insulin sequence, too, was tested in the *trp* system. So two possible routes to insulin were available. Lilly executives opted to develop the two-chain system of Kleid and Goeddel first, because they believed they could bring it to the market faster. Privately, however, they preferred the long-term potential of the proinsulin bugs engineered by Ullrich. The company promptly threw personnel and money at both initiatives.

By that time, Lilly had wholeheartedly lined up behind the project. "There were predictions early on that it was going to take—like everyone else thought—ten years to really bring the technology to fruition, but Genentech started hitting benchmarks *very* rapidly, so that started to make believers out of the people at Lilly," recalls William D. Young, a former Lilly manager and now Genentech's vice-president of manufacturing and process sciences. "They then thought, well, we'd better get some project people in place, begin to think about how we're ultimately going to commercialize this."

At this point, Swanson's scientists ran out of rope. They had lived up to their reputation as astute molecular biologists, but they knew nothing about industrial scale-up. They knew nothing about purification to satisfy human pharmaceutical applications. They knew nothing about animal trials and clinical trials and FDA approval, not to mention process engineering and marketing. Lilly did. By 1980, the insulin handoff was essentially complete. Lilly would shepherd recombinant DNA insulin to the marketplace; Genentech scientists moved on to other projects.

Later that same year, two of the seminal figures in the recombinant DNA revolution (to say nothing of the insulin race) enjoyed coincidentally convergent recognition from the world at large. Around ten in the morning on October 14, 1980, Jeremy Knowles received a telephone call in his office from his faculty colleague Wally Gilbert. Knowles had just finished teaching a nine-o'clock class, had a crowded day ahead of him, and thought that, by prior mutual consent, both scientists were too busy that day to converse. "Wally said hello," he recalls, "and I said, a little irritated, 'We both are very busy today, aren't we?' " Knowles reiterated that it was not a good day to communicate. Gilbert didn't seem to be getting the message. "We both are *very busy* today, aren't we?" Knowles repeated.

In the best of times, Gilbert could be conversationally circumspect, but as Knowles recalls, he hemmed and hawed and managed only to say, "I . . . it is too . . . Jeremy . . . it's rather busy here with the party."

"And with that," Knowles says, "the penny dropped."

The penny, worth $53,000 in this case, was a share of the 1980 Nobel Prize for chemistry. Gilbert had been an odds-on candidate for years, most recently for the development of the DNA sequencing technique with Allan Maxam, although his earlier studies on messenger RNA and the *lac* repressor work added up to a distinguished lifetime achievement in biology. Among admirers, it was only a matter of time before the Nobel committee got around to acknowledging his contributions.

As seems banally inevitable in these cases, Gilbert had learned the news from a reporter shortly after reaching his laboratory that morning; the official notification from Stockholm had been lost in transit and in fact arrived only weeks later. Gilbert shared half the chemistry prize with Frederick Sanger of Cambridge University, in recognition of their independently developed techniques to sequence DNA; the other half went to Paul Berg, professor of biochemistry at Stanford University.

Suitably chastened by his abrupt treatment of a Nobel laureate on the phone, Knowles was all complicity and accommodation when someone in Gilbert's lab called to say, "Jeremy, your place for a party? Three hundred people? Okay." There was a huge celebratory bash, and even some speculation as to whether Gilbert would accept his prize in an orange or purple turtleneck. He did not, of course; before going to Stockholm, however, he made a novel addition to his wardrobe: a dark suit.

At about the same time that Jeremy Knowles was taking Wally Gilbert's phone call that morning, shares in a new stock issue began trading on Wall Street's over-the-counter market. The frenzy that ensued is still talked about today by the investment community. Some 1.1 million shares of Genentech, at a price of $35 per share, were made available to investors on the morning of October 14, 1980. Within twenty minutes, riding the tide of high-tech glamour and pandemonious greed, the price skyrocketed to $89 per share. By the end of the day's trading, it had settled down to 71¼. Nonetheless, it was

295

without question one of the most spectacular Wall Street debuts in history. One analyst told the *New York Times*, "I can't recall any new issue having such a meteoric rise in such a short period of time. . . ."

In the space of several hours, Genentech effortlessly raised $38.5 million for its corporate coffers. All but lost in the flurry of bandwagon buying was the warning on the front page of the prospectus that stated: "The Common Stock offered hereby involves a HIGH DEGREE OF RISK." It was a day of at least financial vindication, too, for Robert Swanson and Herb Boyer. For their initial $500 investment in the enterprise, derided by scientific colleagues and businessmen alike as premature and doomed to failure, each now held 925,000 shares of the company and was worth, on paper, nearly $66 million. All this for a company which, for the first half of 1980, had reported skimpy earnings of $80,000 and had not a single product to offer.

For all the scientists who possessed chunks of stock—which included the early thoroughbreds like Kleid, Goeddel, Crea, Heyneker, and even Richard Scheller—the news was good, too. Scheller, for example, possessed fifteen thousand shares on the basis of his one summer of work making somatostatin DNA for Boyer's company, and looked upon this sudden wealth with bemused nostalgia. He recalls the crucial meeting at Cal Tech back in 1976, when Wally Gilbert dropped in; he recalls Gilbert mentioning blunt-end ligation and Art Riggs shaping the idea into linkers; and he remembers the practicable technology and the company that grew out of that technology. "And we're all still living rather well," he laughs now, "because of it."

To the man in the street, unfamiliar with the previous five feverish years of molecular biology, just what was this outfit known as Genentech? As Robert J. Cole wrote in the next morning's *New York Times*, "Genentech, based in San Francisco, is a four-year-old biotechnology company best known so far for having used gene-splicing to produce human insulin for the treatment of diabetes."

T w e n t y

"There Was Really No Crying Need . . ."

. . . .

Frederick Banting and J.J.R. MacLeod won the 1922 Nobel Prize for their discovery of insulin, but it was Eli Lilly & Co. that saved the University of Toronto researchers considerable embarrassment that same year by figuring out how to make this "wonder cure" quickly available to desperate diabetics. When the scientists in Toronto suddenly lost the touch in isolating and purifying this life-sustaining substance, Lilly's chemists rushed in to save the day. The effort was led by Lilly's legendary research director, H. B. Clowes, who had been perhaps the most enthusiastic listener in the audience on that day in December 1921 when Fred Banting haltingly described the discovery of insulin for the first time at a meeting of the American Physiological Society.

That same kind of scale-up effort was to be reenacted half a century later with genetically engineered insulin, and the task fell, fittingly, to Irving Johnson, who had arrived at Eli Lilly as a cancer researcher in 1953 and reported to, of all people, H. B. Clowes. Like his mentor before him, Johnson faced the unenviable task of making a mountain of insulin out of a molehill of feasibility. The skills to do so—he makes no bones about saying it—resided not in South San Francisco, but in Indianapolis. And only during the scaling-up phase did Genentech's scientists begin to sense not only the magnitude of the economics of the problem, but also the bigger picture of insulin and its market.

Dennis Kleid vividly recalls the moment when some long-cherished conceits about the uniqueness of genetically engineered insulin began to crumble in his own mind. It came during a trip to Indianapolis in December 1979, when the seasoned hands at Lilly showed the visiting upstart cloners through the insulin plant and explained to them how insulin was prepared the old-fashioned way.

"We were under the impression that it was impure and there was crap in it and stuff like that," Kleid remembers. "But that stuff was crystallized *two times* in the process. It was *incredibly* pure stuff. And there were only a few parts per million of contaminants. That was a real eye-opener. They were not making crap. Maybe they used to. But they certainly, in the last few years, weren't making it. And they impressed upon us very, very clearly that this [human insulin] was going to be *no* advantage at all." The newer way of making porcine insulin did not prompt allergic reactions, the biologists were told, and Lilly had enough glands to last for a long time.

As the tour of the facilities continued, the complicated economics of insulin production became clear as well. The Genentech scientists viewed a vault filled with vials of insulin. They were shown a freight car full of frozen pancreases, fresh from the stockyards, 25 cents per gland, shipped by the ton. They were shown where the glands were ground up prior to the extraction and purification of insulin. The recombinant DNA alternative was elegant and less messy, but not necessarily simpler and, more important, not necessarily cheaper, despite all the persistent suggestions, dating back to the ferocious public debates on recombinant DNA, that it might be. And this was the realization that Dennis Kleid had.

"With our stuff," he says, "you had to grow these bugs up. You had to expend energy and media to get your starting material from which you extracted the insulin. Then you have to do all this complicated processing. I mean, Lilly's starting material was essentially made with grass in the sun and cows walking around, and it was a by-product of making beef, so their starting material was even cheaper." This was the kind of rude economic awakening that many of the early entrepreneurial biologists experienced. Medical break-throughs did not necessarily follow from scientific breakthroughs.

The swiftness with which Eli Lilly scaled up the manufacturing process, and zoomed through the regulatory maze of federal agencies, nonetheless surprised many observers. Lilly had clearly warmed to the project as Genen-tech continued to hit its benchmarks. Even as Kleid and Heyneker were polishing off the new improved expression system, Lilly was seeking approval from the NIH's Recombinant DNA Advisory Committee (RAC) to use two-thousand-liter fermenters in their scale-up operation. The request was approved by NIH director Donald Frederickson in April 1980; three months later, Lilly's commitment to recombinant DNA technology was manifest to all when the company announced plans for multimillion-dollar pilot plant facilities in both Indianapolis and Speke, England, a suburb of Liverpool (the total expenditure, according to Lilly officials, amounted to between $70 million and $80 million).

Teams of half a dozen scientists had been assigned to develop the A chain technology, the B chain technology, fermentation, purification, and then putting the two chains back together again. No less than twelve separate tests were employed to ensure that the final product was pure and that no bacterial proteins contaminated the process. "No one knew, in fact, whether you could grow these organisms at large-scale in ten-thousand-gallon fermenters and maintain the genetic stability of the plasmid," recalls Irving Johnson. "It was clear if we were going to be the first—and we were convinced we were—that we'd have regulatory issues to deal with that would be breaking ground for everyone who came behind us."

Johnson and Ronald Chance led the in-house efforts, and human insulin was soon ready for clinical tests. In order for a drug to receive regulatory approval, it must be proved safe in animals and then successfully clear three stages of human testing. Phase I trials test the safety and dosage in healthy individuals, Phase II trials test its efficacy against the targeted illness, and Phase III trials compare this effectiveness against existing treatments for the same illness. It is a process of great expense and considerable uncertainty, which is why each time a drug company enters the process, it represents a huge roll of the corporate dice. As one financial analyst observes, "Most drugs work. The question is: what else do they do that you don't want them to, and how do they compare to drugs already on the market? You don't know until toward the end of clinical trials." By that point, a company may have thrown anywhere from $20 million to $100 million down a hole during development and testing.

From a regulatory point of view, Swanson's instinct to go with insulin back in 1976 proved to be a shrewd choice. More than half a century of data had accumulated on the performance of beef and pork insulins; since 1941, when the University of Toronto's insulin patent lapsed, the FDA began to run standardized tests on each batch of insulin produced by Lilly and the other major U.S. producer, Squibb. Unlike truly novel drugs, whose effects are being measured for the first time, Lilly's new insulin had historical yardsticks against which to be gauged, and this accelerated the process of approval.

Dr. Harry Keen, of the Unit for Metabolic Medicine at Guy's Hospital in London, headed the team that performed the initial clinical tests in 1980. Seventeen male nondiabetic subjects, ranging in age from twenty-four to fifty-six years, served as the human guinea pigs. Keen's team injected Genentech's insulin under the skin to see if it caused allergic reactions. They injected it into folds of the abdominal skin. They injected it directly into the bloodstream. Technicians took blood sugar readings, sometimes every ten minutes; other clinicians slapped thin sheets of cellophane over each and every injection site, meticulously tracing the circumference of welts and bruises. Finally, the human insulin was compared against porcine insulin as it was absorbed into the system. Here its performance, if not representing dramatic improvement, was at least comfortingly similar to that of pork insulin.

As the team noted, in a "Preliminary Communication" published in the British medical journal *The Lancet* on August 23, 1980, "there seems to be little demonstrable difference between the purified porcine and the genetically synthesized human insulins." But perhaps the more important historical point resided in an unrelated comment. "The experiments described in this paper," Keen's team wrote, "report not only the first use of human insulin produced by recombinant DNA but also, to our knowledge, the first use of any recombinant DNA product in man."

One by one, hurdles were surmounted, tests completed, data assembled. By February 1981, there was enough biosynthetic hormone available to begin widespread testing in about two hundred diabetic volunteers in Philadelphia, Detroit, Minneapolis, Seattle, Trenton, and Wichita. By May 14, 1982, Eli Lilly had filed a New Drug Application, or NDA, with the Food and Drug

Administration. Such a filing ordinarily initiates a fifteen-to-twenty-month regulatory slalom through the agency, but there were expectations from the outset that the FDA would act more swiftly on human insulin. Lilly, meanwhile, came up with a name for the first pharmaceutical product of the recombinant DNA era: Humulin.

Any hopes Lilly had of hustling its human insulin into Europe as standard-bearer of the biological revolution, however, met stiff resistance from Novo Industri. The Danish company, which controlled about 60 percent of the European market, had already tested its chemically created human insulin; clinical trials indicated it performed the same as porcine insulin. In early 1982, the governments of France and Austria approved the sale of "Novolin," which meant Novo had the honor of marketing its product as the first "human" insulin. Lilly was thus outgunned in its first marketing battle in Europe.

But the head-to-head battles didn't stop there. In the spring of 1982, Novo formed a partnership with the U.S. pharmaceutical firm E. R. Squibb & Sons, which contracted to market Novo's insulins in the United States. In the meantime, Novo's semi-synthetic human insulin won government approval in Great Britain and West Germany. And, to complete an elliptical research circle, Novo contracted with Biogen in 1981 to develop a human insulin by recombinant DNA means that might be cheaper to make than its chemically clipped version.

On October 29, 1982, roughly six and a half years after Eli Lilly sponsored its symposium on insulin and recombinant DNA technology, unofficially kicking off this scientific and industrial saga, Lilly received approval from the U.S. government to sell Humulin to diabetics. By 1983, it was in the refrigerators of pharmacists. Until October 1985, when Genentech's second product, human growth hormone, was approved for sale, Humulin remained the only recombinant DNA pharmaceutical to reach the U.S. marketplace.

"After all the excitement, all the talk, and all the money," the *Economist* wondered in 1982, "has it been worthwhile?" Did human insulin represent, as Wally Gilbert suggested to the Cambridge City Council back in 1976, "something which is, in a sense, beyond price"?

New drugs, like new movies, face critics in the trade and business press. Humulin, it must be said, did not receive overwhelmingly positive reviews. Almost from the moment it reached the marketplace, there was disagreement as to whether human insulin represented a major therapeutic improvement over the purified pork preparations already available. That debate continues to this day. A sampling of initial reaction shows, however, that Humulin was greeted more as a technological than a medical breakthrough, and that this sentiment was building even before the drug reached pharmacies. As early as 1980, the British magazine *New Scientist* reported, "Other big chemical manufacturers predict that Eli Lilly's massive $40 million investment in two plants to make insulin (one in the U.S. and one in Britain) may be a classic example of backing a loser."

Business Week noted, "Without question, Eli Lilly & Co. has gained a place in manufacturing history by being the first company to produce a drug for humans using recombinant DNA technology." But Dr. Henry Miller, medical officer for the FDA's National Center for Drugs and Biologics, told the *Washington Post* that "the advent of human insulin therapy is not likely to constitute a medical breakthrough for insulin-requiring diabetics." *Biotechnology Newswatch,* an industry newsletter, observed that FDA approval "is being viewed by experts as a giant step for genetic engineering, but only a marginal advance for medical science." And the stuffy *Economist* seemed to answer its own question—was it worthwhile?—by concluding its appraisal: "The first bug-built drug for human use may turn out to be a commercial flop. But the way has now been cleared—and remarkably quickly, too—for biotechnologists with interesting new products to clear the regulatory hurdles and run away with the prizes."

That in fact may be Humulin's great legacy. It is still too early to tell if human insulin will be a big money-maker; early sales trends indicate increasing market penetration, although Lilly may be selling it for less than it would like, in the opinion of physicians. There is little question, however, that work on the hormone cleared the way, politically and scientifically and perhaps even psychologically, for the more exotic products promised for the future—the interferons, the interleukins, and the other anticancer agents. Once one recombinant DNA therapeutic had cleared the invisible threshold of social acceptance, the path for subsequent products was likely to be smoother.

These less than sterling reviews, it must be said, came as a surprise to some of the scientists involved, who are better at predicting scientific consequences than medical ones. Dave Goeddel of Genentech, for example, concedes that he and Dennis Kleid were probably "brainwashed"—his word—by the infectious optimism of Bob Swanson and others who argued *a priori* that human insulin would be significantly better than what was already available. "We didn't hear all the other arguments," Goeddel says now. "We assumed we were going to make a product—or *I* assumed, and probably Dennis did—that we were going to make a product better than the existing one. That was the idea." Kleid demurs: *"I* certainly thought it was well worth doing. No one brainwashes me." Art Riggs makes the most impassioned case for the value of genetically engineered insulin. "Human insulin has a very marginal advantage over porcine insulin, and no advantage over the human insulin made by Novo's semi-synthetic process," he admits. "What is a fact is that approximately ten percent of diabetics become allergic to bovine insulin, and even when they're switched to porcine, they still can be in trouble. In 1978, I was told that it was impossible to supply all the world with porcine insulin. Given our present state of technology, it is medically wrong to start any diabetic on bovine insulin, because there's a ten percent chance of an allergic reaction."

Walter Gilbert's thinking on the subject underwent radical evolution. In 1977, in applying to use MIT's P3 laboratory for their cloning experiments, the insulin researchers noted that a worldwide insulin shortage was anticipated within "three to five years." One year later, in 1978, he says he began to think about the problem differently. "It's unclear that there's a real use for human insulin," he says now, "unless as a hype, public relations use, to give you the real human protein. Because it's unclear that it's anything doctors have really moved toward, or any benefit. There's a *slight* difference in the absorption [compared to porcine insulin], but it's unclear what that has to do with it." Only later, Gilbert says now, did he realize that "there wasn't really a shortage of insulin," that genetically engineered insulin was unlikely to be cheaper, and that it had no particular immunological advantage over purified pork preparations.

Was human insulin, in fact, better than the insulins diabetics were already using?

Consider some of the claims made for human insulin in 1978, when Genentech's achievement was presented to the public, and in 1986, when at least some answers were discernible. When Rachmiel Levine, then deputy director of research at the City of Hope and a diabetes expert of long standing, described the work at the September 1978 press conference, he predicted three major consequences for human insulin. Genetically engineered insulin, he suggested, would prove to be less allergenic and less expensive and should alleviate the problem of potential shortages. He was asked, seven years later, how those predictions panned out.

Is human insulin less allergenic?

"Oh, yes," said Levine. He, too, noted that about 10 percent of all diabetics are allergic to animal insulins, especially beef. In most cases, this allergy amounts to a dermatitis-like rash. That response is eliminated with human insulin, although some patients form antibodies even to the human hormone. In addition, Levine noted that some complications from diabetes may stem from the use of animal insulins, which differ slightly from human insulin. "We do not know whether the severity or the occurrence of the ordinary complications of diabetes may have in time something to do with the difference in chemical structure of the insulins. In time, the more like human insulin you can give, maybe the better it is." That wouldn't become clear, Levine hastened to add, in less than ten or twenty years, but it could represent a substantial improvement.

Is human insulin cheaper?

"No," replied Levine, "that hasn't turned out to be the case yet." At that point, perhaps the only major difference between human insulin by recombinant DNA and purified pork preparations was price: the porcine insulin cost less than Humulin. Not by a whole lot—all that purification processing is not cheap—but usually a couple of dollars less per bottle than Humulin. (Recently, the price differential has vanished, according to diabetologist Jay Skyler, "and purified pork is essentially dead as a product.")

Finally, has human insulin alleviated the problem of supply?

"Well," Levine answered, "we haven't reached the stage where it needs alleviation. But one can never tell."

To hear some tell it, there never was a supply problem with pig pancreases in the first place. "The whole thing was rubbish," insists Paul Haycock, research director at Squibb-Novo. "There never was a shortage of pig pancreases, and there never will be." Haycock blames the scare on a miscalculation by an official who had prepared projections for the Food and Drug Administration—a miscalculation based, ironically, on a mistake in an Eli Lilly training brochure which confused kilograms with pounds. Instead of projecting an insulin shortage by 1982, a revised FDA report predicted adequate insulin supplies through the year 2006. In any event, there is never likely to be a shortage of insulin caused by scarcity of pancreases. A spokesman for Squibb-

Novo confirms that "all of Novo's insulins will be made by recombinant DNA—not overnight, but it will happen." "I think that's the trend," agrees Lilly's Irving Johnson. "When we will reach the point when all the insulins are recombinant—I don't know how to predict that."

"Let me point out to you," Rachmiel Levine said at one point, "that there was really no crying need for human insulin. The refinement of getting human insulin rather than animal insulin has to do with the fact that it differs in chemical structure to a small extent from the ordinarily used pig or beef insulin, and that therefore it has a *slight* antibody situation so that allergies are more common than with human insulin. But really, the incidence of that is not very severe—insulin allergies or antibody phenomena. It's not severe at all. It occurs. But it wasn't something that cried out for doing something drastically."

There are several lessons in all this. One obvious point is that scientists are scientists; they usually are not medical doctors and they almost never are economists. Yet during the pioneering days of biotechnology, the press, the public, and sometimes even sophisticates of the financial community, seduced by the kind of supposition scientists normally loathe to indulge in, invested a great deal of credibility (to say nothing of money) in highly speculative ruminations on the potential efficacy of a genetically engineered drug or the ultimate production costs of a given pharmaceutical. As this one case study shows, insulin has turned out not to be cheaper. Insofar as the price has dropped at all of late, it is primarily due to the competition between Lilly and Novo, not because the new way of making insulin is inherently cheaper. And since neither is human insulin significantly better than the cheaper porcine insulin that was already being used, some physicians have been reluctant to prescribe it, although Lilly claims that "a very significant amount" of all newly diagnosed diabetics start out with human insulin.

A less obvious point, at least from the point of view of basic research, is that it can be short-sighted to hold scientists too accountable for their medical and economic predictions. Human insulin may not have revolutionized the lot of diabetics, just as subsequently interferon, the cancer-killing "magic bullet," has yet to live up to early expectations, but molecular biology has generated so much useful information, even on the way to less-than-blockbuster products, that the value of any particular series of experiments may not lie in its conclusions, but in the refinement of steps along the way. The public often (perhaps too often) thinks of biology solely in terms of the practical medical benefits that come out of research. Scientists themselves think of it as a process during which information—some of it spectacular, much of it obscure—diffuses in a vast, optimistic conspiracy of knowledge and problem-solving. In that sense, the work on recombinant DNA insulin sent ripples to the farthest corners of biology.

There is a final postscript to this story. The work on recombinant DNA insulin never promised to produce a cure for diabetes, only the potential for improved therapy and, perhaps, clues to the origin of the disease. But there continue to be other routes of attack on the problem. Several of them threaten to make insulin a thing of the past.

One promising strategy has been primarily pursued by an insulin researcher named Paul Lacy at Washington University School of Medicine in St. Louis. Lacy attended the famous insulin symposium at Eli Lilly in 1976, although he did not give a talk and certainly was not among the molecular biologists who went after the gene. The more interesting question to Lacy was: is there a way to restore insulin production to a diabetic who has lost it?

In January 1985, Lacy and a surgeon-colleague at Washington University, Dr. David Scharp, provided the first tentative answer to that question. They successfully implanted healthy, functioning islet of Langerhans cells into the spleens of human diabetics. Following surgery, these diabetics were for the first time in years able to produce natural insulin, although they still had to rely on supplementary injections. There is a long, long way to go: only severe diabetics who had previously undergone kidney transplants qualified for the procedure; three out of six patients lost function of the foreign islet cells; and the ability to make insulin lasted no more than one to two months. A seventh patient, with islets from her own diseased pancreas seeded into her liver, nonetheless has enjoyed functioning insulin production for ten years.

There is much more work to be done to prove the safety and effectiveness of the islet transplants, but some biotechnology companies are exploring the concept of enclosing islet cells in permeable man-made membranes that prevent problems of rejection. William Chick, the University of Massachusetts diabetes expert whose rat tumors sparked so much interest in Indianapolis in 1976, is experimenting with semipermeable, veinlike tubes, filled with insulin-producing animal islets, which have been successfully implanted in test animals. The prospects are so promising in 1986 that one financial analyst, Scott

King of Montgomery Securities in San Francisco, predicts that if the rejection problems can be licked, "insulin will be obsolete in ten years."

This is not to diminish all the work—conceptual and technical, basic research and applied—that went into the insulin molecule during the late 1970s. While the marketplace product that emerged in this case may not have revolutionized the care of diabetics, all the techniques discovered and protocols refined and enthusiasm generated by the work have put us well on the way to revolutionizing our understanding of diseases, all diseases, and ultimately of how we may cure them.

Epilogue:
"After All
the Fever . . . ."

Ten years after it all began, ten years after the meeting at Eli Lilly and Co. effectively kicked off the race for human insulin, it is worth indulging in a game of "what if." Gazing back from the mid-1980s, with Genentech now widely regarded as the most successful start-up biotechnology company, one wonders what would have happened if the Harvard team headed by Wally Gilbert or the group at the University of California–San Francisco had expressed human insulin first. What if Biogen, and not Genentech, had enjoyed the tremendous surge of interest, the enormous boost in credibility, that accompanied such an accomplishment?

The initial, and predictable, response at Genentech is to say that it would not have made any difference at all. But with elaboration, this assertion becomes a bit more tentative. "With hindsight," says Arthur Riggs, "I think I could say that it probably wouldn't have changed things much." But he adds, "Let's say if our competitors had got the natural human insulin gene cloned prior to our reaching an agreement with Lilly, then it might have affected the agreement with Lilly, which could have changed the entire picture. Genentech needed those dollars from Lilly."

"The whole thing could have changed," Dennis Kleid concedes, "*if* Gilbert really got it first. I would say that the problem would have been, somebody else getting it first and Lilly signing up with them. Then we would have been stuck with an also-ran, second to Lilly, and a university would have shown again that they knew how to do molecular biology better than a company." Robert Swanson has trouble imagining the company finishing second in this race, but he does admit that if Genentech had lost out to either UCSF or Biogen, "the world would clearly be different."

For what it is worth, the Harvard researchers firmly believe that had they not been forced by the NIH regulations to travel to Porton, the outcome might have been entirely different. "If we hadn't had to go to England, we would have gotten the human clone," says Lydia Villa-Komaroff. "Very quickly, I think." There is an important technical point to make here, though: without the refinement of Art Riggs's link of methionine, the Harvard researchers still would have produced an insulin molecule biochemically welded to a bacterial protein. The Genentech team had figured out a way to harvest those molecules.

The insulin story represents but one bubble in that vast pot of ferment in the mid-to-late-1970s world of molecular biology. Looking back upon events with a decade's perspective, however, some events and decisions appear larger and sharper, while others recede into insignificance. The work on recombinant DNA insulin not only produced classic papers in molecular biology at the time, but it is clear now that the scientists who worked on it continue to find themselves in the vanguard of both basic research and biotechnology. Genentech, Biogen, Chiron, California Biotechnology—all these respected gene-splicing firms in a sense grew out of laboratory work on insulin or growth hormone.

Other events resonate with odd echoes ten years later. Public fears about recombinant DNA work, as they surfaced at the Cambridge City Council hearings, have not totally dissipated, but scientists more than ever are convinced of the safety and value of the techniques. Ten years ago, scientists worried that the commercialization of biology might compromise basic research in academia; now, with federal funding cutbacks threatening virtually every laboratory, scientists are wondering how industry can be coaxed into supporting more of the same kind of academic research that led to the recombinant DNA revolution in the first place. Ten years ago, scientific competition seemed fierce and sometimes cutthroat; now, with a shrinking federal funding pie to share, budget pressures almost certainly will make competition more strained.

Ten years ago, patents on work stemming from basic research were considered unethical by some. A few individuals and companies, according to MIT microbiologist Jonathan King, "took the position that it was okay to take publicly funded research into the commercial sector, privatize it, and sell it back to the public"; this still rankles many investigators, but the value of these patents remains unclear. Some financial analysts predict that, in effect, they won't be worth the paper they're printed on. Nor have biotech businesses necessarily crimped scientific communication, as once feared; Genentech's lab groups, for example, are widely admired in academia for their basic research, and employees often give seminars at academic institutions.

From the scientific point of view, Herb Boyer's decision to employ the synthetic DNA approach for Genentech's first projects gave the company an enormous edge getting out of the starting blocks; conversely, the complex organization of genes in higher organisms, as revealed by discoveries in the mid-1970s, has made the cDNA approach pioneered by the Harvard group and others the preferred way to clone and study genes these days. Both approaches, in a sense, were correct, but for different reasons and with different ramifications.

With a decade's perspective, the uproar over the pBR322 incident at UC–San Francisco seems utterly insignificant from the scientific point of view. What does not seem insignificant, even now, is the reluctance of the scientists involved to inform the public and acknowledge the misunderstanding when it happened. Many scientists believe the take-home lesson from the tumultuous debates of a decade ago is to keep their own counsel. As long as an attitude of self-serving discretion pertains, scientists will be doomed to repeat the same mistakes all over again, with the added risk of diminishing all of science's credibility in the public eye. According to recent press reports, several biotechnology companies have run into precisely those kinds of difficulties because they either neglected to inform local communities of their intentions or interpreted federal regulations in almost comically convenient fashion. The risk to human health and safety, most scientists would agree, is very small indeed, but the risk to the social credibility of scientists could be significant.

Last of all, looking back, perhaps the most impressive aspect of the race for insulin is the sheer energy and optimism and persistent resourcefulness of the biologists themselves. Regardless of the strategic approach, regardless of the ultimate aim (whether basic knowledge or a pharmaceutical product), virtually all the scientists were, to paraphrase Peter Medawar's crucial characterization, artists of the soluble.

In the view of the pharmaceutical industry, the reports on human insulin are still mixed. Lilly has maintained its dominant market share in U.S. insulin sales, but the Squibb-Novo product has slowly encroached on the domestic turf. Some analysts had predicted that by the end of 1985, the Novo share would reach about 25 percent, Lilly slipping to 75 percent; but Lilly fought off the Novo challenge, according to observers, by cutting prices in the hospital market and holding down Novo's share to about 22 percent. Sales of Humulin, according to these same analysts, amounted to only about 6 to 8 percent of the U.S. market, but sales were definitely improving: according to market observers, worldwide sales of Humulin have risen from $8 million in 1983 to an estimated $90-100 million in 1986, with a commensurate jump in royalties for Genentech. Lilly officials believe that as production experience is gained, costs will drop.

The jury is still out, however, on any advantages of human insulin over purified pork preparations. Dr. Mayer Davidson, director of the Diabetes Center at Cedars-Sinai Medical Center in Los Angeles, reflects a general sentiment among clinicians when he says, "So far, I don't think it's been a big breakthrough." Dr. Andrew Drexler, of New York University Medical Center, agrees, saying, "As a general phenomenon, we all see it as a minor, incremental improvement." By 1986, Lilly was well on the way to completing its conversion to the proinsulin-cDNA approach pioneered by Axel Ullrich. Proinsulin is now in advanced clinical trials.

Some observers doubt that the price will ever drop significantly. Why? "It's called an oligopoly," says analyst Scott King.

What has happened to the major players in the insulin story?

Ruth Hubbard, the Harvard biologist who frequently questioned the need for human insulin and the use of recombinant DNA, now belongs to the Boston-based Committee for Responsible Genetics. She believes that some useful things may come out of recombinant DNA technology, and distances herself from the more extreme opposition of Jeremy Rifkin. "But I think," she says, "that we are very skeptical that it is going to be done properly and usefully

311

in a competitive, secretive, capitalist way of doing things, where the control is concentrated in the hands of a few major entrepreneurs who stand to profit from taking shortcuts. So I don't think that the technology can be judged in a vacuum. Given the economic and social realities, I'm very skeptical of it. I think, given quite different social and economic realities, I would want to look at it again. I'm not worried about its offending against God or nature, but I think that in the real world of the late-twentieth-century competitive free enterprise—it's not very free, but that's what it's called—people's health and safety are not going to be the primary concerns."

A prediction Hubbard made at the National Academy of Sciences forum in 1977 bears repeating here. "We know," she said, "that if we produce more insulin, more insulin will be used, whether diabetics need it or not." Those words have an eerie prescience in late 1986 amid reports that a black market has developed around the second product of the genetic engineering revolution: human growth hormone. According to an article in *Science* about the new "cosmetic endocrinology," parents are demanding the drug so that their children will grow not to normal, but *desirable*, height. When told that one of his children would grow to be five feet seven, the story noted, one father replied, "That's absolutely unacceptable!" Easy access to powerful new biological agents presents important dilemmas for the public, and perhaps new, undreamed-of potential for drug abuse.

THE RUTTER-GOODMAN GROUP

John Chirgwin left the Rutter laboratory at the end of 1978, served as an assistant professor of anatomy and neurobiology in the school of medicine at Washington University in St. Louis until 1986, then moved to the University of Texas Health Center in San Antonio as an associate professor in the department of medicine. Raymond Pictet, having worked with Bill Rutter since 1969, left San Francisco in the fall of 1980 and assumed a research position at the Jacques Monod Institute at the University of Paris, where he continues his basic work on insulin. Unusually dour in his retrospective assessment of the rat insulin experiment at UCSF, Pictet says, "I think the originality was in the technique, and probably really Herb Boyer or Stanley Cohen or Paul Berg or David Baltimore has more to do with cloning insulin than the people who actually did it."

John Shine, after returning for several years to his native Australia, joined the entrepreneurial gold rush and re-expatriated himself to California in 1984; he served as scientific director of California Biotechnology, Inc., a leading biotech company in Mountain View, California, until April 1986, when he was promoted to president. Of the period 1975–78, he says, "I doubt that *anything* will ever compare, for me, to those few years in San Francisco. . . . It was a point where a whole area of science had suddenly become possible, and we were there to pioneer it."

Peter Seeburg, upon reporting to Genentech at the end of 1978, immediately got into a race with his former UCSF mentors, Howard Goodman and John Baxter, to clone and express the human growth hormone gene. Genentech was the first to announce the feat, in July 1979, although the UCSF researchers had already submitted a paper to *Science* claiming the same achievement.

It turned out to be an expensive first for Genentech, however. Since Seeburg transferred his growth hormone clones from Goodman's lab to Genentech without prior consent, a dispute ensued and the company arranged a cash settlement with UCSF in June 1980. Genentech's prospectus contains information noting such a transaction amounting to $350,000; Seeburg confirms that this was compensation for the clones he took to Genentech. By the end of 1986, Seeburg expected to quit his position as a senior scientist at Genentech and return to West Germany to head a research lab at the University of Heidelberg.

In 1981, Howard Goodman astounded colleagues in the scientific world by making a unique and far-ranging research deal with the West German chemical and pharmaceutical company Hoechst AG: the company agreed to establish a molecular biology department at Massachusetts General Hospital for Goodman and fund it over a ten-year period to the astronomical tune of $67.3 million, in exchange for which Hoechst would receive first and exclusive licensing rights to develop commercial products out of any of the basic research results achieved by the Goodman group. "It was the insulin work," according to an article in *Science* magazine in 1982, "that first brought Goodman to the attention of Hoechst scientists. . . ."

William Rutter stepped down after thirteen years as chairman of the UCSF biochemistry department in 1982, and resumed research as Hertzstein Professor of Biochemistry. In 1981, motivated in part by his lab's successful cloning of the hepatitis B virus in yeast and the potential development of a vaccine, Rutter became founder and chairman of Chiron Corp., a highly regarded biotechnology company based in Emeryville, California. The firm went public in 1983, and in 1986 Merck Sharp & Dohme received FDA approval to market the hepatitis vaccine developed by Chiron—the first genetically engineered vaccine to receive government certification for human use. Chiron is now working on genetically engineered vaccines for AIDS, malaria, and herpes. Rutter continues to work on pancreatic development at his UCSF laboratory.

Now a senior scientist at Genentech, Axel Ullrich remains one of the foremost cloners of his generation. After successfully isolating and cloning human preproinsulin (the process by which Eli Lilly will make its human insulin), he contributed to the Genentech efforts on interferon, epidermal growth factor, and nerve growth factor; more recently, he cloned the oncogene for human breast cancer, which may help explain the origins of this disease. Contrary to some early dire forecasts, industrial labs such as Ullrich's have consistently produced results of great biological interest. Ullrich's work has helped uncover a curiosity of great interest to cancer researchers, which is that the body's so-called growth factor receptor genes bear a striking similarity with several oncogenes, the genes that apparently trigger the development of cancers.

Bill Rutter and Ullrich found themselves dueling over an important and longstanding basic question several years ago when their respective labs, along with perhaps a dozen other groups, raced to clone the insulin receptor—the slot on the outside surface of the cell into which the insulin molecule fits like a key, permitting glucose to enter cells. Once again it was a close finish. Rutter's twelve-person team reported its results in the April 1985 issue of *Cell*. But Ullrich's fifteen-person team reported its results in the February 28, 1985, issue of *Nature*. Says Alan Permutt, a longtime insulin researcher, "Axel is not only incredibly brilliant, but also incredibly efficient at getting the job done. At least fifteen groups were working on the insulin receptor gene, and they all had head starts. But he was the one to do it."

THE GILBERT GROUP

Forrest Fuller, Wally Gilbert's beleaguered graduate student, received his Ph.D. from the Harvard Bio Labs and now is a staff scientist at California

Biotechnology, where he has studied atrial nitriuretic factor, a hormone that influences heart function. One of Fuller's colleagues there is Karen Talmadge, once voted "most popular" in a *Midnight Hustler* poll. Talmadge briefly studied patent law at Columbia Law School after getting her Harvard Ph.D. and then did her postdoc at Cold Spring Harbor; she now is a chief scientist and patent coordinator at Cal Biotech. Stephanie Broome occupied a laboratory in Biogen's Cambridge headquarters while completing work on her Harvard Ph.D., and is now doing a postdoc at the Scripps Clinic for Experimental Science in California.

Peter Lomedico, after completing his postdoc in the Gilbert lab, became a staff scientist in molecular genetics at the pharmaceutical giant Hoffmann-LaRoche in Nutley, New Jersey; in 1984, Lomedico headed one of several U.S. groups which successfully cloned the gene for interleukin-1, a protein known as a lymphokine, believed to play a unique role in the body's immunological response and to have potential as a cancer-fighting agent. After reluctantly relinquishing editorial control of the *Midnight Hustler,* Allan Maxam moved across the Charles River to head up his own laboratory at the Dana-Farber Cancer Institute in Boston.

Argiris Efstratiadis now leads a research group at Columbia University in New York, where he is a professor and continues to investigate the biology and evolution of the insulin gene. He has published several volumes of translated poetry in Greece, including works by Federico Garcia Lorca, and is currently translating the poems of Sylvia Plath into Greek. His companion from graduate school days, Lydia Villa-Komaroff, left Harvard in 1978 to head a molecular biology laboratory at the University of Massachusetts Medical Center at Worcester (where one of her colleagues was William Chick). She left Worcester in 1985 and now heads a lab group in the Department of Neurobiology at Boston Children's Hospital, studying the molecular biology of brain development.

In 1982, Walter Gilbert gave up his full professorship at Harvard University and abandoned his American Cancer Society chair at the university to run Biogen, the Swiss-based biotechnology company he helped found, on a full-time basis (when Biogen dedicated its new headquarters in Cambridge, former mayor Alfred Vellucci was on hand to snip the ribbon). Biogen's scientific staff has produced a number of significant achievements, beating Genentech in announcing the cloning of human alpha interferon, developing tumor necrosis factor (a potential cancer-fighter), and isolating an important cellular protein known as Mullerian Inhibiting Substance. However, Biogen has not yet duplicated the business success of its cross-continent rival Genentech. When the company went public in March 1983, investor ardor for biotech companies had significantly cooled; the stock, offered at $23 per share, has remained well below that price, although the initial offering raised $57 million and Biogen currently has a market capitalization of $250 million. The company lost $11.6 million in 1983, $13 million in 1984, $19 million in 1985, and more than $28 million in 1986. With expenses far outstripping revenues, Gilbert was under enormous pressure to cut back operations and elected to resign as chairman and chief executive officer of Biogen in December 1984; by the beginning of 1987, Biogen had entered negotiations to sell its Geneva operations, and the company was said by the *Wall Street Journal* to have "faltered." Gilbert, meanwhile quietly resumed research in the Harvard University laboratory of his friend and colleague Jeremy Knowles until 1986, when he rejoined the Harvard faculty as H. H. Timken Professor of Science. He is now back in the Harvard Bio Labs, setting up his new group. He remains on the Scientific Advisory Board of Biogen.

Gilbert's scientific reputation has held up considerably better than his boardroom rating, and he is certainly among the leading theoreticians in molecular biology today. The once controversial terms coined by Gilbert, "introns" and "exons," though perhaps not universally admired, are universally used; more important, Gilbert's theory on "genes in pieces" received dramatic (though still inconclusive) support in research reported by the University of Texas laboratory of Michael S. Brown and Joseph L. Goldstein, winners of the 1985 Nobel Prize. Their research supports the notion, originally advanced by Gilbert, that exons, the expressed portion of genes, may in fact act like minigenes (or genetic cassettes) that code for functional subunits of a protein and can be shuffled around an organism's entire chromosome; Nature somehow manages to move these elements from place to place, in effect creating entirely new combinations of these subunits in a way that dramatically accentuates the impact of mutations. Gilbert, meanwhile, has become one of the leading advocates of perhaps the biggest "Big Science" project ever in biology: the quest to sequence the entire human genome.

The Harvard P-3 laboratory, which sparked the controversial public hearings in Cambridge, finally opened in October 1978, shortly after the Gilbert group returned from the unsuccessful sojourn in England. When the NIH relaxed its guidelines in 1979, the lab became almost instantly obsolete and, after a final cost of nearly $1 million to the university and to taxpayers, was decommissioned in January 1983. It remains the conviction of the Harvard researchers that if they hadn't had to go to Europe, they might well have won the race for human insulin.

THE GENENTECH GROUP

"It's always hard to imagine that," says Robert Swanson, who continues to serve as chief executive officer at Genentech. "What's happened is that essentially every important human pharmaceutical [to come out of recombinant DNA technology] has been done here. Some of the races were closer than others, but we've continued to win those races. We organized ourselves to be first, and we succeeded."

Swanson still occupies his small corner office in the warehouse, but the huge building and eight more like it have been built or converted into a sprawling research, development, and manufacturing center employing about eleven hundred people. By 1985, Genentech had arranged major business collaborations with Corning Glass Works, Hewlett-Packard (later terminated), and Travenol Laboratories, and by 1986 was valued on the market at $2 billion—just as its first home-grown product, Protropin (human growth hormone), reached the market. Genentech, which has reported small but growing profits every year since 1980, is widely considered to be the leading biotechnology company, and Swanson has publicly predicted that it will hit annual sales of $1 billion by the 1990s.

One of the major reasons has been the work of David Goeddel, who has headed Genentech's successful efforts to clone several alpha interferon genes, beta interferon, gamma interferon, human growth hormone, tissue plasminogen activator (TPA), and tumor necrosis factor—virtually all the key proteins in the pipeline as major pharmaceuticals. In heading the team that cloned TPA, Goeddel provided Genentech with a product that appears so successful in dissolving the blood clots associated with heart attacks that it will become Genentech's company-transforming product; the drug is expected to receive regulatory approval in 1987, and analysts predict sales of at least $750 million by 1990. Goeddel became director of Genentech's molecular biology department in 1980, at age twenty-nine. Dennis Kleid, senior scientist and patent agent, led the Genentech research team that produced an animal vaccine for foot-and-mouth disease. The paper reporting that achievement won the American Association for the Advancement of Science's prestigious New-comb-Cleveland Prize in 1982.

Herb Heyneker has never looked back since abandoning a lifetime appointment at the University of Leiden to rejoin Genentech in 1978. Technically Genentech's first scientist, dating back to his days in the Boyer lab, Heyneker is now vice-president for research of a Genentech joint venture with Corning Glass known as Genencor, which is attempting to develop industrial enzymes. Heyneker's cloning colleague, Francisco "Paco" Bolivar, resisted the lure of a job at Genentech and returned to Mexico, where he is a professor of molecular biology at the National University of Mexico and directs the Research Center for Genetic Engineering and Biotechnology in Cuernavaca. Richard Scheller, who worked briefly on the somatostatin project, teaches at Stanford University and currently investigates neurobiology. After heading up Genentech's synthetic DNA department from 1978 to 1981, Roberto Crea left to form his own biotech company, Creative Biomolecules, which is now based in Hopkinton, Massachusetts.

Herbert Boyer, whose confidence in recombinant DNA as a commercial process helped launch the biotechnology revolution in 1976, came to symbolize the academic scientist who sold out to business, and he thus became the symbolic target for criticism about such boardroom biologists. "It wasn't difficult being associated with the enterprise," he said in a 1985 interview. "It was difficult being associated with the controversy." Indeed, colleagues suggest that in response to the enormous crush of attention, culminating in a *Time* magazine cover story in 1981, Boyer retired to his UCSF laboratory, maintained a discreet distance from the day-to-day operations of Genentech, rarely consented to interviews, and sought to rehabilitate his scientific reputation. He began investigating methylation patterns in bacterial DNA, sites which indicate how key cellular proteins (such as repressor molecules or restriction enzymes) interact with the double helix. In a more successful reprise of the "Magnificent Failure" of 1976, Boyer's group worked with John Rosenberg's team at the University of Pittsburgh in discovering the minute, atomic interaction between a protein (*Eco* RI) and DNA; a commentary in *Science* described the work recently as a "milestone" in molecular biology. In 1985, Boyer was elected—an election long overdue, in the view of admirers—to the National Academy of Sciences. There has been frequent published speculation that Boyer's association with a biotechnology company harmed his chances of winning a Nobel Prize.

Of the insulin project, in which he played a self-described "avuncular" role, he says now: "I'm not one to put one value in absolutes. I don't think in terms of a single step or a single experiment. You have to look at the whole trip. But I think it is fair to say that insulin was a very significant step along the way in establishing Genentech as a major force in the genetic engineering community."

On the basis of his work on somatostatin and insulin, Keiichi Itakura now presides over the Department of Molecular Genetics at City of Hope National Medical Center. His ground-breaking wizardry in synthesizing DNA helped pave the way for automated synthesizers, which can now build a small gene in a matter of hours. Arthur Riggs, the biologist who recruited Itakura, is chairman of the biology division at City of Hope. In addition to his administrative duties, he is researching DNA methylation in mammalian cells.

Looking back at the major experiments of the 1975–78 era, the quiet-spoken Riggs now says, "In my mind, somatostatin was the big one. With the *lac* operator, we copied nature. We copied the sequence of the natural *lac* operator, which I think had been determined a year earlier. When we were successful with somatostatin, we didn't copy nature. We just used information that we had extracted from nature. And I don't mind emphasizing it: that's the fundamental thing that seems to have gotten lost in the excitement. The newspaper reporters could care less. Somatostatin might be a new cure for diabetes, you know. We scientists never thought that at all. We were standing there saying, 'Hey, but wait a second, *this is a gene we made ourselves.*' And nobody cared."

People did care, of course, but landmarks quickly subside into prairie when science roars through new territory, as it did in the mid-1970s. In the summer of 1984, one of the guys with his foot habitually on the gas, David Goeddel, sat in his office, talking about the work on insulin. Compared to the early days, he had a new job title (director of molecular biology), but still wore the same old uniform: a plaid shirt, gray corduroy jeans, running shoes. The office was windowless; there were pictures of mountains he had conquered on the wall, a baseball trophy atop a filing cabinet, a basketball on the floor in the corner. How long, he was asked, would it take, utilizing the latest techniques in molecular biology, to repeat the work on insulin that sprawled over continents and consumed many man-years a decade ago. "You could make the DNA in one day," he replied. "You could clone it and express it in one week. You could purify the protein and do the reconstitution in probably a week.

"If you had a good group of people wanting to work hard," Goeddel concluded, in his low and restless yet matter-of-fact voice, "you could probably do it in a little more than two weeks."

A continent away, in his fourteenth-floor office overlooking the Hudson River, still dressed in black but better shaven than in the old days, Argiris Efstratiadis too offered a final thought about the insulin work. "The punchline for all this," he says, "is that after all the fights, after all the publicity, after all the people's behavior, after all the fever, what is left is a beautiful piece of science by all the parties involved."

Interviews

Major interviews were conducted in person and taped. Citations from these interviews attempt to follow the speaker's mode of speech as closely as possible, but some minor editing has been done in the interests of clarity and to eliminate repetition. Some of these interviews were originally conducted during the preparation of an article for *Science 85* magazine.

PHILIP ABELSON, former editor, *Science*, Washington, D.C.: June 13, 1986 (Tel)

KEITH BACKMAN, staff scientist, BioTechnica International, Inc., Cambridge, Mass.: Nov. 18, 1986 (Tel)

LEONOR BALCE-DIRECTO, research technician, Dept. of Biology, City of Hope National Medical Center, Duarte, Cal: March 5, 1985 (Duarte, Cal.); March 7, 1985 (Duarte, Cal.)

DEBORAH BANNISTER, Corporate Communications, Genentech, Inc., South San Francisco, Cal: Oct. 18, 1986 (Tel)

JOHN D. BAXTER, Dept. of Biochemistry, University of California–San Francisco, San Francisco, Cal.: Feb. 17, 1987 (Tel)

M. KATHY BEHRENS, biotechnology analyst, Robertson, Colman, and Stephens, San Francisco, Cal.: May 21, 1984 (Tel)

GRAEME I. BELL, Dept. of Biochemistry, University of Chicago, Chicago, Ill.: Feb. 18, 1987 (Tel)

PATRICIA BERKELEY, biotechnology analyst, Arthur D. Little and Co., Cambridge, Mass.: May 30, 1984 (Cambridge, Mass.)

FRANCISCO BOLIVAR, director, Research Center for Genetic Engineering and Biotechnology, Cuernavaca, and professor of molecular biology, National University of Mexico: June 14, 1986 (Tel)

HERBERT W. BOYER, Dept. of Biochemistry, University of California–San Francisco, Cal., and vice-president, Genentech, Inc.: May 28, 1985 (San Francisco, Cal.); Oct. 23, 1986 (Tel)

STEPHANIE BROOME, postdoctoral fellow, Scripps Clinic for Experimental Science, La Jolla, Cal.: Jan. 18, 1985 (Cambridge, Mass.)

RONALD CAPE, chief executive officer, Cetus Corp., Emeryville, Cal.: March 15, 1985 (Emeryville, Cal.)

SHARON CARLOCK, former employee, Genentech, Inc., South San Francisco, Cal.: Feb. 7, 1986 (Tel)

WILLIAM L. CHICK, Dept. of Biochemistry, University of Massachusetts Medical Center, Worcester, Mass.: Jan. 17, 1985 (Worcester, Mass.); Oct. 30, 1986 (Tel)

JOHN CHIRGWIN, Dept. of Medicine, University of Texas Health Center, San Antonio, Tex.: Jan. 3, 1985 (Tel); Jan. 4, 1985 (Tel); Feb. 11, 1986 (Tel); Oct. 6, 1986 (Tel)

DAVID CLEM, former city councillor, Cambridge, Mass.: Jan. 7, 1987 (Tel)

STANLEY N. COHEN, Dept. of Medicine, Stanford University School of Medicine, Stanford, Cal.: Feb. 27, 1986 (Tel)

ELIZABETH A. CRAIG, Dept. of Physiological Chemistry, University of Wisconsin, Madison, Wis.: July 1, 1986 (Tel)

ROBERTO CREA, vice-president and scientific director, Creative BioMolecules, Inc., Hopkinton, Mass.: July 17, 1984 (South San Francisco, Cal.); March 11, 1985 (South San Francisco, Cal.); June 16, 1986 (Tel)

ROY CURTISS III, Washington University School of Medicine, St. Louis, Mo.: June 3, 1986 (Tel)

MAYER DAVIDSON, director, Diabetes Center, Cedars-Sinai Medical Center, Los Angeles, Cal.: Feb. 20, 1985 (Tel)

FRAN DENOTO, research associate, Dept. of Biochemistry, University of California–San Francisco, San Francisco, Cal.: March 4, 1985 (San Francisco, Cal.)

RICHARD DICKERSON, Dept. of Biochemistry, University of California at Los Angeles, Los Angeles, Cal.: July 23, 1985 (Tel)

ROGER DITZEL, patent officer, University of California–San Francisco, San Francisco, Cal.: Oct. 20, 1986 (Tel)

ANDREW DREXLER, Dept. of Medicine, New York University Medical Center, New York, N.Y.: Oct. 31, 1986 (Tel)

ARGIRIS EFSTRATIADIS, Dept. of Human Genetics and Development, Columbia University, New York, N.Y.: Dec. 2, 1984 (Tel); Dec. 9, 1984 (Tel); Dec. 13, 1984 (New York City); Feb. 15, 1986 (Tel); March 31, 1986 (Tel); April 16, 1986 (New York City); June 21, 1986 (Tel)

STANLEY FALKOW, Stanford University School of Medicine, Stanford, Cal.: June 10, 1986 (Tel)

STUART FEINER, president, INCO Venture Capital Group, New York, N.Y.: Oct. 31, 1986 (Tel)

JOHN FIDDES, vice-president of research, California Biotechnology, Inc., Mountain View, Cal.: June 23, 1986 (Tel)

FORREST FULLER, staff scientist, California Biotechnology, Inc., Mountain View, Cal.: Feb. 27, 1985 (Mountain View, Cal.); March 2, 1985 (Mountain View, Cal.); June 9, 1986 (Tel); June 11, 1986 (Tel)

WILLIAM J. GARTLAND, JR., director, Office of Recombinant DNA Activities, National Institutes of Health, Bethesda, Md.: April 10, 1986 (Tel)

DAVID GELFAND, staff scientist, Cetus Corp., Emeryville, Cal.: June 11, 1986 (Tel)

ROBERT GELFAND, Dept. of Medicine, Yale University School of Medicine, New Haven, Conn.: Feb. 19, 1985 (Tel)

RICHARD and EMMA GILBERT: July 2, 1984 (White Plains, N.Y.)

WALTER GILBERT, H. H. Timken Professor of Science, The Biological Laboratories, Harvard University, Cambridge, Mass.: May 2, 1984 (Geneva); May 31, 1984 (Cambridge, Mass.); June 2, 1984 (Cambridge, Mass.); Oct. 30, 1984 (Cambridge, Mass.)

SHELDON GLASHOW, Dept. of Physics, Harvard University, Cambridge, Mass.: June 27, 1984 (Tel); June 2, 1986 (Tel)

DAVID V. GOEDDEL, director, Department of Molecular Biology, Genentech, Inc., South San Francisco, Cal: July 18, 1984 (South San Francisco, Cal.); Feb. 26, 1985 (South San Francisco, Cal.); June 10, 1986 (Tel); June 11, 1986 (Tel)

GERALD GRODSKY, Dept. of Biochemistry and Biophysics, University of California–San Francisco, San Francisco, Cal.: Feb. 26, 1985 (Tel); Feb. 27, 1985 (San Francisco, Cal.)

ZSOLT HARSANYI, biotechnology analyst, formerly with E. F. Hutton, Washington, D.C.: April 24, 1984 (New York City); Sept. 4, 1984 (Washington, D.C.)

PAUL HAYCOCK, research director, Squibb-Novo, Inc., Princeton, N.J.: Jan. 21, 1986 (Tel); Feb. 18, 1987

HERBERT HEYNEKER, vice-president and director of research, Genencor, Inc., South San Francisco, Cal.: Feb. 28, 1985 (South San Francisco, Cal.); Sept. 19, 1985 (Tel); June 14, 1986 (Tel)

JANET L. HOPSON, science writer, Oakland, Cal.: Jan. 8, 1986 (Tel)

RUTH HUBBARD, The Biological Laboratories, Harvard University, Cambridge, Mass.: Jan. 15, 1985 (Cambridge, Mass.)

KEIICHI ITAKURA, director, Dept. of Molecular Biology, City of Hope National Medical Center, Duarte, Cal.: July 5, 1984 (Duarte, Cal.)

ROBERT IVARIE, Dept. of Genetics, University of Georgia, Athens, Ga.: Oct. 17, 1986 (Tel)

IRVING S. JOHNSON, vice-president of research, Lilly Research Laboratories, Eli Lilly and Co., Indianapolis, Ind.: Feb. 24, 1986 (Tel); Feb. 26, 1986 (Tel); June 12, 1986 (letter to author)

JONATHAN KING, Dept of Microbiology, Massachusetts Institute of Technology, Cambridge, Mass.: June 22, 1984 (Tel); Jan. 18, 1985 (Cambridge, Mass.)

SCOTT R. KING, biotechnology analyst, Montgomery Securities, San Francisco, Cal.: May 22, 1984 (Tel); March 1, 1985 (San Francisco, Cal.); Jan. 7, 1986 (Tel)

DENNIS KLEID, senior scientist and patent coordinator, Genentech, Inc., South San Francisco, Cal.: July 18, 1984 (South San Francisco, Cal.); Feb. 26, 1985 (South San Francisco, Cal.); March 10, 1986 (Tel); June 12, 1986 (Tel); June 13, 1986 (Tel)

JEREMY KNOWLES, Dept. of Biochemistry, Harvard University, Cambridge, Mass.: June 1, 1984 (Cambridge, Mass.); Jan. 16, 1985 (Cambridge, Mass.); June 24, 1986 (Tel)

SHELDON KRIMSKY, Dept. of Urban and Environmental Policy, Tufts University, Boston, Mass.: June 18, 1984 (Tel); Feb. 12, 1985 (Cambridge, Mass.)

PAUL LACY, Dept. of Pathology, Washington University School of Medicine, St. Louis, Mo.: June 13, 1986 (Tel)

C. KEVIN LANDRY, venture capitalist, T. A. Associates, Boston, Mass.: May 30, 1984 (Boston, Mass.)

RACHMIEL LEVINE, former director of research, City of Hope National Medical Center, Duarte, Cal.: March 5, 1985 (Duarte, Cal.)

PETER LOMEDICO, research scientist, Hoffmann–La Roche, Inc., Nutley, N.J.: June 18, 1984 (Tel); Dec. 7, 1984 (Nutley, N.J.); Feb. 20, 1985 (Tel); March 3, 1986 (Tel)

ALLAN MAXAM, Dept. of Molecular Biology, Dana-Farber Cancer Institute, Boston, Mass.: June 1, 1984 (Boston, Mass.); Jan. 16, 1985 (Cambridge, Mass.); Feb. 10, 1985 (Boston, Mass.); June 2, 1986 (Tel)

BRIAN MCCARTHY, Dept. of Molecular Biology, Gladstone Foundation, San Francisco General Hospital, San Francisco, Cal.: May 28, 1985 (San Francisco, Cal.); June 10, 1986 (Tel)

PHIL MCCARTHY, venture capitalist, Chatham, N.J.: May 29, 1984 (Chatham, N.J.)

LINDA I. MILLER, Biotechnology Analyst, PaineWebber, Inc., New York, N.Y.: May 24, 1984 (Tel)

SAMUEL MILSTEIN, biotechnology analyst, Brooklyn, N.Y.: Oct. 24, 1986 (Tel)

JANET MONSON, Dept. of Surgery and Biology, McGill University, and Genetics Unit, Shriner's Hospital, Montreal, Canada: Oct. 19, 1986 (Tel)

PATRICK O'FARRELL, Dept. of Biochemistry, University of California–San Francisco, San Francisco, Cal.: Sept. 29, 1986 (Tel)

DEBRA PEATTIE, Dept. of Tropical Public Health, Harvard Medical School, Boston, Mass.: May 29, 1985 (Stanford, Cal.)

M. ALAN PERMUTT, Washington University School of Medicine, St. Louis, Mo: July 29, 1985 (Tel)

RAYMOND PICTET, research scientist, Institut Jacques Monod, University of Paris, Paris, France: Aug. 6, 1985 (Tel); March 3, 1986 (Tel); April 18, 1986 (Tel); June 17, 1986 (Tel); July 16, 1986 (New York City)

BARRY POLISKY, Dept. of Biology, Indiana University, Bloomington, Ind.: June 20, 1986 (Tel)

ARTHUR D. RIGGS, chairman, Division of Biology, City of Hope National
Medical Center, Duarte, Cal.: July 5, 1984 (Duarte, Cal.); March 5, 1985
(Duarte, Cal.); June 6, 1986 (Tel); June 12, 1986 (Tel)

RAYMOND RODRIGUEZ, Dept. of Biochemistry, University of California at
Davis, Davis, Cal.: Aug. 15, 1985 (Tel)

JOHN ROSENBERG, Dept. of Biological Sciences, University of Pittsburgh,
Pittsburgh, Pa.: Aug. 13, 1985 (Tel)

HENRY ROSOVSKY, former Dean of the Faculty of Arts and Sciences, Harvard
University, Cambridge, Mass.: Jan. 24, 1986 (Tel)

WILLIAM J. RUTTER, Hertzstein Professor of Biochemistry and director of the
Hormone Research Laboratory, University of California–San Francisco,
and chairman of the board, Chiron, Corp., Emeryville, Cal.: July 11,
1984 (San Francisco, Cal.); July 18, 1984 (San Francisco, Cal.); Feb. 27,
1987 (Tel)

RAYMOND SCHAEFER, former vice-president, Venture Capital division, INCO
Ltd., Toronto, Canada: June 20, 1984 (Southbury, Conn.); Aug. 27, 1985
(Tel); June 16, 1986 (Tel)

RICHARD H. SCHELLER, Dept. of Biological Sciences, Stanford University,
Stanford, Cal.: May 29, 1985 (Stanford, Cal.); July 29, 1985 (Tel)

PETER SEEBURG, senior scientist, Genentech, Inc., South San Francisco, Cal.:
Feb. 28, 1985 (South San Francisco, Cal.); June 10, 1986 (Tel)

PHILLIP SHARP, Center for Cancer Research, Massachusetts Institute of Technology, Cambridge, Mass.: June 1, 1984 (Cambridge, Mass.); June 16,
1986 (Tel)

BRIAN SHEEHAN, former vice-president, Genentech, Inc., South San Francisco, Cal.: Jan. 16, 1986 (Tel)

JOHN SHINE, president and chief scientific officer, California Biotechnology,
Inc., Mountain View, Cal.: July 16, 1984 (Mountain View, Cal.); March
13, 1985 (Mountain View, Cal.); Oct. 20, 1986 (Tel)

LOUISE SHIVELY, research associate, Dept. of Biology, City of Hope National
Medical Center, Duarte, Cal.: March 5, 1985 (Duarte, Cal.); March 7,
1985 (Duarte, Cal.)

NINA SIEGLER, biotechnology analyst, Prudential-Bache, New York, N.Y.:
May 21, 1984 (New York City)

JAY S. SKYLER, University of Miami Medical Center, Miami, Fla.: Feb. 25,
1987 (Tel)

STEPHEN STAHL, staff scientist, Biogen S.A., Geneva, Switzerland: May 2,
1984 (Geneva, Switz.)

DONALD F. STEINER, Dept. of Biochemistry, University of Chicago, Chicago,
Ill.: Dec. 21, 1984 (Chicago, Ill.)

J. GREGOR SUTCLIFFE, research scientist, Scripps Clinic for Experimental Science, La Jolla, Cal.: Dec. 6, 1984 (Tel)

ROBERT A. SWANSON, chief executive officer, Genentech, Inc., South San Francisco, Cal.: May 28, 1985 (South San Francisco, Cal.)

BERNARD TALBOT, National Institutes of Health, Bethesda, Md.: April 10, 1986 (Tel)

KAREN TALMADGE, chief scientist and patent coordinator, California Biotechnology, Inc., Mountain View, Cal.: July 16, 1984 (Mountain View, Cal.); Feb. 27, 1985 (Palo Alto, Cal.); June 15, 1986 (Tel)

EDMUND TISCHER, research associate, Dept. of Molecular Biology, Massachusetts General Hospital, Boston, Mass.: Jan. 16, 1985 (Boston, Mass.)

RICHARD TIZARD, staff scientist, Biogen, Inc., Cambridge, Mass.: Jan. 18, 1985 (Cambridge, Mass.)

AXEL ULLRICH, senior scientist, Genentech, Inc., South San Francisco, Cal.: July 19, 1984 (South San Francisco, Cal.); Feb. 28, 1985 (South San Francisco, Cal.); June 26, 1986 (Tel); Oct. 19, 1986 (Tel); Oct. 20, 1986 (Tel); Oct. 22, 1986 (Tel)

ALFRED VELLUCCI, former mayor, Cambridge, Mass.: Jan. 31, 1986 (Tel)

LYDIA VILLA-KOMAROFF, Dept. of Neurobiology, Boston Children's Hospital, Boston, Mass.: July 2, 1984 (Tel); Sept. 22, 1984 (Boston, Mass.); Feb. 9, 1985 (Cambridge, Mass.); March 3, 1986 (Tel); March 13, 1986 (Tel); June 15, 1986 (Tel); June 21, 1986 (Tel)

NICHOLAS WADE, editorial writer, *New York Times*, New York, N.Y.: March 6, 1986 (Tel)

RONALD WETZEL, Dept. of Catalyses, Genentech, Inc., South San Francisco, Cal.: March 15, 1985 (South San Francisco, Cal.); Letters to author, March 19, 1985 and Oct. 18, 1985

THOMAS WHITE, director of research, Cetus Corp., Emeryville, Cal.: March 15, 1985 (Emeryville)

DANIEL YANSURA, staff scientist, Genentech, Inc., South San Francisco, Cal.: July 19, 1984 (South San Francisco, Cal.)

WILLIAM D. YOUNG, vice-president for manufacturing and process sciences, Genentech, Inc., South San Francisco, Cal.: March 15, 1985 (South San Francisco, Cal.)

Several former postdocs and members of the Biochemistry Dept. faculty at the University of California–San Francisco agreed to be interviewed only on the condition of anonymity.

Index

About the Author

A native of Cleveland, Ohio, Stephen S. Hall earned a B.A. in English literature from Beloit College. He has written for several publications, including *Geo, Smithsonian,* and *The New York Times Magazine.* He is currently a contributing editor of *Hippocrates* magazine. **INVISIBLE FRONTIERS** is his first book.

Other Titles from Tempus Books

The World of Mathematics
A Small Library of the Literature of Mathematics from A'h-mosé the Scribe to Albert Einstein, Presented with Commentaries and Notes by James R. Newman

Published in 1956 to critical acclaim and with sales of more than 200,000 copies, THE WORLD OF MATHEMATICS is now available for a new generation of readers. Out of print for many years, this four-volume anthology is a rich collection of 133 articles. The essays provide first-person windows on the concepts of pure math, the laws of probability, statistics, puzzles, and paradoxes, and the role of mathematics in economics, art, music, and literature. Each article is prefaced by Newman's insightful commentary that places it in historical perspective and makes even the most difficult concepts accessible to a wide range of readers. The selections range from hard-to-find classical pieces by Archimedes, Galileo, and Mendel to works of twentieth-century thinkers, including John Maynard Keynes, Bertrand Russell, and A.M. Turing.

James R. Newman, 2800 pages, four volumes $50.00 softcover, #86-96544, $99.95 cloth, #86-96593

Thursday's Universe
A Report from the Frontier on the Origin, Nature, and Destiny of the Universe

"Marcia Bartusiak entered the universe of research at the frontiers of astronomy and cosmology and returned with a gem of a book. I recommend it."

> Heinz Pagels, author of *The Cosmic Code*

How did the universe begin? How will it end? What populates the endless expanse of space? Cited by the *New York Times* as one of the best science books of 1987, THURSDAY'S UNIVERSE explores current ideas about the moment of creation, including the birth and death of stars, quasars, and galaxies, the composition of black holes and neutrinos, and the puzzle of the universe's missing mass. Based on interviews with leaders in the fields of astronomy and astrophysics, THURSDAY'S UNIVERSE is a fascinating voyage to the outer reaches of the cosmos.

Marcia Bartusiak, 336 pages, $8.95 softcover, #86-96627

The New Wizard War
How the Soviets Steal U.S. High Technology — And How We Give It Away

Is the United States giving the Soviet Union an unprecedented technological edge in the name of free trade? How easy is it for Soviet spies to smuggle high-technology equipment and information to the Soviet Union? How adept is the U.S. government at controlling our technology exports? And are our allies abetting the Soviet acquisition enterprise? THE NEW WIZARD WAR is a riveting and timely look at the legal and illegal transfer of high technology from the U.S. to the U.S.S.R. The case studies in the book — mixing international intrigue, military secrets, million-dollar payoffs, and bureaucratic snafus — read like scenes from a fast-paced spy thriller. Much more than a good read, THE NEW WIZARD WAR is an eye-opening, provocative report on controversial and urgent election-year issues: the loss of U.S. technological leadership, the question of national security, the wisdom of U.S.-Soviet scientific cooperation, Japan's role in technology transfer, the Gorbachev agenda, and the balance between government and private control over the flow of technology.

Robyn Shotwell Metcalfe, 256 pages, $17.95 cloth, #86-96338

Time
The Familiar Stranger

"Fraser is perhaps the leading intellectual authority in the world on the study of time. Now he has given us a book that is both stimulating and provocative."

Jeremy Rifkin, author of *Time Wars*

This wide-ranging and learned book surveys the enormous variety of our understandings of time, both in the everyday world and in the sciences and humanities. From the visions of time and the timeless in religion, to those in contemporary physics and cosmology and the more common conceptions of time, J.T. Fraser offers the general reader a fascinating history of the idea and experience of time. Included are such issues as the beginning and end of the universe; the biology of aging and death; human perception of time; dreaming *vs* waking reality; expectation and memory; calendars and chronologies; and the ways in which technological rhythms control our lives.

J.T. Fraser, 400 pages, $9.95 softcover, #86-96809
Available in November 1988

Mathematics
Queen and Servant of Science

"Bell is a lively, stimulating writer, with a good sense of historical circumstance...a sound grasp of the entire mathematical scene, and a gift for clear and orderly explanation."

James R. Newman, editor of THE WORLD OF MATHEMATICS

Here — from a modern-day master of mathematics literature — is a fascinating and lively survey of the developments of pure and applied mathematics. Bell explores mathematical theories from those of Pythagoras and Euclid to those of Einstein and Gödel with vigor, intelligence, and wit that are unsurpassed. He covers a broad range of traditional subjects — algebra, geometry, calculus, topology, logic, rings, and groups — making even the most abstruse subjects come alive. Published in cooperation with the Mathematical Association of America.

Eric Temple Bell, 432 pages, $11.95 softcover, #86-96825
Available in November 1988

The Tomorrow Makers
A Brave New World of Living-Brain Machines

"Talespinners have nothing on the hard-core science freaks. This nonfiction book has enough new ideas for 16 Star Trek sequels. And better dialogue."

Rudy Rucker, *San Francisco Chronicle*, author of *Infinity and the Mind*

THE TOMORROW MAKERS is a spellbinding account of visionary researchers and scientists and their modern-day quest to engineer immortality. Award-winning science writer Grant Fjermedal details the astounding work being done in robotics and artificial intelligence today on organic molecular computers smaller than a grain of sand; on lifelike androids that are intelligent and charming; and — most extraordinary of all — on the process of downloading human minds into computerized robots that will live forever. Cited by the American Library Association as one of 1987's most notable books.

Grant Fjermedal, 288 pages, $8.95 softcover, #86-96247

Nobel Dreams

Power, Deceit, and the Ultimate Experiment

"...one of those rare science books that tell about science in the course of telling about the human comedy."

> Lee Dembart
> *Los Angeles Times*

An intimate and colorful look at Carlo Rubbia's quest for the 1985 Nobel Prize in Physics. It is a vivid saga of big science, big ego, frenetic competition, and hardball politics that demonstrates the chaotic and chancy nature of scientific research. Taubes' narrative is a very readable layperson's primer on high-energy physics, as well as a captivating modern-day adventure story.

Gary Taubes, 288 pages, $8.95 softcover, #86-96239

Inventors at Work

Interviews with 16 Notable American Inventors

"Invention becomes an art in these accounts of serendipitous associations and 'lateral thinking.'"

> *Publishers Weekly*

A critically acclaimed collection of 16 engaging and illuminating interviews with the most notable inventors of our time—from professional R and D specialists such as NASA's Maxime Faget, Hughes Aircraft's Harold Rosen, and Xerox's Bob Gundlach to independents such as entrepreneur Stanford Ovshinsky and artificial intelligence expert Raymond Kurzweil. Their accomplishments—the laser, the microprocessor, the man-powered airplane, the implantable pacemaker, the Apple II computer, the plastic soda bottle, and many others—represent both the boldly significant and the subtly brilliant. Along with fascinating individual stories, INVENTORS AT WORK reveals instructive glimpses into the creative process, thoughts on the personal challenges and institutional roadblocks an inventor faces today, a look at invention as art and invention by committee, and discussions of the impact of United States patent laws and a capitalist economy on the inventive spirit.

Kenneth A. Brown, 408 pages, $9.95 softcover, #86-96080, $17.95 cloth, #86-96312

Computer Lib/Dream Machines

"If you want a vision at once backwards towards the dawning of the personal computer age and forward beyond the limits of the machines we use today, take a look at COMPUTER LIB."

> Jim Seymour
> syndicated columnist

First published in 1974, COMPUTER LIB became the first cult book of the computer generation, predicting the major issues of today: design of easy-to-use computer systems, image synthesis, artificial intelligence, and computer-assisted instruction. Ted Nelson's vision of a nonsequential way of storing data—hypertext—is particularly relevant today with the emergence of CD ROM. Republished by Tempus Books, COMPUTER LIB's wildly Utopian introduction to computers is now available to a new generation. Included is new material from Ted Nelson—commentaries, insights, updates, and reconsiderations—all in his characteristically opinionated, uplifting, and irreverent style.

Theodor Nelson, 336 pages, $18.95 softcover, #86-96031

The Pursuit of Growth

The Challenges, Opportunities, and Dangers of Managing and Investing in Today's Economy

Success stories of phenomenal business growth — IBM, DuPont, Proctor and Gamble — have been well publicized, as have spectacular growth failures — Atari, Bendix, People Express. What is this peculiar obsession with growth at any cost? When does it serve the best interests of managers, entrepreneurs, politicians, and the community? And when is it a formula for disaster? The authors provide a highly readable and insightful analysis of growth as an underlying tenet of business and government.

G. Ray Funkhouser and Robert R. Rothberg, 274 pages, $17.95 hardcover, #86-96098

Machinery of the Mind

Inside the New Science of Artificial Intelligence

"An ideal presentation of what artificial intelligence is all about."

Douglas Hofstadter, author of *Gödel, Escher, Bach*

Focusing on the work of giants in the artificial intelligence field — including Marvin Minsky, Roger Schank, and Edward Feigenbaum — George Johnson gives us an intimate look at the state of AI today. We see how machines are beginning to understand English, discover scientific theories, and create original works of art; how research in AI is helping us to understand the human mind; and how AI affects our lives today. Captivating reading for anyone with an interest in science and technology.

George Johnson, 352 pages, $9.95 softcover, #86-96072